Technology and the Virtues

TECHNOLOGY AND THE VIRTUES

A Philosophical Guide to a Future Worth Wanting

SHANNON VALLOR

OXFORD
UNIVERSITY PRESS

Oxford University Press is a department of the University of Oxford. It furthers the University's objective of excellence in research, scholarship, and education by publishing worldwide. Oxford is a registered trade mark of Oxford University Press in the UK and certain other countries.

Published in the United States of America by Oxford University Press
198 Madison Avenue, New York, NY 10016, United States of America.

Library of Congress Cataloging-in-Publication Data
Names: Vallor, Shannon, author.
Title: Technology and the virtues : a philosophical guide
to a future worth wanting / Shannon Vallor.
Description: New York : Oxford University Press, 2016. |
Includes bibliographical references and index.
Identifiers: LCCN 2016004301 | ISBN 9780190498511 (hardcover : alk. paper) |
ISBN 9780190498528 (ebook (updf)) | ISBN 9780190498535 (ebook (epub)) |
ISBN 9780190498542 (online content)
Subjects: LCSH: Technology—Moral and ethical aspects. | Virtues.
Classification: LCC BJ59 .V34 2016 | DDC 179/.9—dc23
LC record available at https://lccn.loc.gov/2016004301

Contents

Acknowledgements

THIS WORK IS indebted to far too many persons and organizations to count, or name.

I am deeply grateful to the Markkula Center for Applied Ethics and others at Santa Clara University who provided financial support for earlier phases of this project. I owe an immense debt to the participants in the 2007 workshop at the University of Twente, Netherlands on "The Good Life in a Technological Age," where this project was born; in particular I wish to thank Charles Ess for his encouragement, intellectual engagement, and inspiration then and ever since. This book is about humanity's need for more people with the exemplary moral character and practical wisdom to help us navigate the uncertain and rapidly shifting technological and social terrain ahead; in Charles, at least, we have one.

I also thank the Arnold L. and Lois S. Graves Foundation for funding of early research for the book. That research was assisted by the remarkable hospitality and intellectual generosity of Philip. J. Ivanhoe and his family and colleagues, who welcomed me to Hong Kong in 2010 and helped to deepen my understanding of the Confucian virtue tradition.

The ideas in this book, at least the worthwhile ones, owe their mature shape to the fellowship of my peers in the field of philosophy and ethics of emerging technologies, with whom I have enjoyed countless stimulating conversations, arguments, and whiskies. Though I should thank many more, I am especially indebted to Don Howard, Patrick Lin, Evan Selinger, and John Sullins for their invaluable contributions to my thinking and their enduring friendship.

To my editor Lucy Randall and others at OUP, thank you for bringing this work into the light and making it better every step of the way.

Thanks to my grandfather Howard, who taught me at a very early age to love science, truth, argument, and by example, virtue; to my parents Kathleen and William Bainbridge for their boundless support, love, and patience; and finally, to Dan, the deepest well of goodness, brilliance, and love that I have ever known.

Technology and the Virtues

Introduction

ENVISIONING THE GOOD LIFE IN THE 21ST CENTURY AND BEYOND

IN MAY 2014, cosmologist Stephen Hawking, computer scientist Stuart Russell, and physicists Max Tegmark and Frank Wilczek published an open letter in the UK news outlet *The Independent*, sounding the alarm about the grave risks to humanity posed by emerging technologies of artificial intelligence. They invited readers to imagine these technologies "outsmarting financial markets, out-inventing human researchers, out-manipulating human leaders, and developing weapons we cannot even understand."[1] The authors note that while the successful creation of artificial intelligence (AI) has the potential to bring "huge benefits" to our world, and would undoubtedly be "the biggest event in human history . . . it might also be the last." Hawking echoed the warning later that year, telling the BBC that unrestricted AI development "could spell the end of the human race." While some AI enthusiasts dismiss such warnings as fearmongering hype, celebrated high-tech inventors Elon Musk, Steve Wozniak, Bill Gates, and thousands of AI and robotics researchers have joined the chorus of voices calling for wiser and more effective human oversight of these new technologies.[2]

How worried should we be? More importantly: what should we *do?*

AI is only one of many emerging technologies—from genome editing and 3D printing to a globally networked "Internet of Things"—shaping a future unparalleled in human history in its promise *and* its peril. Are we up to the challenge this future presents? If not, *how can we get there*? How can humans hope to live well in a world made increasingly more complex and unpredictable by emerging technologies? Though it will require the remainder of the book to fully respond to that question, in essence my answer is this: we need to cultivate in ourselves, collectively, a special kind of moral character, one that expresses what I will call the *technomoral virtues.*

What do I mean by *technomoral* virtue? To explain this concept will require introducing some ideas in moral philosophy, the study of ethics. At its most basic, ethics is about what the ancient Greek philosopher Socrates called the "good life": the kind of life that is most worthy of a human being, the kind of life worth choosing from among all the different ways we might live. While there are many kinds of lives worth choosing, most of us would agree that there are also some kinds of lives *not* worth choosing, since we have better alternatives. For example, a life filled mostly with willful ignorance, cruelty, fear, pain, selfishness, and hatred might still have some value, but it would not be a kind of life worth *choosing* for ourselves or our loved ones, since there are far happier choices available to us—better and more virtuous ways that one can live, for ourselves and everyone around us. But what does ethics or moral philosophy have to do with *technology*?

In reality, human social practices, including our moral practices, have always been intertwined with our technologies.[3] Technological practices—everything from agriculture and masonry to markets and writing—have shaped the social, political, economic, and educational histories of human beings. Today, we depend upon global systems of electronic communication, digital computation, transportation, mass manufacturing, banking, agricultural production, and health care so heavily that most of us barely notice the extent to which our daily lives are technologically conditioned. Yet even our earliest ancestors used technology, from handaxes and spears to hammers and needles, and their tools shaped how they dealt with one another—how they divided their labor, shared their resources and living spaces, and managed their conflicts. Among our primate cousins, female chimpanzees have been observed to stop fights among males through technological disarmament—repeatedly confiscating stones from an aggressor's hand.[4]

Ethics and technology are connected because technologies invite or *afford* specific patterns of thought, behavior, and valuing; they open up new possibilities for human action and foreclose or obscure others. For example, the invention of the bow and arrow afforded us the possibility of killing an animal from a safe distance—*or* doing the same to a human rival, a new affordance that changed the social and moral landscape. Today's technologies open their own new social and moral possibilities for action. Indeed, human technological activity has now begun to reshape the very planetary conditions that make life possible. Thus 21st century decisions about how to live well—that is, about *ethics*—are not simply moral choices. They are *technomoral* choices, for they depend on the evolving affordances of the technological systems that we rely upon to support and mediate our lives in ways and to degrees never before witnessed.

While ethics has always been embedded in technological contexts, humans have, until very recently, been the primary authors of their moral choices, and the consequences of those choices were usually restricted to impacts on individual or local group welfare. Today, however, our aggregated moral choices in technological contexts routinely impact the well-being of people on the other side of the planet, a staggering number of other species, and whole generations not yet born. Meanwhile it is increasingly less clear how much of the future moral labor of our species will be performed by human individuals. Driverless cars are already being programmed to make 'ethical' driving decisions on our behalf while we relax and daydream, even as other cars roll out of the factory programmed to commit the *unethical* act of cheating on their innocent owners' emissions tests.[5] High-frequency trading algorithms now direct the global flow of vital goods and wealth at speeds and scales no human observer can follow. Artificially intelligent life coach apps are here to 'nudge' us when we need to lower our voices, call our mothers, or write nicer emails to our employees. Advanced algorithms inscrutable to human inspection increasingly do the work of labeling us as combatant or civilian, good loan risk or future deadbeat, likely or unlikely criminal, hireable or unhireable.

For these reasons, a contemporary theory of ethics—that is, a theory of what counts as a *good life* for human beings—must include an explicit conception of how to live well with technologies, especially those which are still emerging and have yet to become settled, seamlessly embedded features of the human environment. Robotics and artificial intelligence, new social media and communications technologies, digital surveillance, and biomedical enhancement technologies are among those emerging innovations that will radically change the kinds of lives from which humans are able to choose in the 21st century and beyond. How can we choose wisely from the apparently endless options that emerging technologies offer? The choices we make will shape the future for our children, our societies, our species, and others who share our planet, in ways never before possible. Are we prepared to choose *well*?

This question involves the future, but what it really asks about is our readiness to make choices in the *present.* The 21st century is entering its adolescence, a time of great excitement, confusion, and intense anxiety, an age both wildly hopeful and deeply troubled. As with many adolescents, our era is also deeply self-absorbed. In popular and scholarly media, we find both historical consciousness and the 'long view' of humanity giving way to an obsessive quest to define the distinctive identity of the present age, an identity almost always framed in technological terms. Whether we claim to be living in the 'Age of Information,' the 'Mobile Era,' the 'New Media Age,' or the 'Robot Age,' we seem to think that defining the technological essence of our era will allow us to better fathom the course of its future—*our* future.

Yet in one of those cruel paradoxes of adolescence, all our ruminations and fevered speculations about the mature shape of life in this century seem only to make the picture *more* opaque and unsettled, like a stream bottom kicked up by shuffling feet. Among all the contingencies pondered by philosophers, scientists, novelists, and armchair futurists, the possibilities presented by emerging technology have proved to be the most enticing to the imagination—and the most difficult to successfully predict. Of course, early visions of a postindustrial technological society were strikingly prescient in many respects. Debates about today's emerging technologies echo many of the utopian and dystopian motifs of 20th century science fiction: fears *and* hopes of a 'brave new world' of bioengineered humans constructed by exquisite design rather than evolutionary chance; of humans working side-by-side with intelligent robotic caregivers, surgeons, and soldiers; of digitally-enabled 'Big Brothers' recording and analyzing our every act; and of the rise of a globally networked hive mind in the 'cloud' that radically transforms the nature of human communication, productivity, creativity, and sociality.

Still, we cannot help but smile wistfully at the lacunae of even our most far-seeing science fiction visionaries. In the classic Ray Bradbury tale 'The Veldt,' first published in 1950, we encounter the existential and moral dilemma of the Hadley family, whose complete surrender to the technological comforts of the 'Happy-life Home' has stripped their lives of labor, but also of joy, purpose, and filial love. In a present marked by the increasingly sophisticated design of 'smart homes,' Bradbury's story resonates still. It may have taken a few decades longer than he expected, but affluent modern families can now, just like the Hadleys, enjoy a home that anticipates their every personal preference for lighting, room temperature, music, and a perfectly brewed cup of coffee—and the 'smart homes' of the future will even more closely approximate Bradbury's vision. We also recognize all too well the Hadleys' parental anxiety and regret when their children, irretrievably spoiled by the virtual world of their interactive playroom, fly into an incandescent rage at the thought of having their electronic amusements removed.

Yet today we can only laugh or cry when Lydia, the children's mother, complains that her surrender to domestic technology has left her without "enough to do," and too much "time to think." No technologically-savvy 21st century parent can identify with Lydia Hadley's existential plight.[6] Rather, the promised land of unlimited technological leisure has given way to a reality of electronic overstimulation and hypersaturation, a 24-hour news cycle, and smartphones on which your boss texts you from the 18th hole in Dubai while you sit at the dinner table wolfing down take-out, supervising your child's Web research on whale sharks, feverishly trying to get caught up on your email, responding to your Facebook invitations, and updating the spreadsheet figures your colleagues need for their afternoon presentation in Seoul. Leisure is one thing our age does *not* afford most modern technology consumers, who struggle each night to ignore the incoming status updates on their bedside devices so that they may grab a few precious hours of sleep before rejoining that electronic day that knows neither dusk nor dawn.

Indeed, the contemporary human situation is far more complicated, dynamic, and unstable than any of the worlds depicted in our first imaginings of a high-tech future. Today, exponential leaps in technological prowess and productivity are coeval with widespread economic stagnation, terrestrial resource depletion, and rising ecological instability. A global information society enabled by a massive electronic communications network of unprecedented bandwidth and computing power has indeed emerged; but far from enabling a 'new world order' of a utopian *or* dystopian sort, the information age heralds an increasingly *dis*-ordered geopolitics and widening fractures in the public commons. The rapid amplification of consumerism by converging innovations and ever-shorter product marketing cycles continues apace; yet far from ensuring the oft-predicted rise of technocratic states ruled by scientific experts, the relationship between science, governance, and public trust is increasingly contentious and unsettled.[7] Paradoxically, such tensions appear to be greatest where scientific and technical power have been most successfully consolidated and embedded into our way of life; consider that the nation that gave birth to Apple, Microsoft, Google, Intel, Amazon, and other tech behemoths has slashed federal funding for basic science research, struggled with declining scientific literacy and technical competence among its population, and adopted increasingly ambivalent and politicized science policy—even as it continues to shower the tech industry with tax loopholes and political access.[8]

Such complexities remind us that predicting the general shape of tomorrow's innovations is not, in fact, our biggest challenge: far harder, and more significant, is the job of figuring out what we will *do* with these technologies once we have them, and what they will do with *us*. This cannot be done without attending to a host of interrelated political, cultural, economic, environmental, and historical factors that co-direct human innovation and practice. Indeed, a futurist's true aim is not to envision the technological future but our techno*social* future—a future defined not by which gadgets we invent, but by how our evolving technological powers become embedded in co-evolving social practices, values, and institutions. Yet by this standard, our present condition seems not only to defy confident predictions about where we are heading, but even to defy the construction of a coherent narrative about where exactly we *are*. Has the short history of digital culture been one of overall human improvement, or decline? On a developmental curve, are we approaching the next dizzying explosion of technosocial progress as some believe, or teetering on a precipice awaiting a calamitous fall, as others would have it?[9]

Should it matter whether our future can be envisioned with any degree of confidence? Of course we might *want* to know where we are and where we are heading, but humans characteristically want a lot of things, and not all of these are necessary or even objectively worthwhile. Could it be that our understandable adolescent curiosity about what awaits us in our century's adulthood is, in the grand scheme of things, unimportant to satisfy? Let us imagine for the sake of argument that given certain efforts, we could better predict the future shape of life in this century. Other than idle curiosity, what reason would we have to make such efforts? Why not just take the future as it comes? Why strain to see any better through the fog of technosocial contingencies presently obscuring our view? There is a simple answer. Our growing technosocial blindness, a condition that I will call *acute technosocial opacity*, makes it increasingly difficult to identify, seek, and secure the ultimate goal of ethics—a life worth choosing; a life lived *well*.

Ethics, defined broadly as reflective inquiry into the good life, is among the oldest, most universal, and culturally significant intellectual preoccupations of human beings. Few would deny that humans have always and generally preferred to live well rather than badly, and have sought useful guidance in meeting this desire. Yet the phenomenon of acute technosocial opacity is a serious problem for ethics—and a relatively new one.[10] The founders of the most enduring classical traditions of ethics—Plato, Aristotle, Aquinas, Confucius, the Buddha—had the luxury of assuming that the practical conditions under which they and their cohorts lived would be, if not wholly static, at least relatively stable. While they knew that unprecedented political developments or natural calamities *might* at any time redefine the ethical landscape, the safest bet for a moral sage of premodern times would be that he, his fellows, and their children would confront essentially similar moral opportunities and challenges over the course of their lives.

Without this modest degree of foresight, ethical norms would seem to have little if any power to guide our actions. For even a timeless and universally binding ethical principle presupposes that we can imagine how adopting that principle today is likely to sustain or enrich the quality of our lives *tomorrow*. Few are moved by an ethical norm or ideal until we have been able to envision its concrete expression in a future form of life that is possible for us, one that we recognize as relevantly similar to, but qualitatively better than, our current one. When our future is opaque, it is harder to envision the specific conditions of life we will face tomorrow that can be improved by following an ethical principle or rule today, and such ideals may then fail to motivate us.

While philosophical ethics first emerged in Greece and Asia in the 6th–4th centuries BCE, the need for ethical guidance as we face our future applies equally to modern systems of ethics. Yet modern ethical frameworks often provide *fewer* resources for mitigating the difficulty posed by an uncertain future than do classical traditions. For example, the ethical framework of 18th century German philosopher Immanuel Kant supplied a single moral principle, known as the *categorical imperative*, which is supposed to be able to resolve any ethical dilemma. It simply asks a person to consider whether she could will the principle upon which she is about to act in her particular case to be universally obeyed by all other persons in relevantly similar cases.[11] If she *can't* will her own 'subjective' principle of action to function as a universal rule for everyone to follow, then her act is morally wrong. So if I cannot will a world in which *everyone* lies whenever it would spare them trouble, then it cannot be right for me to lie.

Although it can be applied to any situation, the rule itself is highly abstract and general. It tells us nothing specific about the shape of moral life in 18th century Europe, nor that of any other time or place. At first we might think this makes the principle *more* useful to us today, since it is so broad that it can apply to any future scenario we might imagine. Yet this intuition is mistaken. Consider the dutiful Kantian today, who must ask herself whether she can will a future in which *all* our actions are recorded by pervasive surveillance tools, or a future where we *all* share our lives with social robots, or a future in which *all* humans use biomedical technology to radically transform their genes, minds, and bodies. How can any of these possible worlds be envisioned with enough clarity to inform a person's will? To envision a world of pervasive and constant surveillance, you need to know what will be *done* with the recordings, *who* might control them, and *how* they would be accessed or shared. To know whether to will a future full of social robots, you would first need to know what *roles* such robots would play in our lives, and how they might transform human interactions. To will a world where all humans enhance their own bodies with technology, won't you first need to know which *parts* of ourselves we would enhance, in what ways, and what those changes would *do* to us in the long run, for example, whether we would end up improving or degrading our own ability to reason morally? Once even a fraction of the possible paths of technosocial development are considered, the practical uncertainties will swamp the cognitive powers of any Kantian agent, paralyzing her attempt to choose in a rational and universally consistent manner.

Modern utilitarian ethics of the sort promoted by 19th-century British philosophers Jeremy Bentham and John Stuart Mill fares little better by telling us that we may secure the good life simply by choosing, among the available courses of action, that which promises the greatest happiness for all those affected. The problem of discerning *which* course of action promises the greatest overall happiness or the least harm—among all the novel paths of biomedical, mechanical, and computational development open to us—is simply incalculable. The technological potentials are too opaque, and too many, to assign reliable probabilities of specific outcomes. Moreover, technology often involves effects on humanity created by the aggregate choices of *many* groups and individuals. When we factor in the interaction effects between converging technologies, social practices, and institutions, the difficulty becomes intractable.

In their book *Unfit for the Future*, philosophers Ingmar Persson and Julian Savulescu note that the technological and scientific advances of the 20th century have further destabilized the traditional moral calculus by granting humans an unprecedented power to bring about "Ultimate Harm," namely, "making worthwhile life *forever* impossible on this planet."[12] We might destroy ourselves with a bioengineered virus for which we have no natural defenses. Carbon dioxide, nitrogen, and phosphorus from large-scale industry and agriculture may acidify our oceans and poison our waterways beyond repair. Or we might unleash a global nuclear holocaust, a risk that experts warn is once again on the rise.[13] How can existential risks such as these, scenarios that would ruin *any* future possibility for happiness, possibly be factored into the calculation?

Moreover, emerging technologies such as nanomedicine and geoengineering in theory have the potential to forestall 'Ultimate Harm' to humanity *or* to cause it; and not enough is known to reliably calculate the odds of either scenario. Add to this the fact that engineers and scientists are constantly envisioning new and untested avenues of technological development and the insolubility of the moral calculus becomes even more obvious. John Stuart Mill himself noted that the practicality of utilitarian ethics relies heavily upon our collective inheritance of centuries of accumulated moral wisdom about how to maximize utility in the *known* human environment.[14] Even on the timescale of our own lives, this environment is increasingly unstable and unpredictable, and it is not clear how much of our accumulated wisdom still applies.

Given this unprecedented degree of technosocial opacity, how can humans continue to do ethics in any serious and useful way? The question compels an answer; to abandon the philosophical project of ethics in the face of these conditions would not only amplify the risk of 'Ultimate Harm,' it would violate a deep-seated human impulse. Consider once again Ray Bradbury, whose stories are still among the most widely read and appreciated in the tradition of science fiction. What drives the imagination of a storyteller like Bradbury, and what makes his stories resonate with so many? Reading his most lauded works *Fahrenheit 451*, *The Martian Chronicles*, and the collection *The Illustrated Man* (which leads with 'The Veldt'), one notices how closely Bradbury's vision tracked human beings of a future Earth, or human descendants of Earth. Why this anthropological fidelity in a writer hardly wanting for imaginative horsepower?

Even the Martians in Bradbury's stories serve as literary foils who expose and reflect upon the distinctive powers, obsessions, and weaknesses of human beings.

And why is the human future usually envisioned on a time scale of fifty years, or a hundred and fifty? Why not a thousand years, or ten thousand? Why do so many of Bradbury's tales have a patently *ethical* arc, driven less by saintly heroes and diabolical villains than by ordinary, flawed humans working out for themselves how well or how poorly their lives in an era defined by rockets, robots, and 'televisors' have gone? Here is one plausible answer: Bradbury seemed compelled to imagine how human beings *more or less like himself*, and those he cared about, would fare in the not so distant technological future—to envision the possibilities for *us* living well with emerging technologies, and more often, the possibilities for our *failing* to live well.

All of this is meant to suggest that the ethical dilemmas we face as 21st century humans are not 'business as usual,' and require a novel approach. Now, it is a common habit of many academics to roll their eyes at the first hint of a suggestion that the human situation has entered some radically new phase. As a prophylactic against overwrought claims of this kind, these sober-minded individuals keep on hand an emergency intellectual toolkit (which perhaps should be labeled 'Break Glass In Case of Moral Panic') from which they can readily draw a litany of examples of any given assertion of transformative social change being trumpeted just as loudly a century ago, or five, or ten. This impulse is often well-motivated: libraries worldwide are stocked with dusty treatises by those who, either from a lack of historical perspective or an intemperate desire to sell books, falsely asserted some massive seismic shift in human history that supposedly warranted great cultural alarm.

Yet sometimes things really *do* change in ways that we would be remiss to ignore, and which demand that we loosen up our scripted cultural patterns of response. At risk of inviting the scorn of the keepers of academic dispassion, I suggest that this is one of those times. The technologies that have emerged in the last half century have led to the unprecedented economic and physical interdependence of nations and peoples and an equally unprecedented transmissibility of information, norms, ideas, and values. A great many intellectual and cultural scripts are being rewritten as a result—scripts about modern state power, about socioeconomic development, labor and human progress, and about our relationship with our environment, to offer just a few examples. The conventional scripts of philosophical ethics must be rewritten as well. While an irreducible plurality of ethical narratives is both inevitable and desirable in a world as culturally diverse as ours, we need a common framework in which these narratives can be situated if humans are going to be able to address these emerging problems of *collective* technosocial action wisely and well. This framework must facilitate not only a shared moral dialogue, but also a global commitment to the cultivation of the specific technomoral habits and virtues required to meet this challenge.

Fortunately for us, a tradition already exists in philosophy that can provide such a framework. That tradition is *virtue ethics,* a way of thinking about the good life as achievable through specific moral traits and capacities that humans can actively cultivate in themselves.[15] Part I of this book explains the distinctive advantages of a virtue-driven approach to emerging technology ethics, and anticipates some of the challenges this project may face. Part II develops the theoretical foundations for our approach. Here we explore the rich conceptual resources of the classical virtue traditions of Aristotelian, Confucian, and Buddhist ethics, from which we construct a contemporary framework of *technomoral virtues* explicitly designed to foster human capacities for flourishing with new technologies. Part III applies the framework to four domains of emerging technology (social media, surveillance, robotics, and biomedical enhancement technology) that are likely to reshape human existence in the next one hundred years, assuming that we are fortunate and prudent enough to make it to the 22nd century.

No ethical framework can cut through the general constraints of technosocial opacity. The contingencies that obscure a clear vision of even the next few decades of technological and scientific development are simply far too numerous to resolve—in fact, given accelerating physical, geopolitical, and cultural changes in our present environment, these contingencies and their obscuring effects are likely to multiply rather than diminish. What this book offers is not an ethical *solution* to technosocial opacity, but an ethical *strategy* for cultivating the type of moral character that can aid us in coping, and even flourishing, under such challenging conditions.

The framework developed in the following chapters adapts Aristotelian, Confucian, and Buddhist reflections on moral development and virtue to our need for a profile of *technomoral* virtues for 21st century life.[16] These will not be radically new traits of character, for they must remain consistent with the basic moral psychology of our species. Rather the technomoral virtues are new alignments of our existing moral capacities, adapted to a rapidly changing environment that increasingly calls for collective moral wisdom on a global scale. In these challenging circumstances, the technomoral virtues offer the philosophical equivalent of a blind man's cane. While we face a future that remains cloaked in a technosocial fog, this need not mean that we go into it unprepared or ill-equipped, especially when it comes to matters of ethical life. The technomoral virtues, cultivated through the practices and habits of moral self-cultivation that we can learn from the classical virtue traditions examined in this book, are humanity's best chance to cope and even thrive in the midst of the great uncertainties and vicissitudes of technosocial life that lie ahead.[17] This hope will only be realized, however, if these virtues are more consciously cultivated in our families, schools, and communities, supported and actively encouraged by our local and global institutions, and exercised not only individually but *together*, in acts of collective human wisdom. This is a tall order; but not beyond our capabilities.

There is, however, what philosophers call a 'bootstrapping problem.' Our hope of flourishing in this and coming centuries—or even of securing our continued existence in the face of species-level threats created by our present lack of technomoral wisdom—requires us to act very soon to commit significant educational and cultural resources to the local *and* global cultivation of such wisdom. The framework articulated in this book, which draws strength from multiple cultural sources, can help us accomplish just that. Yet our existing technomoral vices, along with the normal human range of cognitive biases and limitations, impede many of us from grasping the depth, scope, or immediacy of the threats to human flourishing now confronting us. Even among those who recognize the dangers, many fail to grasp that the solution must be an ethical one. We cannot lift ourselves out of the hole we are in simply by creating more and newer technologies, so long as these continue to be designed, marketed, distributed, and used by humans every bit as deficient in technomoral wisdom as the generations that used *their* vast new technological powers to dig the hole in the first place!

While the first step out of the hole requires reallocating individual, local, and global resources to technomoral education and practice, we can and must make wise and creative use of technology to aid in the effort. Each of the emerging technologies explored in the book has the potential to be designed and used in ways that reinforce, rather than impede, our efforts to become wiser and more virtuous technological citizens. Thus our way out of the hole is a *recursive* procedure, in which traditional philosophical and educational techniques for cultivating virtue are used to generate the motivation to design and adopt new technological practices that shape our moral habits in more constructive ways. These in turn can reinforce our efforts of moral self-cultivation, forming a virtuous circle that makes us even more ethically discerning in technosocial contexts as a result of increasing moral practice in those domains. This growing moral expertise can enable the development of still better, more ethical, and more sustainable technologies. Used as alternating and mutually reinforcing handholds, this interweaving of moral and technological expertise is a practical and powerful strategy for cultivating technomoral selves: human beings with the virtues needed to flourish together in the 21st century and beyond.

The Motivation of the Book

I was driven to write this book by a deep moral concern for the future of human character, one that arose over many years of watching my own moral and intellectual habits, and those of my students, be gradually yet profoundly transformed by ever new waves of emerging digital technology. Far from regarding my initial classroom forays into this topic as silly technophobia, my students responded with overwhelming gratitude, even desperation, for a chance to talk about how their own happiness, health, security, and moral character were being shaped by their new technological habits in ways that often bypassed their understanding or conscious choice.

These concerns will be familiar to readers of popular writing on digital culture. Nicholas Carr, Evgeny Morozov, and Jaron Lanier are just a few of the prominent cultural critics who have recently expressed alarm at the possibility, even likelihood, that our mediatized digital culture is undermining core human values, capacities, and virtues. Carr's *The Shallows* warns us of deleterious cognitive and moral effects that our new digital consumption habits may be having on our brains. Morozov's *The Net Delusion* and *To Save Everything, Click Here* challenge our unreflective faith in technocratic 'solutionism.' From Lanier, a computer scientist and pioneer innovator of virtual reality technology, came the widely read humanistic manifesto *You Are Not a Gadget*, which laments the domination of contemporary technosocial life by the increasingly libertarian and antihumanistic values celebrated by many Silicon Valley technologists: unrestrained capitalism, consumerism, and reductive efficiency.

This book shares with these critics a deeply humanistic and explicitly moralized conception of value. It assumes that the 'good life,' by which we mean a human future worth seeking, choosing, building, and enjoying, must be a life lived *by* and *with* persons who have cultivated some degree of ethical character. It assumes that this is the *only* kind of human life that is truly worth choosing, despite the perpetual challenges we encounter in building and sustaining such lives. It also holds that a good and choiceworthy life has never been attained in any great measure by isolated individuals, but only by persons who were fortunate enough to enjoy some degree of care, cooperation, and support from other humans, and who were highly motivated to give the same. This book is therefore fundamentally inconsistent with antihumanistic and neoliberal philosophies, and if Lanier is right, inconsistent with the philosophy of many of those driving the emerging technological developments it proposes to examine.

Yet the reader will also find in this book a resolute hope for the future of human flourishing *with*, not without or in spite of, the technosocial innovations that will continue to shape and enrich our lives for as long as human culture endures. As a scholar who chose out of all possible specialties the philosophy of science and technology, who as a young girl wrote adventure games in BASIC for her Commodore PET and eschewed the Barbie Dream'Vette in favor of Star Wars AT-AT and X-Wing toys, it is simply impossible for me to be antitechnology, personally or philosophically. Indeed, to be antitechnology is in some sense to be antihuman, for we are what we do and make, and humans have always engineered our worlds as mirrors of our distinctive needs, desires, values, and beliefs. Of course we are not alone—increasingly, researchers find other intelligent animals such as birds, elephants, and cephalopods reshaping their environments and practices in surprisingly skillful and creative ways. Perhaps to be antitechnology is also to be antilife, or antisentience. But however widely we share this part of ourselves with other creatures, humanity without technology is not a desirable proposition—it is not even a meaningful one. The only meaningful questions are: *which* technologies shall we create, with what knowledge and designs, affording what, shared with whom, for whose benefit, and *to what greater ends?* These are the larger questions driving this book. Yet humans lacking the technomoral habits and virtues described within its pages could, I think, never hope to answer them. Let us not surrender that hope.

PART I

Foundations for a Technomoral Virtue Ethic

1

Virtue Ethics, Technology, and Human Flourishing

TO SOME, THE title of this chapter may seem faintly anachronistic. In popular moral discourse, the term 'virtue' often retains lingering connotations of Victorian sexual mores, or other historical associations with religious conceptions of morality that focus narrowly on ideals of piety, obedience, and chastity. Outside of the moral context, contemporary use of the term 'virtue' expresses something roughly synonymous with 'advantage' (e.g., "the virtue of this engineering approach is that it more effectively limits cost overruns"). Neither use captures the special meaning of 'virtue' in the context of philosophical ethics. So what does virtue mean in this context? How does it relate to cultivating moral character? And why should virtue, a concept rooted in philosophical theories of the good life dating back to the 5th century BCE, occupy the central place in a book about how 21st century humans can seek to live well with emerging technologies?

The term 'virtue' has its etymological roots in the Latin *virtus,* itself linked to the ancient Greek term *arête,* meaning 'excellence.' In its broadest sense, the Greek concept of virtue refers to any stable trait that allows its possessor to excel in fulfilling its distinctive function: for example, a primary virtue of a knife would be the sharpness that enables it to cut well. Yet philosophical discussions of ethics by Plato and Aristotle in the 5th and 4th centuries BCE reveal a growing theoretical concern with distinctly human forms of *arête,* and here the concept acquires an explicitly *moral* sense entailing excellence of character.[1] A distinct but related term *de* (德) appears in classical Chinese ethics from approximately the same period. *De* originally meant a characteristic 'power' or influence, but in Confucian thought it acquired the sense of a distinctly *ethical* power of the exemplary person, one that fosters 'uprightness' or 'right-seeing.'[2] Buddhist ethics makes use of a comparable concept, *śīla,* implying character that coordinates and upholds right conduct.[3] The perfection of moral character (*śīla pāramitā*) in Buddhism expresses a sense of cultivated personal excellence akin to other ethical conceptions of virtue. Thus the concept of 'virtue' as a descriptor of moral excellence has for millennia occupied a central place in various *normative* theories of human action—that is, theories that aim to prescribe certain kinds of human action as right or good.

In the Western philosophical tradition, the most influential account of virtue is Aristotle's, articulated most fully in his *Nicomachean Ethics* (~350 BCE). Other notable accounts of virtue in the West include those of the Stoics, St. Thomas Aquinas, David Hume, Francis Hutcheson, and Friedrich Nietzsche. Yet Aristotle remains the dominant influence on the conceptual profile of virtue most commonly engaged by contemporary ethicists, and this profile will be our starting point. While cultural and philosophical limitations of the Aristotelian model will lead us to extend and modify this profile in subsequent chapters, its basic practical commitments will remain largely intact.

Moral virtues are understood by Aristotle to be states of a person's *character*: stable dispositions such as honesty, courage, moderation, and patience that promote their possessor's reliable performance of right or excellent actions. Such actions, when the result of genuine virtue, are not only praiseworthy in themselves but imply the praiseworthiness of the person performing them. In human beings, genuine virtues of character are not gifts of birth or passive circumstance, nor can they be taught in any simple sense. They are states that the person must cultivate in herself, and that once cultivated, lead to deliberate, effective, and reasoned choices of the good.[4] The virtuous state emerges gradually from habitual and committed practice and study of right actions. Thus one builds the virtue of courage only by repeatedly performing courageous acts; first by patterning one's behavior after exemplary social models of human courage, and later by activating one's acquired ability to see for oneself what courage calls for in a given situation. Virtue implies an alignment of the agent's feelings, beliefs, desires, and perceptions in ways that are appropriate to the varied practical arenas and circumstances in which the person is called to act.[5] Moral virtues are conceived as personal excellences in their own right; their value is therefore not exhausted in the good actions or consequences they promote. When properly integrated, individual virtues contribute to a person's character *writ large*; that is, they motivate us to describe such a person as *virtuous,* rather than merely noting their embodiment of a particular virtue such as courage, honesty, or justice. States of character contrary to virtue are *vices*, and a person whose character is dominated by these traits is therefore *vicious*—broadly incapable of living well.

Most understandings of virtue ethics make room for something like what Aristotle called *phronēsis,* variously translated as prudence, prudential reason, or practical wisdom.[6] This virtue directs, modulates, and integrates the enactment of a person's individual moral virtues, adjusting their habitual expression to the unique moral demands of each situation. A fully virtuous person, then, is never blindly or reactively courageous or benevolent—rather, her virtues are expressed *intelligently,* in a manner that is both harmonious with her overall character and appropriate to the concrete situation with which she is confronted.[7] Virtues enable their possessor to strike the mean between an excessive and a deficient response, which varies by circumstance. The honest person is not the one who mindlessly spills everyone's secrets, but the one who knows *how much* truth it is right to tell, and *when* and *where* to tell it, to *whom*, and in what *manner*.[8] Reasoning is therefore central to virtue ethics. Yet unlike theories of morality that hinge on rationality alone, such as Kant's, here reason must work *with* rather than against or independently of the agent's habits, emotions, and desires. The virtuous person not only tends to *think* and *act* rightly, but also to *feel* and *want* rightly.[9]

A virtuous person is not merely conceived *as* good, they are also understood to be moving toward the accomplishment of a good *life*; that is, they are *living well*. In most cases, they enjoy a life of the sort that others recognize as admirable, desirable, and worthy of being chosen. Of course not every life that *appears* desirable or admirable is, in fact, so. Conversely, a virtuous person with the misfortune to live among the vicious is unlikely to be widely admired, although this does nothing to diminish the fact of their living well. The active flourishing of the virtuous person is not a subjective appearance; virtue just *is* the activity of living well. This means that while virtue ethics can allow for many different types of flourishing lives, it is incompatible with moral relativism. There are certain biological, psychological, and social facts about human persons that constrain what it can mean for us to flourish, just as a nutrient-starved, drought-parched lawn fails to flourish whether or not anyone notices its poor condition. While the cultivation of virtue is not egoistic, since it does not aim at securing the agent's own good *independently* of the good of others, a virtuous character is conceptually inseparable from the possibility of a good life for the agent.[10] This is why Aristotle describes the virtuous person as objectively *happy*; even in misfortune they will retain more of their happiness than the vicious would.[11] Although it is widely recognized that the Greek term *eudaimonia*, which we translate as 'happiness,' is far richer than the modern, psychological sense of that term (an issue to which we will return later in this book), it will serve our preliminary analysis well to note that the classical virtue ethical tradition regards virtue as a *necessary,* if not sufficient, condition for living well and happily.[12]

If thinkers in this tradition are correct, then *just as in every previous human era, living well in the 21st century will demand the successful cultivation of moral virtue*. Yet given what was noted at the beginning of this chapter—namely, that the popular understanding of virtue is largely divorced from the philosophical teachings of virtue traditions—we have to ask: how can we possibly reconnect popular ideas about living well with technology to a robust discourse about the moral virtues actually needed to achieve that end? While a satisfactory answer to this question cannot be given until later in this book, it may be helpful to briefly examine the circumstances that have led to the revival of contemporary *philosophical* discourse about the moral virtues and their role in the good life.

1.1 The Contemporary Renewal of Virtue Ethics

Ethical theories in which the concept of virtue plays an essential and central role are collectively known as theories of *virtue ethics*. Such theories treat virtue and character as more fundamental to ethics than moral rules or principles. Advocates of other types of ethical theory generally see virtues as playing a lesser and more derivative role in morality; these include the two approaches that previously dominated philosophical ethics in the modern West: *consequentialist* ethics (for example, Jeremy Bentham and John Stuart Mill's utilitarianism) and *deontological* or rule-based ethics (such as Immanuel Kant's categorical imperative).[13]

Compared with these alternatives, virtue ethics stood in general disfavor in the West for much of the 19th and 20th centuries. Reasons for the relative neglect of virtue ethics in this period include its historical roots in tightly knit, premodern societies, which appeared to make the approach incompatible with Enlightenment ideals of modern cosmopolitanism. Thanks to the medieval philosopher St. Thomas Aquinas's use of Aristotelian ideas throughout his writings, virtue ethics had also acquired strong associations with the Thomistic moral theology of the Catholic Church. This made it an even less obvious candidate for a universal and secular ethic. Virtue ethics was seen as incompatible with evolutionary science, which denied what Aristotle and many other virtue ethicists had assumed—that human lives are naturally guided toward a *telos*, a single fixed goal or final purpose. Virtue ethics' emphasis on habit and emotion was also seen as undermining rationality and moral objectivity; its focus on moral persons rather than moral acts was often conflated with egoism. Finally, virtue ethics' eschewing of universal and fixed moral rules was thought by some to render it incapable of issuing reliable moral guidance.[14]

The contemporary reversal of the fortunes of virtue ethics began with the publication of G.E.M. Anscombe's 1958 essay "Modern Moral Philosophy," in which she sharply criticized modern deontological and utilitarian frameworks for their narrow preoccupations with law, duty, obligation, and right to the exclusion of considerations of character, human flourishing, and the *good*. Anscombe also claimed that modern moral theories of right and wrong, having detached themselves from their conceptual origins in religious law, were now crippled by vacuity or incoherence, supplying poor foundations for secular ethics. Her proposal that moral philosophers abandon such theories and revisit the conceptual foundations of virtue was the guiding inspiration for a new generation of thinkers, whose diverse works have restored the philosophical reputation of virtue ethics as a serious competitor to Kantianism, utilitarianism, and other rule- or principle-based theories of morality.[15] Among Western philosophers, scholarly interest in virtue ethics continues to grow today thanks to the prominent work of neo-Aristotelian thinkers such as Alasdair MacIntyre, John McDowell, Martha Nussbaum, Rosalind Hursthouse, and Julia Annas, to name just a few.

Yet Anscombe was clear that Aristotelian virtue theory was not a satisfactory modern ethic. Even contemporary virtue ethicists who identify as neo-Aristotelian typically disavow one or more of Aristotle's theoretical commitments, such as his view of human nature as having a natural *telos* or purpose, or his claims about the biological and moral inferiority of women and non-Greeks. No contemporary virtue ethicist can deny that there are significant problems, ambiguities and lacunae in Aristotle's account; whether these can be amended, clarified, and filled in without destroying the integrity or contemporary value of his framework is a matter of ongoing discussion. As a consequence, contemporary Western virtue ethics represents not a single theoretical framework but a diverse range of approaches. Many remain neo-Aristotelian, while others are Thomistic, Stoic, Nietzschean, or Humean in inspiration, and some offer radically new theoretical foundations for moral virtue.[16]

In addition to internal disagreements, the contemporary renewal of virtue ethics has met with external resistance from critics who challenge the moral psychology of character upon which virtue theories rely. Using evidence from familiar studies such as the Milgram and Stanford prison experiments, along with more recent variations, these critics argue that moral behavior is determined not by stable character traits of individuals, but by the concrete situations in which moral agents find themselves.[17] Fortunately, virtue ethicists have been able to respond to this 'situationist' challenge. First, the impact of unconscious situational influences, blind spots, and cognitive biases on moral behavior is entirely compatible with virtue ethics, which already regards human moral judgments as imperfect and contextually variable. Moreover, unconscious biases can, once discovered, be mitigated by a range of compensating moral and social techniques.[18] Perhaps the most powerful response to the situationists is that robust moral virtue is by definition exemplary rather than typical; indeed, the experiments most often used as evidence *against* the existence of virtue consistently reveal substantial minorities of subjects who respond with exemplary moral resistance to situational pressure—*exactly* what virtue ethics predicts.[19] Thus despite its critics, the contemporary renewal of virtue ethics as a compelling alternative to principle- and rule-based ethics shows no sign of losing steam; if anything, intensified critical scrutiny is a healthy indicator of virtue ethics' returning philosophical strength.

While a survey of contemporary virtue ethics in the West might stop here, it would be dangerously provincial and chauvinistic to ignore the equally rich virtue ethical traditions of East and Southeast Asia, especially Confucian and Buddhist virtue ethics. While there is important contemporary work being done in this area, relatively few Anglo-American virtue ethicists have acknowledged or attempted to engage this work.[20] This is a substantial loss. To ignore the content of active and longstanding virtue traditions with related, but very distinct, conceptions of human flourishing is to forgo an opportunity to gain a deeper critical perspective on the admittedly narrow preoccupations of Aristotelian virtue theory.[21]

As we move beyond the realm of theory and into the domain of *applied* virtue ethics, Western provincialism becomes entirely unsustainable; for applied ethics—which tackles real-world moral problems through the lens of philosophy—is increasingly confronted with problems of global and collective action. Environmental ethics offers the starkest selection of practical problems demanding global cooperation and coordinated human responses that reach across national, philosophical, and ethnic lines, but this is hardly an isolated case. The expansion of global markets for new technologies is having profound and systemic moral impacts on the entire human community—primarily by strengthening the shared economic, cultural, and physical networks upon which our existence and flourishing increasingly depend. If we look at the spread of global information and communications systems, unmanned weapons systems, consumer robotics, or genetic engineering, we see that emerging technologies and their effects do not respect the cultural and philosophical boundaries that separate capitalists from socialists, Buddhists from Christians, or neo-Confucians from neo-Aristotelians.

It may be, for example, that liberal Europeans value personal privacy in a manner that is, on theoretical grounds, quite distinct from the way in which privacy values are framed in traditional Chinese society, where boundaries between self and community are far less sharp.[22] Yet if Google aims to connect us all, and we want Google to act ethically with respect to our privacy concerns, then it cannot *in practical terms* be true that these distinct privacy values have nothing to do with one another—as illustrated by the cultural complexities of the dispute over the European Court of Justice's 2014 decision that EU citizens have a fundamental human "right to be forgotten" by search engines. Likewise, applied ethicists must increasingly attend to cultural differences between Japanese, South Korean, and American attitudes toward robots precisely because of emerging market incentives to develop robotic solutions to various demographic and military challenges these nations share.[23] And if international tribunals of justice and human rights are to have any continuing legitimacy, they will have to find ways of successfully framing the ethical stakes of technologies such as autonomous weapons that effect significant change upon *global* habits of military and political practice. In the midst of such developments, ethical discourse that speaks only to the concerns of a particular moral or philosophical 'tribe' will be helpless to confront the ethical impact of the technosocial realities that increasingly address humans *collectively*—realities which demand the effective cultivation and application of some measure of cooperative human wisdom.

1.2 Virtue Ethics and Philosophy of Technology

In the Introduction, we identified the emerging conditions of acute technosocial opacity: rapid technological, sociopolitical, and environmental change accompanied by existential risks that make the ethical pursuit of the good life in the 21st century extraordinarily challenging and fraught with uncertainty. I suggested that our current patterns of thinking about ethics and the good life may well prove ineffective, deleterious, or even catastrophic if we do not adapt them to these new technosocial realities. I also claimed that new moral resources for meeting this challenge with grace and wisdom can be found in the philosophical tradition of *virtue ethics*. Having now outlined the general significance of the term 'virtue ethics' in classical and contemporary philosophy, and the cultural scope of our interest in virtue ethical frameworks, let us turn to the important task of explaining more fully why this type of ethical approach, more so than the obvious alternatives, offers a uniquely helpful scaffold for a new technosocial ethics.

Let me begin by explaining further why we should be dissatisfied with the primary alternatives. Each is vulnerable to several well-known objections. Religious laws and norms speak only to their believers, and thus are poor candidates for a global technosocial ethic. Utilitarianism, which employs a universal moral calculus designed to maximize the greatest good for all concerned, is often criticized for decoupling the moral worth of acts from the moral worth of persons, and for legitimizing choices in which the lives or well-being of a minority may be deliberately sacrificed for a greater overall yield of happiness. Kantian deontology, which supplies a single *categorical imperative* mandating universal rational consistency in moral action, is criticized for treating the rational consistency of the agent as more important to morality than natural human bonds of care and concern.[24]

Kantian and utilitarian ethics have also been thought to be overly demanding, for example, in asking agents to be impartial in weighing the competing interests of strangers and loved ones. Such considerations lead many to conclude that these accounts stray too far from common moral intuitions.[25] Still, the above criticisms can be difficult to prosecute without begging the question in favor of a virtue- or character-based account. Fortunately, there is a more fundamental and compelling reason to prefer virtue ethics, one made increasingly relevant by the conditions of growing technosocial opacity to which 21st century humans are subject.

A well-known weakness of virtue ethics' competitors is their adherence to rules or fixed principles as the final arbiters of sound moral judgment. Consider Kant's categorical imperative and its famous blanket injunction against lying, even to the "inquiring murderer" at our door who wishes to know if we are sheltering the innocent person he wishes to kill. Or recall the equally familiar challenge to utilitarian morality in which it is noted that the principle of greatest happiness can seemingly be used to justify the brutal execution of a randomly chosen scapegoat. When such apparent counterexamples cannot be explained away, defenders of rule or principle-based ethics are generally left with two options: admit limited exceptions to the moral principle, or accept and try to defend these morally counterintuitive implications. By rendering the theory unpalatable or offensive to many if not most reasonable people, the latter strategy preserves the integrity of the theory at the expense of its motivating force. The former strategy, however, tends to produce *ad hoc* and apparently arbitrary exceptions, undermining confidence in the principle upon which the whole theory rests.

Virtue ethicists, on the other hand, avoid this dilemma by denying that right action is ever captured by fixed principles or rules, and claiming instead that moral principles simply codify, in very general and defeasible ways, patterns of reasoning typically exhibited by virtuous persons. On this view, moral expertise does not come from fixed moral principles, but is *reflected* in them; and imperfectly at that. Consider an adult who can easily understand and follow rules, but is wholly lacking in practical wisdom—a cultivated ability of her own to recognize and independently solve moral problems. There is no moral rule or set of rules that one could hand such a person that would, from then on, steer her reliably and safely to the destination of a life lived well. If the rules are well chosen, she *might* do morally better in life with them than she would without them, but inevitably she will fail to recognize situations to which the rules are supposed to apply, and be caught short by moral situations for which the rules do not seem to fit or where their proper application is not obvious. It is not simply that virtue—that is, excellence in practical reasoning—is needed to correctly apply abstract moral principles such as those of Kantian or utilitarian ethics.[26] For the virtue of practical wisdom or *phronēsis* encompasses considerations of universal rationality *as well as* considerations of an irreducibly contextual, embodied, relational, and emotional nature—considerations that Kant and others have erroneously regarded as irrelevant to morality. From the totality of these considerations the virtuous person must make moral sense of each concrete situation encountered, and give an appropriate response. A successful moral response is distinguishable from a failed or inappropriate response *in practice,* and the reasons behind the success of that response can always be articulated after the fact. But the difference between moral success and moral failure can rarely, if ever, be deduced in advance from *a priori* principles.

Who, for example, could have deduced *in advance* from universal moral principles that on December 20, 1943, in the midst of the ferocity and desperation of a World War II battle over Bremen, Lt. Franz Stigler, a Luftwaffe ace fighter pilot, should choose to spare the life of Charles Brown, the American pilot of a crippled B-17, and his helpless, injured crew, by escorting them safely out of German airspace?[27] Other morally salient considerations and principles that spoke against such an act of mercy, such as loyalty and duty, might have bound Stigler, and did at many other moments in the war. This unexpected act, the virtue of which is nevertheless easy to explain in hindsight, expressed the total and singular moral sense he was able to make of the complex demands of *that* worldly moment confronting him.

Not all moral acts appear so singular or unexpected in their general form; yet virtue is expressed even in the distinctive *way* one moral person chooses to do things that every moral person must: deliver a painful but necessary truth, express one's deep gratitude, or forgive a petty slight. To fully appreciate this, we need a deeper explanation of the virtue called *practical wisdom.* Practical wisdom is the kind of excellence we find in moral experts, persons whose moral lives are guided by appropriate feeling and intelligence, rather than mindless habit or rote compulsion to follow fixed moral scripts provided by religious, political, or cultural institutions. As noted by the moral philosopher Kongzi (commonly known by his Latinized name, Confucius), the acts of a virtuous person are made noble not simply by their correct content—though that content *will* typically respect important moral conventions—but by the singular and authentic moral *style* in which that person chooses to express their virtue. It is this aesthetic mode of personally expressing a moral convention, rule, or script that embodies and presents one's virtue. The person who enacts fixed moral rules 'correctly' but rigidly—without style, feeling, thought, or flexibility—is, on this view, a shallow parody of virtue, what the Confucian tradition refers to derisively as the "village honest man."[28]

Even reliably pro-social habits such as following laws and telling the truth fail to guarantee virtue. For while the virtuous person will certainly have such habits, moral intelligence is required to ensure that these habits do not produce acts that violate the moral sense of the situation—for example, mindless obedience to a lawful but profoundly immoral and indefensible order. Actions issuing from the moral habits of a virtuous person—that is, a person with practical wisdom—are properly attuned to the unique and changing demands of each concrete moral situation. In contrast, a person who is prone to thoughtless and unmodulated action is likely to go wrong as often as not. Thus moral virtue presupposes knowledge or understanding. Yet unlike theoretical knowledge, the kind of knowledge required for moral virtue is not satisfied by a grasp of universal principles, but requires recognition of the relevant and operative practical conditions.

Moral expertise thus entails a kind of knowledge extending well beyond a cognitive grasp of rules and principles to include emotional and social intelligence: keen awareness of the motivations, feelings, beliefs, and desires of others; a sensitivity to the morally salient features of particular situations; and a creative knack for devising appropriate practical responses to those situations, especially where they involve novel or dynamically unstable circumstances. For example, the famous 'doctrine of the mean' embedded in classical virtue theories entails that the morally wise agent has a quasi-perceptual ability to *see* how an emergent moral situation requires a spontaneous and often unprecedented realignment of conventional moral behaviors.[29] Even if it is, as a rule, morally wrong to touch naked strangers without their consent, I had better not hesitate to give CPR to my neighbor's naked, unconscious body—if the situation calls for it, which depends on many factors, such as whether I have adequate training. Nor can the full content of all such tacit and embodied moral knowledge ever be captured in explicit and fixed decision procedures.[30] As Aristotle took pains to note, matters of practical ethics by nature "exhibit much variety and fluctuation," requiring a distinctive kind of reasoning that displays an understanding of changing particulars as well as fixed universals.[31] On Aristotle's view, while it is true that rational principles are part of ethics, it is the virtue of practical wisdom that *establishes* the correct moral principle for a given case, rather than wisdom being defined *by* its correspondence with a prior principle.[32]

Now we are in a position to understand why, if our aim is to learn how to live well with emerging technologies, a virtue ethics approach will generally be more useful than one that relies upon consequentialist or deontological principles. If the practical conditions of ethical life in the 4th century BCE already displayed too much variety and flux for us to rely upon a principle-based ethics, requiring instead an account that articulates the specific virtues of persons who judge wisely and well under dynamic conditions, then the practical uncertainties and cultural instabilities produced by emerging technologies of the 21st century would seem to make the contemporary case for virtue ethics that much stronger.[33]

A key phenomenon accelerating the acute technosocial opacity that defines our age is that of *technological convergence*: discrete technologies merging synergistically in ways that greatly magnify their scope and power to alter lives and institutions, while also amplifying the complexity and unpredictability of technosocial change. The technologies most commonly identified as convergent are the fields of applied technoscience referred to as 'NBIC' technologies: nanotechnology, biotechnology, information technology and cognitive science.[34] The specific ethical challenges and opportunities presented by many of these technologies are discussed in Part III of this book. Yet consider just briefly the impact of their convergence on the emerging markets for brain implants, cybernetic prosthetics, replacement organs, lab-grown meat, 'smart' drugs, 'lie-detecting' or 'mind-reading' brain scanners, and artificially intelligent robots—*and* the panoply of new ethical dilemmas already being generated by these innovations. Now ask which practical strategy is more likely to serve humans best in dealing with these unprecedented moral questions: a stronger commitment to adhere strictly to fixed rules and moral principles (whether Kantian or utilitarian)? Or stronger and more widely cultivated habits of moral virtue, guided by excellence in practical and context-adaptive moral reasoning? I hope I have given the reader cause to entertain the latter conclusion, which should become increasingly plausible as this book continues.

We can already enhance the plausibility of this claim by noticing an emerging asymmetry between the moral dilemmas presented by today's converging technologies and the topics that still dominate most applied ethics courses and their textbooks. These textbooks have sections devoted to weighing the ethics of abortion, capital punishment, torture, eating meat, and so on. In each case, it seems reasonable to frame the relevant moral question as, "Is *x* (where *x* is an act or practice from the above list) right or wrong?" Of course, these questions may or may not have definitive answers, and to get an answer one may need to specify the conditions under which the act is being considered—for example, whether alternative sources of nutrition are available to the meat eater. Still, such questions make sense, and we can see how applying various moral principles might lead a person to concrete answers. Compare this with the following question: "Is Twitter right or wrong?" Or: "Are social robots right or wrong?" There is something plainly ill-formed about such questions.

At this point, the reader will likely object that we are asking about the rightness or wrongness of technologies rather than of acts, and that this is the primary source of our confusion. But notice that it does not actually help things to reform our questions in action-terms such as, "Is *tweeting* wrong?" or "Is it wrong to *develop* a social robot?" The asymmetry is of a different nature. It is not even that one set of problems involves technology and another does not; after all, technology is heavily implicated in modern practices of abortion and capital punishment. The problem is that *emerging* technologies like social networking software, social robotics, global surveillance networks, and biomedical human enhancement are not yet sufficiently developed to be assignable to specific practices with clear consequences for definite stakeholders. They present open developmental *possibilities* for human culture as a whole, rather than fixed options from which to choose. The kind of deliberation they require, then, is entirely different from the kind of deliberation involved in the former set of problems.

Of course, the line is not a bright one, as emerging technologies also impact long-standing practices where fixed moral principles retain considerable normative force, such as data privacy and copyright protection. Yet it remains the case that very often, the answers for which questions about emerging technology beg are simply *not* of the 'yes/no' or 'right/wrong' sort. Instead, they are questions of this sort: 'How might interacting with social robots help, hurt, or change us?'; 'What can tweeting do to, or for, our capacities to enjoy and benefit from information and discourse?'; 'What would count as a 'better,' 'enhanced' human being?' It should be clear to the reader by now that these questions invite answers that address the nature of human flourishing, character, and excellence—*precisely the subject matter of virtue ethics.*

That said, philosophical recognition of virtue ethics as an appropriate, even ideal framework for thinking about ethics and technology has emerged only recently.[35] While reflections on technology and the good life are found in Plato, Aristotle, Sir Francis Bacon, J.J. Rousseau, G.W. Hegel, Karl Marx, John Dewey, and José Ortega y Gasset, to name just a few, the 'philosophy of technology' came into focus only with the post-World War II writings of thinkers such as Martin Heidegger, Jacques Ellul, Herbert Marcuse, and Lewis Mumford.[36] Whether fairly or not, the latter group's works are often described as pessimistic, even fatalistic philosophies in which technology is a monolithic force that invariably undermines human freedom, condemning us to narrow lives of consumption, exploitation, and mindless efficiency.[37] Once we surrender to this thesis of 'technological determinism,' there would seem to be little left to say with regard to ethics, a study that presupposes our human freedom to choose a good life.

Yet by the end of the 20th century a second wave of philosophers, while still strongly influenced by the first, had found sufficient reason to hope for a robust ethics of technology. Foremost among them is Hans Jonas, who in the 1970s and 1980s developed the first explicit call for a new technosocial ethics. Though continuing to portray technology as a monolithic "colossus" whose advance threatens to bring about the "obsolescence of man," Jonas claimed that if we can somehow manage to survive the species-level threats to our existence posed by technology, we will need a new ethics of technology to confront the profound challenges that these emerging innovations pose to the human image, and to our conceptions of the good life.[38] Yet he reminds us that "we need wisdom most when we believe in it least"—in an era when the foundations of all moral theories have been challenged, when ethical relativism is a default stance for many, and when religious principles are no longer widely accepted as starting points for ethics, where are we to start looking for the technosocial wisdom we so desperately need?[39] This is why Jonas famously remarks that philosophy is "sadly unprepared for this, its first cosmic task."[40]

Jonas recognized that traditional moral systems and principles would offer little help—primarily because such systems were historically designed to resolve moral dilemmas that confront an *individual*, where the ethically relevant consequences of his or her action unfold within a foreseeable present shared by all other moral stakeholders.[41] On the contrary, as we have already indicated, emerging technologies 1) present new problems of *collective* moral action; 2) are likely to impact *future* persons, groups and systems as much or more so than present stakeholders; and 3) have unpredictable consequences that unfold on an open-ended time-horizon. The general happiness problem becomes incalculable for the utilitarian, and the Kantian imperative fails to capture the ethical stakes that lie beyond the "abstract compatibility" of my private choice with my rational will.[42] Though Jonas seems more favorably disposed to virtue ethics, speaking often of virtues like wisdom and humility, he claims that the traditional Aristotelian view falls short of what we need. This is for two reasons; first, because Aristotle's notion of what is good for a human life rests on "presumed invariables of man's nature and condition" (such as basic human biology) which, thanks to biomedical technologies like germline engineering, are no longer invariable. Second, Aristotle's account of practical wisdom involves a person's ability to perceive and respond to the morally salient features of his *own* local context—yet as Jonas knew, and as we have said, the choices made by 21st century humans will impact the flourishing of the entire species (and not *only* our species), up to and including future generations of life on this planet.[43] It is clear to Jonas, then, that what is needed is a genuinely *new* ethics.[44]

Unfortunately, Jonas's constructive proposal for a new ethics still remained closely wedded to the pessimistic tone and metaphysical essentialism of first-wave philosophers of technology. Like Heidegger and those who followed in his footsteps, Jonas framed technology as a monolithic threat to the 'essence of Man.'[45] Jonas's proposal limits us to thinking of a new technosocial ethic as a way to head off largely dystopian scenarios, using what Jonas calls the "heuristics of fear."[46] Those who see emerging technologies as presenting not only grave threats but also many constructive opportunities for humanity, and who think that a blanket fear of technology breeds more problems than it solves, will find Jonas's account deeply inhospitable to their view.[47]

Albert Borgmann is another second-wave philosopher of technology who touches on the question of a new technosocial ethics and the potential contributions of Aristotelian thought to this enterprise. Unlike Jonas, Borgmann's writings on technology do not directly address the problem of finding an appropriate theoretical scaffold upon which to hang a technosocial ethics. Instead, Borgmann's reflections on modern technology and the good life merge a broader interest in virtue and practical wisdom with a concern for human freedom in the face of technological compulsion.[48] Borgmann takes the 'pursuit of excellence'—understood as the cultivation of social, physical, and intellectual capacities or virtues—as a revealing measure of our connection to the good life.

Yet Borgmann retains a largely, though not entirely, pessimistic view of the relation between human morality and emerging technologies; by his account, the dominant thrust of modern technological life is to *subvert* the cultivation of moral excellence in favor of a life of mindless consumption—what he calls the 'device paradigm.'[49] He notes that we are not wholly powerless to resist this paradigm—nor must we abandon technology in order to do so. Instead, his proposal for effective reform of technological society asks us to recognize that the ultimate end for humans is a life of *engagement,* facilitated by any of a great number of practices, such as running or gourmet cooking, which not only cultivate personal excellences but can be used to honor "focal things" that center family and community life and give it meaning. Using this definition of the good life, which he takes to be Aristotelian in inspiration, Borgmann calls us to reconfigure our technologies so that they facilitate rather than impede a life of engagement.

While the general concept of an engaged life sounds promising as a candidate for 'a life lived well,' Borgmann's account is ambiguous and incomplete. How are we to determine what counts as facilitating engagement and what counts as its subversion? What *specific* excellences are most worth developing in focal practices? Are some forms of personal excellence, or some focal things and practices, more important to protect from technological subversion than others? More importantly for our purposes, how can a narrow focus on reform help us with those emerging technologies whose impact on our practices is not yet manifest? Furthermore, Borgmann's proposal for reform falls short of resolving the problems of collective moral action with which emerging technologies confront us. He does address the need for public discourse about the good life, as a means of politically negotiating and enacting meaningful technosocial reforms.[50] But within his account, such discourse seems to aim primarily at preserving opportunities for the *private* pursuit of engaged living, rather than the collective and long-term flourishing of the global human family. And while he notes that technosocial reform presupposes "shared and public" affirmations of certain types of civic behavior needed to facilitate public discourse, he is vague about the content of such affirmations, mentioning only briefly the value of social expectations of "politeness, sociability, or civility."[51] We need, at the very least, a more complete and explicit account of the specific virtues that will facilitate improved public discourse about technology and the good life. Ideally, we should also form an idea of how to encourage the global cultivation of these virtues, as a precondition for the emergence of collective, rather than strictly personal and private, technosocial wisdom.

While the most influential 20th century philosophers of technology each had a unique way of conceptualizing the ethical issues concerning technology, they *all* tended to describe an ethics of technology as a response to a singular problem or phenomenon, whether conceived as the human 'enframing' of reality as a resource to be manipulated (Martin Heidegger); the oppressive mandate of technological efficiency (Jacques Ellul); the spread of 'one-dimensional thought' enslaved to technopolitical interests (Herbert Marcuse); the relentless technological cycle of manufactured needs and desires (Jonas); or the suffocating rule of the 'device paradigm'(Borgmann). Notice also that each presents humans as somewhat passive subjects of the singular technological 'problem,' who must somehow reclaim their freedom with respect to technology. In contrast, the early 21st century has seen a fundamental shift in how philosophers approach the ethics of technology. A new generation of thinkers has consciously moved away from essentialist conceptions of technology *versus* humanity, and toward the analysis of individual technolog*ies* as features of specific human contexts; away from globalizing characterizations of the 'problem with technology' to more neutral and localized descriptions of a diverse multitude of ethical issues raised by particular technologies and their uses; and finally, away from metaphysical concerns about human freedom and essence and toward more empirically-grounded accounts of human-technology relations.[52]

This shift, commonly referred to as "the empirical turn," has done much to revitalize the philosophy of technology and to expand the scope of inquiry into technology's ethical implications. Articles in scholarly journals such as *Ethics and Information Technology* now employ a variety of empirical and philosophical approaches to explore a wide range of localized technosocial phenomena, without adhering to any one metanarrative about technology and its relation to the good life. Largely thanks to the critical force of the empirical turn, the study of ethics and technology has at last developed a range of theoretical perspectives and commitments sufficiently diverse to constitute a genuinely philosophical *field*. Yet this turn, as productive and welcome as it has been, has its price. The idea that contemporary technosocial life might require a new ethical framework has not been wholly abandoned; but it is increasingly challenged by the sheer diversity and dynamics of the technosocial phenomena being uncovered.[53] This difficulty is further magnified by the lingering critical reaction against totalizing, essentialist narratives, with which the call for a new ethics of technology remains associated. Yet unless we are willing to confront the broader significance of contemporary technosocial relations, the ethics of technology risks being reduced to an endless catalogue of seemingly unrelated phenomena.[54]

Once again, this dilemma reinforces the viability of virtue ethics as a framework for a new ethics of technology. For as we have noted, it is the only theoretical structure designed for maximum flexibility to accommodate the diverse and changing particulars of ethical life. On many points we must concede to the critics of traditional ethical theories, including Aristotelian virtue ethics, yet this does not exclude virtue ethics as a suitable framework. Rather, our new ethical horizon of collective human action in the face of growing technosocial opacity invites us to consider the possibility of the global cultivation of virtues newly adapted to our present need. Aristotelian, Confucian, and Buddhist virtue ethics each articulated qualities of individuals embodying moral wisdom in their own times, regions, or cultures. Does it not then follow that a contemporary virtue ethic could describe characteristic qualities of those who live wisely and well in the globally networked environment of the 21st century? Granted, this cohort is of a size and diversity unprecedented in the history of moral communities. Yet it is a group that shares a need to make collective and morally wise technosocial decisions, even if attuned to more open-ended and culturally complex horizons than the sages of the classical world could have envisioned. But perhaps the reader thinks it implausible that such persons or qualities exist. *Are* there habits of character ideally suited to our new technosocial circumstances?

Certainly it seems there are people who are especially *ill*-equipped to flourish in such circumstances. Consider the person who: 1) is characteristically incapable of empathizing with or giving moral consideration to others beyond their local circle; 2) is unable to communicate or deliberate well with others, especially with those holding different metaphysical and value commitments; 3) reasons unusually poorly in circumstances involving great uncertainty and risk; and 4) has concerns for the good life that rarely if ever extend beyond maximizing gains in the present and immediately foreseeable future. If we can recognize such a character profile in people we know (and I think most readers will), then it seems plausible to say that a person who fits this description would, in the context of any new *technosocial* ethics, be describable as generally vicious.[55] If we can identify technosocial vice, then it seems entirely plausible that with some effort we can identify and further articulate the nature of technosocial *virtue*. This is the effort that culminates in chapter 6 of this book.

Virtue ethics is a uniquely attractive candidate for framing many of the broader normative implications of emerging technologies in a way that can motivate constructive proposals for improving technosocial systems and human participation in them. It also allows us to avoid the perils of essentialism, overgeneralization, and abstraction that weakened first- and second-wave attempts to frame an ethics of technology. Virtue ethics is ideally suited for adaptation to the open-ended and varied encounters with particular technologies that will shape the human condition in this and coming centuries. Virtue ethical traditions privilege the spontaneity and flexibility of practical wisdom over rigid adherence to fixed rules of conduct—a great advantage for those confronting complex, novel, and constantly evolving moral challenges such as those generated by the disruptive effects of new technologies. Moreover, a turn to virtue ethics doesn't mean that we need to abandon the fruits of the 'empirical turn' in philosophy of technology; in fact, the empirical turn can *feed* practical wisdom by opening up ever more avenues for gathering vital contextual information about new and emerging technosocial realities.

Furthermore, as we will explore in chapter 2, virtue ethics treats persons not as atomistic individuals confronting narrowly circumscribed choices, but as beings whose actions are always informed by a particular social context of concrete roles, relationships, and responsibilities to others. This approach allows us to expand traditional understandings of the ways in which our moral choices and obligations bind and connect us to one another, as the networks of relationships upon which we depend for our personal and collective flourishing continue to grow in scale and complexity. Finally, while visions of the good life and of right action vary considerably across virtue traditions, we will see in Part II that the basic structure and tools of moral self-cultivation they employ manifest a great deal of cross-cultural similarity and conceptual overlap. A technosocial virtue ethic thus has the potential to resonate widely with a culturally diverse and global network of persons confronting the need for collective moral wisdom in dealing with the uncertainties, risks, and opportunities generated by new technologies.

Similar considerations have already guided virtue theory's adoption as a promising model for other areas of applied ethics, including environmental ethics, bioethics, media ethics, and business ethics.[56] It is time to add technosocial ethics to this growing list. Yet to consider our project here as simply outlining one more region of applied virtue ethics would be too modest. As we have seen, emerging science and technology now condition virtually *every* aspect of life. Every area of applied ethics, from bioethics to media ethics to business ethics to environmental ethics, now operates within a technosocial context of global scope. This context reaches with ever more intimacy and pervasiveness into our homes, cars, schools, cafes, sports fields, parks, and libraries. Arguably, a technosocial virtue ethic is an applied ethic of contemporary life *writ large*.

Yet before such an ethic can be developed and used, we must establish the theoretical foundations upon which it will stand. This foundation will become clear in Part II, where we see how core concepts and practices of classical virtue ethics can be remolded into a powerful framework for living well in the 21st century and beyond. Yet even before this, we must verify that the project of framing a *global technosocial virtue ethic* is philosophically coherent and practically sound. To do this, it is important to establish that the framework we aim to build can withstand the significant tensions that arise from its historically and culturally diverse origins. This task is our focus in chapter 2.

2

The Case for a Global Technomoral Virtue Ethic

IN CHAPTER 1 WE saw that *virtue ethics* offers the most promising framework for living wisely and well with emerging technologies. In Part II, we will find rich resources for contemporary technomoral life in the classical virtue traditions of Aristotelian, Confucian, and Buddhist ethics. Yet any attempt to develop a contemporary virtue ethic by drawing upon these diverse traditions must address their individual integrity, for each is rooted in cultural and historical understandings of the good life that are irreducibly distinct from those of other moral traditions. Even Confucian and Buddhist practices, which have historically intermingled in many East Asian cultures (while also interacting with other systems of thought such as Daoism, Shintoism, Christianity, and Islam) reflect very different perspectives on virtue and the good life. Likewise, while many features of the classical Greek vision of human excellence have been adapted to Stoic, Christian, and Islamic thought as well as to modern secular liberalism, the vision of the good life that Aristotle expressed in the 4th century BCE remains incommensurable in many respects with its later adaptations.[1] Thus before we go any further, we need to know how a conceptually coherent framework for contemporary technomoral life could possibly emerge from these diverse classical sources.

Moreover, there is good reason to think that certain commitments of these classical traditions are simply *wrong*, not just because they fail to speak to other cultures or to contemporary ways of life but because they are based on objectively false claims about the subject of ethics (i.e., people). Classical views about the intellectual and moral capacities of women, slaves, and 'barbarians' are oft-discussed examples of such factual errors.[2] Or consider Aristotle's inattention to the fundamental biological vulnerability and social dependence of the human animal.[3] Since it took time and distance to reveal the blind spots and weaknesses in traditional virtue ethics that are plain to us now, we cannot assume that more won't reveal themselves. So *even if* the resources we intend to draw from classical virtue traditions can be made compatible and relevant to contemporary forms of life, we will still have to ask what degree of confidence in the action-guiding content of these traditions is warranted.

On what grounds, then, does it make sense to pursue anything like the goal of this book, namely, a global virtue ethic for contemporary technosocial life that draws upon the resources of these culturally diverse classical traditions? To answer that question we must take a closer look at the traditions themselves, to see what conceptual unity licenses us to synthesize their resources in service of a global virtue ethic for contemporary technosocial life. Once we have established that this project is conceptually coherent, we will turn to the reasons why its development is needed, *and* within practical reach. In service of our goals in this chapter, and as a preliminary step towards the deeper and more targeted analyses in Part II, let us review in very broad strokes the central ethical teachings of the three classical traditions drawn from in this book.

2.1 Classical Virtue Traditions: Aristotelian, Confucian, and Buddhist Ethics

2.1.1 Aristotelian Ethics

The ancient Greek philosopher Aristotle (384–322 BCE) greatly extended and modified the ethical teachings of his Athenian predecessors Socrates and Plato in a definitive series of lectures on the good life, the notes of which were collected in a volume titled *Nicomachean Ethics*.[4] With the notable exception of its enduring influence upon the Catholic moral tradition (largely owed to the writings of St. Thomas Aquinas), Aristotelian ethics largely fell out of favor in the modern West until the mid-20th century, when it was returned to contemporary relevance by a growing community of neo-Aristotelian virtue ethicists. As noted in chapter 1, its current revival is generating new scholarly interest in the applicability of Aristotelian ethics to contemporary moral problems.

Many key features of Aristotle's ethics were sketched in chapter 1. However, it will help to summarize the core commitments of his approach. Aristotle claimed that the highest good of a human life, that for the sake of which all voluntary human action occurs, is *eudaimonia* (variously translated as 'happiness' or 'human flourishing'). He claimed that *eudaimonia* is constituted by a complete life of virtuous activity, defined as excellence in the active fulfillment of our unique function (*ergon*) as human beings. Believing this function to be the exercise of our capacity to reason, Aristotle argued that its fulfillment in a life of happiness with others presupposes the self-cultivation of various moral excellences or 'virtues' (*arêtes*) of character, such as courage (*andreia*), honesty (*aletheia*), patience (*praotes*), friendliness (*philia*), justice (*dikaiosunē*), and moderation (*sôphrosunê*), along with the unifying intellectual virtue of practical wisdom (*phronēsis*).

A virtuous person who successfully cultivates these excellences is able to perceive the right end or goal to seek in a wide range of practical situations. To be right, the end must actually be achievable by the person in that situation, *and* likely to promote the ultimate aim of all human action: human flourishing or living *well* (which for social animals like us, always means living well *with others*). A virtuous person reliably discerns and employs effective practical means to achieve these ends, as appropriate to the specific circumstances in which he finds himself. So, for example, we can imagine that in one situation a virtuous person may best promote the moral end by fiercely protecting an important secret; yet in a different time and place, the *same* person might rightly judge that promoting the moral end requires him to expose that secret.

In Aristotle's view, this discerning skill in living rightly is cultivated through a process of habituation and gradual refinement of one's character, in which one's repeated practice of moral actions is guided and encouraged by a combination of proper laws, moral education, and the presence of noble human models (called *phronimoi*: 'practically wise persons') who exemplify and inspire virtuous living. In a fully cultivated or virtuous person (a *phronimos*), one's habits, thoughts, and emotions have been refined and harmonized to such a degree that virtuous actions are consistent, produced with spontaneity and pleasurable ease, infused with appropriate moral feeling and belief, and above all, intelligently guided by practical wisdom or *phronēsis*. While Aristotle reserves the very highest and rarest form of *eudaimonia* for those suited to a life of theoretical reason—that is, a life spent contemplating eternal and divine realities—he claims that any human being equipped with practical reason and a basic level of material security can achieve a flourishing and happy life through political activity and civic friendship (*philia politikē*). This counts as *eudaimonia* because, like philosophical contemplation, living well in community with others *also* requires the constant exercise of our distinctive capacity for higher reasoning. For Aristotle, expressing this function (*ergon*) of our particular animal kind is what makes our lives *humanly* good.

2.1.2 Confucian Ethics

Kongzi (551–479 BCE) or 'Master Kong' (Latinized as 'Confucius') was a teacher and local government minister in the Chinese state of Lu. His moral and political philosophy is collected in a handful of classic works, the best known of which today is the posthumously assembled *Analects*. Confucian moral philosophy, also known as 'Ruism,' was greatly extended and enriched by disciples and later followers of Kongzi, among the best known of whom are Mengzi (also known as Mencius) and Xunzi. From the time of the Han Dynasty (206 BCE–220 AD) to the present day, Confucian thought has exerted a powerful influence on Chinese culture and the ideology of the Chinese state, though its classical form has been repeatedly altered over the centuries by interactions with other moral and political schools of thought, most notably Chinese Legalism, Buddhism, Daoism, and Communism.[5]

As with all rich moral traditions, Confucianism is marked by doctrinal disputes over the 'correct' version of the original account. Among the best known of these is the dispute between Mengzi and Xunzi over whether human nature is fundamentally good or evil. Yet classical Confucian thought maintains a stable core, centered on the need for persons to cultivate in themselves the kind of moral virtues that enable the flourishing of relationships within the family—virtues that are then gradually extended outward to other relationships to promote broader political flourishing.[6] The Confucian self is not an isolated, autonomous individual but a being defined by relationships and reciprocal obligations to others. A life that enables familial and political flourishing is understood by Confucians as a life in harmony with the Way (*Dao*), which is seen as a timeless ideal for the functioning of human societies.

To become the sort of person whose life engages this ideal pattern, one must allow one's character to be shaped by lifelong study of moral tradition and the practice of accepted moral rituals or 'rites' (*li*). These rituals express and foster attitudes of moral respect and deference in everything from the treatment of one's family members, to the conduct of public ceremonies, to one's personal dress and bodily gestures. Yet as we will see, ritual practice must never be allowed to become rote, mechanical, or rigid; it must embody a deep cognitive and emotional sensitivity to the particular roles, circumstances, and human needs that determine the proper form of ritual expression called for in each situation. The practice of moral self-cultivation aims to gradually refine this intelligent sensitivity.

As with Aristotelian virtue ethics, Confucianism is shaped by a philosophical belief that the ultimate goal of a human life (here, a life lived in accordance with the Way/*Dao*) is timeless and fixed by nature, transcending cultures and history. As did Aristotle, Confucians also rely heavily on the use of human models to illustrate and inspire virtuous living; while 'adherence to the Way' is the final standard of a good life, individuals draw from observation or historical accounts of the behavior of particular 'exemplary persons' (*junzi*) to learn what adherence to the Way looks like. As do Aristotelians, Confucians describe complete virtue as a form of habituated practical wisdom or "intelligent awareness" that harmonizes cognitive, perceptual, and affective motivations and guides their unified and appropriate expression in particular moral contexts.[7] The cultivated person reliably discerns the 'middle way'—the 'mean state' or "due measure and degree" of response called for by a given moral situation—and enacts that response with spontaneous ease and an elegant, authentic style.[8] Thus in both traditions, the rigid following of established moral rules and the maximizing of moral utility are subservient and conditional aspects of virtuous living at best, and mere 'semblances' of virtue at worst.[9] This stands in stark contrast to modern deontological or consequentialist moral theories that regard virtue as defined by strict adherence to such principles.

Yet we must not elide important differences between Confucian and Aristotelian conceptions of the good life. For one thing, Confucians stress the logical and natural priority of family virtue, while Aristotle privileges the flourishing of the political state or *polis* and sees family virtue as guided by and ultimately in service to the state.[10] Nor do individual Confucian virtues map neatly onto Aristotelian ones. Though significant resonances and overlaps can be discerned, for example between *zhi* (wisdom or intelligent awareness) and Aristotelian *phronēsis* (practical wisdom),[11] even virtues with nominal counterparts in Aristotle, such as courage (*yong*), are framed very differently by Confucians.[12] Among the Confucian virtues having no direct counterpart in Aristotle are benevolence or humanity (*ren*), ritual propriety (*li*), appropriateness (*yi*), and empathic reciprocity (*shu*). Finally, nothing in Confucian thought replicates the tension found in Aristotle (and in Buddhism) between active political flourishing and a life of contemplative and philosophical well-being; for Confucians, *all* human flourishing is embodied in family and political life.

2.1.3 Buddhist Ethics

The religious and philosophical practice of Buddhism dates back to its origins on the Indian subcontinent in the 5th or perhaps 6th century BCE. Buddhism grew out of a wider social phenomenon of spiritual wanderers devoted to challenging the orthodox beliefs and practices of the Vedic tradition upon which modern Hinduism is founded. The historical individual credited with Buddhism's founding is Siddhārtha Gautama (Sanskrit; in Pāli script: Siddhattha Gotama), more commonly known as the Buddha (or the 'awakened one'). Born to a royal family in what is now Nepal, as a young man Gautama renounced conventional social and religious life and took up a personal quest for new spiritual and moral wisdom. As traditional accounts have it, he spent six years on his quest before becoming enlightened or 'awakened' at the age of thirty-five, after which he spent the remainder of his life spreading his religious and ethical teachings throughout northeastern India.[13]

Following his passing at the reported age of eighty, Buddhist teachings and practices expanded, splintered, and were widely disseminated throughout Southeast and East Asia. While the highest concentrations of Buddhist practitioners remain on the Asian continent, 20th century Buddhism is a global phenomenon with adherents numbering in the hundreds of millions—over a billion on some accounts. Two major traditions of Buddhist thought are distinguished today, embodying significant divergences among their accepted teachings and practices while retaining a common core of basic concepts and values. The oldest, Theravada Buddhism, remains dominant in Southeast Asia, while Mahāyāna Buddhism is prevalent in East Asia where it takes a variety of cultural forms (including Pure Land, Zen, Tibetan, and other schools of Mahāyāna practice).

The core teachings of Buddhism are heavily rooted in traditional Vedic metaphysics, incorporating its central concepts of *karma* (spiritually significant action), *samsāra* (cycle of rebirths), *Dharma* (sacred duty or doctrine), and *yoga* (spiritual discipline). However, classical Buddhism departs from Vedic philosophy and Hinduism by denying the substantive reality of a self (*ātman*), a doctrine known as *anātman* (no-self). On this view, the worldly self that appears stable, unified, and enduring is really no more than a collection of transient mental and physical phenomena of various types (the *skandhas* or 'aggregates'); nor is there any 'deeper' unified self beneath the worldly one.[14] Another key metaphysical principle is that of *pratītya-samutpāda*, or 'dependent co-arising.' Many of Buddhism's ethical norms follow from this view of all beings as causally interconnected, in a manner that surpasses local spatiotemporal boundaries. Buddhism also posits that suffering (*duhkha*), which it holds as fundamentally characterizing worldly existence within the cycle of rebirths, can be radically transcended by reaching a state of enlightened liberation (*nirvāna*). Such a state, in which one achieves the same 'awakened' condition as did the original Buddha, is in principle attainable within the lifetime of any individual by following the Noble Eightfold Path, a developmental practice which jointly cultivates various practical forms of spiritual knowledge (*prajñā*), ethical conduct (*śīla*), and concentrated awareness (*samādhi*).

While the moral teachings of Buddhism are sometimes described narrowly in terms of *śīla*—that is, adherence to the various 'precepts' or rules of ethical conduct—a fuller understanding of Buddhist ethics conceives of *śīla* as one aspect of complete virtue, which must be integrated with other virtuous dimensions of the Eightfold Path. That is, just as Confucian ritual action (*li*) and Aristotelian moral habit (*hexis*) lead to genuine virtue only when integrated with appropriate thought, perception, and feeling, Buddhist self-cultivation requires that fully ethical conduct inform, and be informed by, right belief, right intention, and mental and emotional discipline.[15] Such enrichment enables the enlightened person to modulate his or her expression of the conventional moral precepts, or even in rare cases adopt 'skillful means' (*upāya kaushalya*) to suspend them, as called for by the morally relevant features of the particular situation. In this way, as with other virtue traditions, Buddhist ethics takes a holistic, flexible, and contextual approach to moral action that is fundamentally distinct from either rule-based (deontological) or consequentialist models of ethical life.

Buddhism's resonances with other classical virtue traditions do not end here. As with the central role granted by Confucian and Aristotelian ethics to 'exemplary persons' (the *junzi* and *phronimoi* respectively), *bodhisattvas* (persons actively seeking enlightenment) generally receive direction to or assistance on the path of self-cultivation from the community of exemplary persons to which they have access. In Buddhism this is the monastic community and lay members of the *Sangha*, those spiritual adepts who, having attained a certain degree of enlightenment, are able to give spiritual and moral direction to others. Indeed, the virtuous community of the *Sangha* represent one of the 'Three Jewels' of Buddhist teaching, the other two being the sacred teachings of *Dharma* and the perfected being of the Buddha nature itself.

As with Confucian and Aristotelian ethics, the moral vision of Buddhism is motivated by a teleology that regards a certain sort of life or nature (here, the Buddha-nature) as a timeless and nonnegotiable ideal for human beings. As do the other virtue traditions, classical Buddhism also stresses the role of habituation in moral self-cultivation; enlightenment is typically sought through devotional practices (*yoga*) that function to gradually shape one's mental/emotional dispositions into a highly refined and disciplined form. Thus at its core, Buddhist ethics, as with all virtue ethics, is about gradually transforming oneself into a certain sort of person—one who can live in a way that is worthy of human aspiration.[16] Right rules or precepts of action are only guideposts on the path to that destination, not the destination itself, nor even the entire path.

Still, there are deep disagreements between Buddhism and other virtue traditions, and among various Buddhist schools, about the good life and the path leading to it. First and foremost, while individual points of contact can be found between the expansive catalog of Buddhist virtues and the far more limited Aristotelian and Confucian ones, the overall character profiles of the exemplary person appear quite different among all three. In particular, the Buddhist virtues of humility, detached equanimity, and expansive compassion find no clear parallels, and even direct opposition, in the Aristotelian model which encourages great men to cultivate a character marked by warranted pride, appropriate ambition, and righteous indignation.[17]

Comparing the profiles of the Buddhist and Confucian virtues may at first seem to reveal closer affinities; but even here, virtues that seem nominally similar turn out to have very different meanings in their full moral contexts. For example, one might try to pair the virtue of Buddhist generosity (*dāna*) with Confucian benevolence (*ren*), or Buddhist compassion (*karunā*) with Confucian empathy (*shu*). Yet such an impulse obscures the fact that while Buddhists regard traditional kinship loyalties as resulting from deluded ignorance of our common nature and our moral obligation to reduce suffering for all creatures, the Confucian worldview emphasizes graded love and compassion, with those to whom we are related and to whom we have incurred reciprocal social obligations having a far stronger claim to our generosity and concern than do strangers. Faced with a dilemma pitting important interests of one's father or ruler against the interests of a suffering multitude from a neighboring kingdom, a virtuous Buddhist and a virtuous Confucian might make very different choices.[18]

Additionally, devout Buddhists' relative disregard for material goods and political status would be incomprehensible to Aristotle and Kongzi, who each regarded poverty and political disenfranchisement as significant if not insuperable obstacles to human flourishing. Finally, the state of human flourishing at which Buddhists aim, namely *nirvāna*, implies a transcendence of local and worldly striving that is wholly incompatible with Confucian ethics or with Aristotle's life of political happiness, and distinct even from Aristotle's highest ideal, a life of rational philosophical activity.

2.2 The Shared Commitments of Virtue Traditions

These comparisons yield a general outline of a life that leads to virtue. It is a life that ideally begins with proper moral habituation into social roles and responsibilities—guided by established precepts, examples, laws, and rituals—but under the right conditions and with sufficient personal effort becomes increasingly self-directed, reflective, and intelligent. Through this process of moral self-cultivation, a virtuous person attains a level of ethical mastery in which both the ultimate aims of moral living and the appropriate means of attaining them in particular contexts and situations are understood and correctly valued. Such a person does the right thing with relative ease and joy. Moreover, she expresses an authentic and appealing moral style of acting in the world that inspires and provides an exemplar for others who wish to do the same. A virtuous person need not be infallible or wholly incapable of moral weakness, but he or she lives in a fully human way—that is, as a person who embodies for others the best moral possibilities of the species.

However attractive this picture may be, one quickly realizes that it is notably lacking in concrete moral content. This is because as we have noted above, there is much variation among classical virtue traditions in the specific action-guiding principles and norms they employ. Herein lies the problem for contemporary defenders of virtue ethics. Given the robust differences between classical visions of the good life, is there any justification for the assertion that the Aristotelian, Confucian, and Buddhist traditions *are* all members of a single family of ethical theories called 'virtue ethics'? We might throw other classical conceptions of virtue into the mix, such as those of ancient Homeric or Nordic societies.[19] But an exhaustive account of such traditions is both impractical and unnecessary for our purposes. If we can show that there *is* a shared conceptual core even among the three widely influential traditions we have sketched above, one that offers useful action-guidance *even* for contemporary humans in a globally networked world, this will be sufficient to rescue the promise of a technomoral virtue ethic of global reach.

Let us start with the question of whether these three traditions represent a single class of moral theory, namely *virtue ethics*. While Aristotelian ethics is regarded in the West as a paradigmatic virtue theory, scholars of comparative philosophy have not been in universal agreement about whether Confucian or Buddhist ethics belong under the same theoretical umbrella. A few have held that Confucian ethics, with its emphasis on ritual conduct, or Buddhist ethics, with its clear precepts for moral action, are best classified as *deontological* or rule-based moral theories. Others have claimed that Buddhism, which aims at the comprehensive reduction of suffering by sentient creatures, is really a *consequentialist* or utilitarian framework. Still other scholars have claimed that Confucianism or Buddhism are not *ethical* theories at all, but mere systems of political or religious thought. However, such views are in the minority.[20] Still, even among the majority of comparative scholars who regard Confucian and Buddhist thought as more akin to virtue ethics than to any other Western philosophical approach, not all are convinced that they share with Aristotelian virtue ethics any robust conceptual core.

Yet there are compelling grounds for the claim that they do. One common strategy for defending this claim is to distinguish between 'thick' and 'thin' moral concepts. *Thin* concepts supply only the essential structure or skeleton of an idea, whereas *thick* concepts flesh out that idea in richer detail. Perhaps, then, culturally distinct virtue traditions share a thin core of basic theoretical commitments that are filled out very differently by their respective thick contents. These visions of human virtue would rest on very similar conceptual foundations, yet diverge noticeably in their detailed recommendations for how to live.[21] Comparative philosopher Bryan Van Norden has suggested that there are at least four thin commitments shared by virtue traditions, which I articulate (with some modifications) below:[22]

1. *A conception of the 'highest human good'* or flourishing that serves as the aim of ethics; this 'thin' concept may then take diverse thick forms: a life of excellent rational activity (Aristotelianism), a life in harmony with the *Dao* (Confucianism), the attainment of *nirvāna* (Buddhism), or some other concrete vision of human flourishing.
2. *A conception of moral virtues as cultivated states of character, manifested by those exemplary persons* in one's community who have come closest to achieving the highest human good. The thick form of this shared concept displays the diverse ideals embodied by the *phronimoi, junzi, Sangha,* or other concrete exemplars.
3. *A conception of the practical path of moral self-cultivation*, that is, an account of how the virtues can be cultivated by those who do not already possess them. Different cultures may have different 'thick' accounts of the method of self-cultivation.
4. *A conception of what human beings are generally like*, and the way in which their nature typically conditions their ability to become cultivated. Again, we would expect the 'thick' concepts of human nature and potential to differ between traditions, or even within sub-schools of the same tradition.

In every case, the 'thick' content that fills out each shared 'thin' commitment is shaped by the unique social, historical, and political configurations of the particular culture in which that tradition or its schools is rooted. Still, I will argue in chapter 3 that with respect to items (2), (3), and (4) above, the shared content among these traditions is even 'thicker' than many comparative thinkers have noted, enough to provide significant action-guiding content for a global technosocial ethic. It is important to note the features of classical virtue traditions *not* included in the thin commitments. While all three traditions are goal-directed in the very broad sense noted under (1), a virtue ethic need not define its goal in terms of a distinctive human function (*ergon*) of the sort Aristotle invoked, or metaphysical commitments such as those which inform the Buddhist conception of *nirvāna*, or a belief in a fixed universal order (*Dao*) of the sort that Confucians hold. Furthermore, while a virtue ethic must be naturalistic in the limited sense required by (4), it need not be reductively so, nor need it subscribe to any essentialist account of a universal, fixed human nature. This is important because it leaves room for a virtue ethic that is *pluralistic* (open to more than one mode of expression of human flourishing) and *malleable* (adapted to the needs and affordances of the present human condition and environment). A contemporary technosocial virtue ethic need not invoke any single, ahistorical, and universal image of the good life for a human being. It need only hold that we can identify, within the limits of precision and accuracy afforded by practical wisdom, certain ways of living that plausibly count as flourishing *in the present human context of technosocial life*, and others that do not.

The framework of 'thick' versus 'thin' concepts is not the only well-known philosophical strategy for identifying what virtue traditions have in common. Alasdair MacIntyre has also claimed that virtue traditions share a conceptual core, even as he rightly rejects the idea that we can extract from these diverse traditions, with their disparate practices and conceptions of the good, any single list of universal moral virtues. Once we examine MacIntyre's account of the conceptual structure that unifies virtue traditions, we will use his analysis to test the coherence of our own endeavor.[23]

For MacIntyre, a given reference to a virtue is only meaningful:

1. Within the context of a recognized human *practice* dedicated to securing moral goods internal to that practice;
2. Where that practice is embedded in a coherent *narrative* concerning a whole human life;
3. Where that life is itself understood as participating in a shared moral *tradition* of seeking the highest good for a human being.[24]

On MacIntyre's view, this tripartite standard is sufficient to capture the conceptual unity behind virtue ethics' diverse and incommensurable manifestations, and explains how one virtue tradition can successfully adapt the resources of others to its own practices. I will not subject his influential account to a critical inspection here; while I believe that the commonalities among classical virtue traditions are even richer than MacIntyre acknowledges, his analysis remains insightful and compelling. Luckily for us, a technomoral virtue ethic of the kind developed in this book can satisfy these three requirements for a virtue tradition. This is because emerging technomoral practices have the potential to create a new global narrative concerning the good life for human beings, a narrative for which many classical virtue resources from diverse cultures can be usefully adapted. In order to see how, it will help to break down and examine MacIntyre's requirements one by one, beginning with the notion of a *practice*.

MacIntyre defines a practice as a "cooperative human activity through which goods internal to that form of activity" are sought and realized as part of the effort to *excel* at that activity.[25] He offers music, farming, chess, architecture, and football as examples, and of course philosophy itself would count as such a practice. All of these practices aim to produce certain goods, such as wisdom, beauty, shelter, physical sustenance, and so on. We say the goods are *internal* to the practice when, unlike external goods (say, a salary that one might get paid for doing such things), the internal good is produced purely from excelling with others in the practice. A well-coordinated and highly skilled orchestra can produce its internal good—beautiful, stirring sound—even if their activity goes unpaid or unappreciated.

Now consider the increasingly global reach of information, communication, biomedical and robotic technologies, and in particular, the ways in which these technologies have enabled unprecedented global mobility and "cooperative human activity." The emergence of electronic trading and purchasing enabled a 24-hour global marketplace within which the economic fates of nations and cultures are now inextricably intertwined, defining multinational corporations as new global powers in the process. New technoscientific capacities have fostered global research efforts such as the Large Hadron Collider, which required the economic and scientific cooperation of 111 different nations. New military and surveillance technologies entering a multinational arms market, and resisted by underground electronic networks of 'hacktivists' working to limit state power, are rapidly transforming how nations, groups, and individuals collaborate to advance competing visions of global security. New information, communications, and mobile technologies are driving unprecedented coordination of activities by individuals and groups devoted to the shared pursuit of human health, environmental well-being, and social and economic justice around the world. Just this brief survey of modern technological society makes it evident how much humans are now relying on *cooperative technosocial activity* as a means of seeking the good life, individually *and* collectively. This is not to say that fervent nationalism and provincialism are dead—far from it. It is only to say that technology has enabled new global human agencies to be added to the mix.

What, then, are the unique 'goods' internal to this global technosocial activity? There are a number of goods we can identify as goals of technosocial activity considered broadly: efficiency, speed, mobility, analytical power, knowledge, pleasure, friendship, and creativity are just a few of the goods that have been celebrated as being fostered by new technosocial practices.[26] But are these *internal* or *external* goods by MacIntyre's standard? Unlike wealth or power, internal goods do not typically accrue to individuals in a zero-sum fashion; they enrich the individual and the "whole relevant community" at once.[27] Most of the goods above could be said to satisfy this requirement. Yet there is no coherent technosocial practice that pursues *all* of these goods. Furthermore, each of these goods can be sought through many other sorts of practices that do not involve modern technology. MacIntyre is clear that an 'internal good' is a good uniquely associated with a particular practice, or another "of that specific kind."[28]

Yet there *are* distinctive internal goods associated with specific kinds of technomoral practice, goods that are directly relevant to our concerns about humanity's ability to successfully navigate a profoundly uncertain and risk-filled future.

First, the notion of *global community* is itself such a good. But with what sort of technomoral practice is it uniquely associated? Consider that prior to the information revolution, the goal of a genuinely 'global village' could be pursued only in extremely limited ways, and only by means of practices that were themselves dependent on technologies such as trading ships, telegraphs, and printing presses. These technologies enabled cooperative activities of communication and exchange relevantly similar in structure, if not in speed or scale, to those that increasingly dominate contemporary technosocial life. Most such activities continue to be employed for purely private enrichment rather than for moral goods of the sort with which we are concerned. Yet we have also seen the emergence of the idea of a new 'global commons,' and a variety of coordinated attempts to cultivate new technomoral practices to realize this goal. Such efforts range from the open source, 'Net Neutrality,' and digital equality movements to experiments with global Internet currencies such as Bitcoin, to attempts (however noble or reckless you judge them to be) by Wikileaks, Anonymous, and other hacktivists to 'make information free' from national, economic, and cultural hegemonies.

Similarly, *intercultural understanding* and *global justice* are moral goods being pursued through modern technosocial practice in unique and unprecedented ways. Social activists, relief workers, and educators have seized upon the Internet's ability to cheaply transmit live images of violent political or religious conflict around the world, to give marginalized or disenfranchised persons a global voice through social media, and to cultivate and share expertise in using these new technosocial means to successfully marshal global financial, logistical, or popular support for local causes. There simply *were* no practices prior to the information revolution through which locally marginalized persons or groups could reliably solicit global attention and concern. Other than their humanity, what can a pregnant Sudanese convert to Christianity facing a death sentence, a transgender American teenager imprisoned without charge in solitary confinement, a Pakistani schoolgirl challenging the Taliban, and an Egyptian marketing executive for Google in Dubai have in common? The answer: a networked information commons that can raise local injustices to political awareness and action on a global scale. What we are seeing is a global social experiment with coordinated technosocial activity aimed at securing moral goods of the sort that persons committed to this practice regard as worthy of lifelong pursuit for a human being: goods such as global religious freedom, gender equality, and other forms of political, economic, and criminal justice. This fully satisfies MacIntyre's requirements for the emergence of a moral practice.

Human security is another internal good that may be considered distinctively realizable through global technomoral practices. By 'human security' I do not mean the security of individual living humans or groups, but the security of the present and future flourishing of the human species from potential threats such as global climate shifts, the emergence and spread of new pathogens, or the rise of robust artificial intelligence. It is virtually impossible to pursue this goal without technologies of global communication and coordination; without new technomoral practices built around such technologies such a good could only be *hoped* for, as a delivery of blind luck, divine providence, or the aggregate yet uncoordinated result of individual and local acts of human wisdom. Effective, deliberate, and intelligent pursuit of this good through a human *practice* demands the massive coordination of international will and resources that only new information and communications technologies (ICT's) make sustainable.

Finally, while *collective human wisdom* has always been pursued through regional and partisan human affiliations, and occasionally envisioned as a global hope, only with modern ICT's such as ubiquitous translation software and streaming video does it become possible to design practices that aim at genuinely *global* edification (as opposed to, say, the edification of literate Latin- or English-speaking populaces taken to represent the whole). It is likely that further goods internal to global technomoral practice can be identified, and others will emerge in time; but we have already done enough to establish the possibility, even the inchoate beginnings of, global technomoral practices that aim at internal goods of the sort associated with virtue traditions.

Of course, each of these goods can be sought only with great difficulty and demands particularly noble motivations, and none have yet been consistently or broadly realized in practice. Yet this is the case with human practices *generally*. For example, the distinctive goods of novel prediction and theoretical unification internal to scientific practice have *always* been realized imperfectly, inconsistently, and only with great effort. The same goes for the distinctive goods realized by religious or political practice. Furthermore, technomoral practices on a global scale remain extremely immature from a developmental perspective, as a result of how recently these new technological affordances for global action have emerged. It would be unreasonable to expect *any* human practice to be consistently realizing its highest internal goods in its relative infancy. As a corollary, it is not surprising that general rules and shared "standards of excellence" for pursuing global technomoral goods are still evolving; we would expect to find such stable guidelines only in mature moral practices.[29]

Our account of global technomoral practice also easily satisfies the second and third core requirements of MacIntyre's analysis: that actions which are part of a genuine moral practice be part of a personal narrative driven by the agent's coherent intentions and beliefs, *and* that this individual story make sense as part of a broader historical tradition, in which I and others are implicated and can hold each other mutually accountable for our success or failure in pursuing the good life for a human being.[30]

As MacIntyre emphasizes, a moral practice is irreducibly a social enterprise, not just the sum total of individual efforts to live well. If we consider the internal goods of global community, intercultural understanding, global justice, human security, and collective wisdom, it becomes immediately evident that each of these readily makes sense as part of a personal narrative in which an individual's beliefs and intentions are directed toward the joint achievement of these goods. While there are certainly those who find themselves supporting such goals accidentally or without any motivational integrity (the person who 'likes' the Facebook page of a social justice movement because it happens to be 'trending,' giving it no further thought), most humans will not engage in *sustained* and *coordinated* activity to promote these goals without a set of beliefs and intentions directed at becoming a certain sort of person, one whose life has been spent with others in pursuit of values that one regards as ultimately *good* and affirming of a life lived *well.*

What about the third requirement, that this personal narrative be subsumed within a broader community of practitioners who hold each other accountable for the success or failure of their efforts, and whose collective activity constitutes a historical *tradition* of pursuing what they agree to be the highest human goods? It is difficult to envision the person today who successfully engages in the enduring pursuit of global community, understanding, justice, security, or wisdom without ever coordinating her efforts with others who share the belief that pursuit of these particular goods is the *best* way to pursue human flourishing in our present circumstances. The efforts of legacy NGOs such as OXFAM, Greenpeace, and Amnesty International as well as 21st century networks such as 350.org, EFF (Electronic Frontier Foundation), and ICRAC (International Committee for Robot Arms Control) testify to the fact that accountability for success or failure in achieving global technomoral goods will inevitably fall to persons in community rather than isolated individuals.

2.3 The Need for a Global Technomoral Virtue Ethic

Most importantly, our account of technomoral practice passes one more crucial test of MacIntyre's: if something is a practice with its own internal goods, then there will be distinctive *virtues* "the possession and exercise of which tends to enable us to achieve those goods,"[31] and without which we are prevented from achieving them. Is it at all plausible that goods such as global community, intercultural understanding, global justice, human security, or collective wisdom can be successfully achieved by persons lacking in virtues such as courage, civility, or caring? Or is it clear that along with nonmoral excellences such as creativity, analytical intelligence, and perseverance, successful technosocial practice will indeed require the cultivation of distinctly *moral* excellences of character, adapted to function in this new context as *technomoral virtues?*

Many of these virtues are also associated with classical traditions, despite those moral worldviews remaining incommensurable in many ways with contemporary technosocial goods and practices. A single virtue of character will usually be conducive to a diverse range of incommensurable goods and practices, and its visible aspect will vary accordingly. While a soldier, a physician, and a politician all need courage to pursue their respective goods in ways that appropriately channel and modulate fear and hope, courageous action looks very different in these various practical contexts. The meaning of a virtue is historically and culturally contingent as well—the courageous soldier in the age of Homer would have expressed his resolve differently than did the courageous Japanese pilot in World War II, and as we move into an era of automated remote warfare, courage in military practice may acquire a radically new visage.

Still, it would be a mistake to be a thoroughgoing antirealist or cultural relativist about the virtues. For while the practices to which they are adapted change, as do their concrete expressions, the virtues themselves are rooted in cognitive, emotional, and perceptual capacities and vulnerabilities that are deeply rooted in the human organism, and that evolve far more slowly than do cultures. Humans in ancient Greece were subject to fear and its ability to cloud moral judgment, and hence needed courage to flourish in fearful circumstances. So were humans in Kongzi's time and place, and so are humans today.[32] Humans in all times and places have needed cooperative social bonds of family, friendship, and community in order to flourish; this is simply a fact of our biological and environmental dependence. Traits of caring, honesty, and civility promote such bonds and the goods they secure, hence their cross-cultural and historical resonance. Indeed, researchers have found a remarkable degree of such resonance in empirical investigations of the overlap among classical and contemporary virtue traditions.[33] What matters is that any given virtue be adapted to and explicable in terms of *some* coherent moral practice, narrative, and tradition, and that we be careful to employ as our interpretive standard for defining a virtue that set of internal goods that we are presently seeking to achieve within our contemporary moral practices, rather than those of a tradition that is no longer living for us.

Thus a framework of technomoral virtues must not be an incoherent mélange pieced together from parts of disparate, incommensurable traditions. Rather, following MacIntyre, we start from the guiding concept of a virtue as a character trait that enables us to cooperatively achieve, through excellent activity, shared goods that are internal to our practices. Next, we recognize in the current human environment the emergence of a new kind of coordinated technosocial practice with its own distinctive moral goods. The third step is to grasp that the chances of humans flourishing in a century where conditions for life are increasingly interdependent will depend in large part on how effective we become at coordinating our decisions and activities in order to secure these moral goods. The final step is to realize that to have any hope of realizing these goods to a significant degree, we will need to begin to cultivate in ourselves that set of moral virtues most likely to foster their realization.

In identifying this set we must not be afraid to adapt, extend or qualify traditional virtue concepts. Our decisions will be driven not by historical fidelity but by the unique needs of the human present, as set out in the introduction of this book. This also mitigates those concerns raised at the start of this chapter about the *reliability* of existing virtue traditions and their action-guiding content, for we have no reason to carry over and endorse every action-guiding ideal, ritual, or character trait celebrated by those traditions as conducive to a good life—unless they just so happen to *also* prove conducive to securing those technomoral goods which are our present concern.

2.3.1 The Cultural Challenge Revisited

A contemporary technomoral virtue ethic can, then, be conceptually coherent. Yet we still need to overcome another sort of worry about the incommensurability of cultures and traditions. This is the worry that, even within the contemporary human context, there are insurmountable cultural obstacles to the emergence of a global technomoral virtue ethic of the sort I have proposed. This worry can be articulated as two distinct objections:

1: Not all of the goods you have proposed as goals of this emerging ethical practice (*global community, intercultural understanding, global justice, human security, and collective human wisdom*) are universally recognized today as essential human goods.

2: Any of those goods on the list that *are* universally recognized as such will be understood, from different cultural perspectives, to mean incommensurably different things in practice. Under that circumstance, no coherent global standards of excellent moral practice can emerge, and thus there can be no meaningful global articulation of the virtues that will promote these goods.

These are serious objections requiring careful response. The first is easiest to deal with. It is a mistake to hold that a virtue tradition must be universally attractive to, or adopted by, all human beings in order to be meaningful or effective. Every virtue tradition that has ever existed has had nonadherents and competing value systems, even within the same ethnic or religious culture. Aristotelian virtue ethics emerged within a diverse Greek world in which Spartan and other value systems were direct and robust competitors. The Confucian value system had opponents in Mohism, Daoism, and Legalism, to name just a few. Indeed that period is often referred to as the time of conflict among the 'Hundred Schools' (*zhūzǐ bǎijiā*).[34] Buddhist ethics has always had cultural competitors in Hinduism and Jainism.

Hence it is a given that a contemporary technosocial virtue ethic will not be attractive or compelling to everyone, and that the goods that it seeks to promote will not be universally accepted as the highest goods to which a contemporary human being can devote her life efforts. Furthermore, there will inevitably remain many who are indifferent to consideration of any 'highest good' for a human life, including those who live largely without moral reflection, who regard morality as a mere matter of convention or power relations, or who accept religious or legal duty as exhaustive of their moral obligations.

However, the objection is not wholly without force. A contemporary technosocial virtue ethic must, in order to effect any one of its internal goods, resonate broadly enough to motivate significant social cooperation on a global scale. An ethical narrative that has motivational force only for liberal Anglo-American philosophers, for example, will fail to suffice. The problems of global justice, security, and community confronting the human species in the 21st century cut across cultural, political, and religious lines, and any technosocial ethic which does not do the same is doomed. So let us consider the plausible scope of our proposed internal goods of technosocial practice, one by one.

Consider *global community*. Why should this draw broad support as a goal of moral technosocial practice? First, let us clarify what this term means; it refers not to a 'world government' or any other particular institution, but to a global *moral* community of active reciprocal concern and respect. Is this a notion that has broad cultural resonance or not? Some scholars argue that humans have expanded their circles of moral concern over time, such that early tribal distinctions between 'us' and 'them' have gradually opened up to encompass today, in many people's minds, the notion of the 'human family.' Further widening of the circle of concern is proposed to include other sentient creatures or the biosphere, perhaps even the universe itself.[35]

Of course, many of these claims occur within Anglo-American and European perspectives informed by the cosmopolitan leanings of the modern liberal tradition. Yet we must not forget that each of the classical virtue traditions make reference to a similar conception of global community; Aristotle acknowledges the friendship (however faint its obligations) that binds all human beings; Kongzi repeatedly states that the civilizing force of virtue reaches out even to barbarian tribes; and even a passing knowledge of Buddhism confirms that the notion of a global moral community is no European liberal invention.[36] There is no reason to doubt, then, that the attractiveness of this notion cuts across cultural, political, and religious lines.

Recognition of the goods of *intercultural understanding* and *global justice* follows naturally from any commitment to foster global community, for these goods are essential conditions for the latter. Any community needs its members to understand each other to some significant degree in order for that community to function; likewise, a global community of active reciprocal concern and respect is impossible if global injustices are routinely tolerated or ignored. Finally, recognition of the good of *human security* is necessarily implied by commitment to global human community; the human race cannot *be* a community if it no longer exists, or if it can no longer flourish in any meaningful sense.

Finally, collective human wisdom is a good that nearly anyone can appreciate, for the following reason: The first two decades of the 21st century have made it *abundantly* clear that the social, political, economic, and environmental conditions under which any one of us must live, along with those we most care about, are increasingly subject to great disturbance by failures of collective human wisdom. The global financial crisis of 2008 and the failure of international efforts to mitigate the effects of climate change supply striking examples, but many smaller scale failures can have similarly profound effects. Indeed, no one on the planet today is fully insulated from the failures of human beings to jointly and wisely deliberate about the collective impact of their actions. We need only consider the environmental dimension of this looming problem to see that those humans still blissfully unaware of their exposure will, unfortunately, learn of it all too soon. Thus confidence in the cross-cultural resonance of each of these goods is magnified by the growing worldwide recognition of political, economic and environmental problems that require massive human coordination to solve; by these lights, commitments to foster global human community, understanding, justice, security, and wisdom appear not only morally desirable, but morally *necessary* for the good life in the 21st century—and possibly for any meaningful life at all.

2.3.2 Ethical Pluralism as a Response to the Cultural Challenge

Here we turn our attention to the second, more daunting objection—for it is simply true that different cultures and groups will have different understandings of what a global community should look like, what 'understanding' someone from a different cultural standpoint amounts to, what constitutes 'justice' or 'wisdom,' or what would count as a 'secure' existence of the human species. If the adherents of a moral practice cannot agree on a substantive description of the goods they collectively seek, then they will not be able to effectively coordinate their activity to achieve those goods, nor will they agree about when these goods have been successfully secured.

In order to develop an adequate response to this objection it will help to draw upon the work of philosophers of technology who have extensively considered the prospects for an intercultural or global ethics of technology, and who have weighed the force of similar objections.[37] Among them, I highlight here the work of philosopher Charles Ess, who has arguably devoted the greatest share of his work to this question. As Ess has argued in various scholarly fora, there is a plain and compelling need for a global ethic that addresses the growing role of technology in our lives. He notes that just as the global network of ICTs requires a shared body of *technical* standards in order to function, so also its functioning will increasingly require certain shared *ethical standards* among the networked citizens of this new and diverse "global city."[38]

Yet as we have noted, the plurality of seemingly incommensurable moral traditions and cultures is a daunting obstacle for a global technosocial ethic. Ess explores the prospects for a form of ethical pluralism that can overcome this obstacle, while avoiding the unworkable alternatives of ethical absolutism or ethical relativism. He rejects as unsatisfying those philosophical accounts of ethical pluralism that call for *mere* tolerance or appreciation from a distance of ethical differences, as well as accounts that would ultimately elide such differences by appealing to a common moral identity or single set of standards that underlies them. Instead, Ess advocates a delicate middle position inspired by Plato and Aristotelian metaphysics, what he describes as an "interpretive pluralism" in which agents from diverse moral traditions can recognize the Other as someone *connected to*, but not identified with, their own moral perspective.[39] The nature of their connection is their diverse interpretations of the Good. Just as Aristotle acknowledged that we talk about existence in many incompatible ways that nevertheless each point 'toward one' (*pros hen*) phenomenon, Ess argues for the possibility of a "*pros hen* pluralism."[40] The idea here is that while we each encounter others in our world talking about the good life in diverse, even incompatible ways, we can still recognize that such talk is all pointing 'toward one' (*pros hen*) overarching aim that we have in common—finding some way to live well with one another.

One might think this common ground a mighty thin ledge on which to stand. Yet the cultural diversity of virtue traditions, when combined with their overarching commitments to moral self-cultivation, habitual practice, discerning judgment, and flourishing relationships, offers strong prospects for a pluralistic ethical discourse about living well with emerging technologies. Ess supports this claim by showing that even in the culturally and legally contentious domains of online trust and information privacy, agents having diverse and incommensurable cultural interpretations of what is right and good in these matters nevertheless display a certain ethical *resonance* in their mutual expectation that agents operating in these arenas will exhibit certain virtues. These include honesty, respect, empathy and, most of all, a discerning judgment of the moral situation. The latter is the virtue that Aristotle calls *phronēsis* or practical wisdom, arguably the top candidate for a universal moral virtue if there is one. Still, what the appropriate display of any one of these virtues looks like in a given circumstance will depend upon the cultural and relational context in which one is operating, as well as one's own particular identity in that context. There will certainly, then, be unavoidable disagreements between agents from different cultural and relational contexts about what ought to be done in any given case. When the impact of the agent's conduct is purely local, these disputes usually can be safely tolerated, or recognized through intercultural dialogue as complementary and instructive. For example, Ess notes that East Asian youth's attitudes about privacy and the self that are rooted in Confucian and Buddhist values seem to be taking cues from Western narratives on these subjects and vice versa.[41]

At other times, more robustly cooperative global deliberations are needed in order to secure a moral result. If we are dealing with matters of Facebook privacy options, Twitter policies for shielding users from sexist or racist abuse, or protocols for verification in e-commerce, we are confronting choices that impact users networked together across the globe and that can't be left up to individuals to independently decide for themselves. When we debate whether to fund the development of a radically new and potentially dangerous biomedical technology, whether to decommission or build more nuclear reactors, or whether to employ, restrict or ban automated lethal systems in our military campaigns, the stakeholders of the debate simply cannot assume that their own culturally or locally situated interpretations of ethics will decide the issue. Yet if this means that ethical considerations have to be set aside, then it looks like such debates will simply be decided by considerations of economic, political, or kinetic *might* rather than *right*. The cynical realist will note that this is precisely how such debates *are* settled. The moral philosopher will reply that the human race and state of the Earth are almost certainly the worse today for it, and that continuing this particular practice for the next one hundred years will result in a state of global human flourishing that is at best gravely suboptimal, and at worst, nonexistent.

Fortunately, the intercultural resonance of virtue ethics suggests an alternative. As I argue in the next chapter, diverse virtue traditions appeal to strikingly similar *structures and methods* of moral self-cultivation, and this provides robust resources for bringing ethical considerations into global technomoral debates in a productive way, one not doomed to the global imposition of the values of dominant actors *or* to parties talking entirely past one another. Instead, we can begin such debates by appealing to common hopes or expectations that the relevant actors will pattern their inquiry and deliberations after the broadly recognizable habits of virtuous or 'morally cultivated' persons. These common patterns of activity are articulated in Part II.

But that is not all. We can reasonably ask that actors who are party to the debate agree to protect and promote goods critical to global human flourishing *when* those are determined to be at stake and within practical reach. Of course, there will still be disagreement among the parties about how those goods are defined. So it may seem that the second objection has come around to bite us once more. Fortunately, the wound is not fatal. For even robust and long-standing virtue traditions make room for disagreement about how to define their common goods; Aristotelians offer different accounts of happiness or *eudaimonia*; Mengzi and Xunzi argued about whether the cultivated state of the *junzi* was a development or a reformation of human nature; and a host of Buddhist schools developed different views of what constitutes enlightenment. Moreover, such disagreements have profoundly *enriched* the content of these traditions, making them more nuanced, active, and responsive to criticism. If perfect, universal agreement upon a fine-grained description of the goals to be reached is necessary for a meaningful virtue ethic, then *there has never been a meaningful virtue ethic*. Assuming this *reductio* is accepted as yielding a false conclusion, then we need only consider what level of agreement is necessary in order for members of a virtue tradition to effectively coordinate their moral activities with broadly salutary results.

Finally, as we determined in our analysis of the internal goods of global technomoral practice, the ability of human actors to adequately and reliably secure such goods in coordinated action with others will depend on our cultivation of the particular *technomoral virtues* likely to be conducive to such success. Why should we think that such cultivation is practically possible? Consider that the alternatives are to surrender any hope for continued human flourishing, to place all our hope in an extended string of dumb cosmic luck, or to pray for a divine salvation that we can do nothing to earn. I am confident I am not alone in finding these alternatives unappealing.

Moreover, if Aristotle had even some success in fostering the cultivation of civic virtues, if Buddhism has encouraged any more compassion and tolerance, or Confucianism more filial care and loyalty, what is to prevent a new tradition from emerging around the technomoral virtues needed for human flourishing today? As Ess notes, what we need is something like a renewed account of the German notion of *Bildung*; that is, a guide to the personal and social cultivation of those virtues conducive to a more humane and enlightened technological society.[42] A fuller articulation of twelve of these virtues will be given in chapter 6, along with practical suggestions for encouraging and strengthening their global cultivation. First, however, chapters 3, 4, and 5 will extend our comparison of the classical virtue traditions, in order to develop from their accounts of moral self-cultivation an idea of how 21st century humans might acquire the technomoral virtues we so desperately need.

PART II

Cultivating the Technomoral Self

Classical Virtue Traditions as a Contemporary Guide

3

The Practice of Moral Self-Cultivation in Classical Virtue Traditions

WE HAVE SEEN why we need a globally accessible account of technomoral virtue. But what would that actually *look* like? The first three chapters in this section draw upon Aristotelian, Confucian, and Buddhist theories of moral self-cultivation to construct a framework of habits and practices that any person or institution can use to support the cultivation of technomoral character. Chapter 6, the final chapter of this section, identifies the specific technomoral virtues that make up this character—traits that 21st century humans must cultivate more successfully if we want to be able to live well with emerging technologies. Our tour of technomoral virtue will then be complete, and we will be ready to apply it to the emerging technologies that are the subject of Part III.

As the reader, you may wonder why it is necessary to undertake the deeper comparative analyses of classical virtue traditions that constitute chapters 3, 4, and 5. Knowing our larger task, why not simply reveal our list of the technomoral virtues and get on with the business of determining how they can be usefully applied?

3.1 Learning from Other Virtue Traditions

For one thing, we would be foolish to neglect the rich resources for our task already available to us within the classical virtue traditions of Aristotelian, Confucian, and Buddhist ethics. These resources are of more than historical interest, as the shared conceptual pillars of these systems will turn out to be critical to the successful management of our own present condition.

We should also recognize that every human understanding of moral excellence bears the traces of one or more past traditions, and humans have always sought insight from the moral sages of history, who likewise sought to lift themselves and their own communities out of shortsighted, base, ignorant, and destructive patterns of activity. As Alasdair MacIntyre notes, when inheritors of one moral tradition survey the resources of another, nothing stands in the way of learning from that rival view, either by finding in it previously underappreciated aspects of their own tradition, or by encountering values or moral insights "which by their own standards they ought to have entertained, but did not."[1] Our present understanding of how to realize the goods of technomoral practice is still profoundly immature, as noted in chapter 2, and almost certainly inadequate. If we want it to reach maturity, we should seize any opportunity to learn from other moral traditions how we might strengthen its foundations and refine its insights.

Additionally, the diversity of norms embodied by classical virtue traditions should remind us that identifying the good life is not, as most of these traditions have assumed, a matter of discovering the one eternally 'true' way to live well. The principle of charity suggests that we should assume that each of the distinct paths to flourishing articulated by Aristotelian, Confucian, and Buddhist ethics more or less *worked,* or continue to work, in those cultural and historical settings where they find purchase. In Aristotle's Athens, a life of civic justice, friendship, and engaged political wisdom would have likely been a good one—far better than most—but was it really the best, or even the next best sort of life possible for a human being? After all, if Aristotle really was mistaken about the rational and moral capacities of women and non-Greeks, then clearly his vision of human flourishing in political society left considerable room for improvement.

We should also be willing to believe that at various times and places, including some places in the present, a good life—far better than most—has been achievable through a Confucian ethic of filial love and piety, or the Buddhist practice of mindful and compassionate detachment from the causes of suffering. Yet these ways of life are no less legitimate subjects of critical reflection and refinement than was Aristotle's. Moreover, each of these distinct conceptions of human flourishing is embedded in a nonfungible cultural and historical context. What sense would it make to claim that a 4th century BCE Chinese politician, *or* a modern one, would have lived better if he had mirrored the conduct of the *phronimoi* of Athens? Or that the *phronimoi* would have flourished more, had they adopted the practices and precepts of the monastic *Sangha*?

Thus we will not attempt to assess the relative merits of these classical visions of the good life, a project of highly dubious value. Fortunately, we have no need for such an assessment. We are not asking which classical vision of virtue should guide us in our present technomoral condition. None of them fully can, for they did not develop out of this condition. Instead we ask what conceptual and practical resources they *can* provide for the journey we need to make from our contemporary condition toward the good life for *us*, where 'us' refers to the increasingly interdependent technomoral community now being constituted on a global scale.[2] Given the enduring and widespread influence of all three traditions, it is reasonable to expect that the resources they can offer us will be significant, allowing our new technomoral ethic to establish links with many existing moral cultures and communities around the globe.

Chapters 3, 4, and 5 unfold the core of these resources: a set of remarkably similar claims, not about *what* moral excellence or a good life is, but about *how* human beings become excellent. As we recover this shared wisdom, we should expect these classical virtue traditions to occasionally challenge our own contemporary biases and dogmas about moral development and the good life. As we will see, living well in the 21st century may in fact require the reactivation of some classical habits and virtues that have suffered from modern neglect. Paradoxically, while humanity today faces challenges of a wholly unprecedented sort, a close look at how premodern traditions understood moral cultivation might in fact be our best preparation for what lies ahead.

3.2 Cultivating the Technomoral Self

As we saw in chapter 2, diverse virtue traditions share a deep conceptual unity accounted for by MacIntyre's notion of a virtuous *practice*. Yet his analysis of that unity assumes that the content of any particular virtuous practice is always relative to a distinctive form of life, one embedded in a specific cultural and historical context that is in key respects incommensurable with that of other traditions. While this is right insofar as it goes, there is at least one moral practice that is claimed as central by most, perhaps all, classical virtue traditions, however diverse these may be in their moral worldviews. This is the practice of moral self-cultivation *itself*. Moreover, the different versions of this practice are linked by significant structural commonalities and conceptual resonances worthy of exploration in light of our present needs.[3]

As we saw in chapter 1, moral self-cultivation is understood by virtue ethicists as a developmental practice that, if pursued properly and under favorable social and environmental conditions, will lead a person to a virtuous and good life. It is a voluntary practice, one freely embraced in order to slowly and painstakingly transform oneself into the sort of human being that one aspires to be, one who is capable of living well. As Stanley Cavell states, drawing upon the words of Ralph Waldo Emerson, moral self-cultivation is a fundamentally human quest for my "further, next, unattained but attainable, self."[4] It presupposes key facts about human beings: that we can freely choose an aspirational path of moral development; that such a choice can be effective in shaping one's moral character; and that we are unlikely to acquire virtuous character without a self-directed, habitual, and enduring effort to improve ourselves. Aristotelian, Confucian, and Buddhist ethics each articulate strikingly similar (while not identical) practices of moral self-cultivation informed by a philosophical anthropology of this sort. It stands to reason that these similarities do not result from arbitrary coincidence, but are rooted in basic truths about what it takes for human beings to become virtuous. If this basic anthropology is correct, its implications reach beyond classical systems of ethics to inform our own project, as they entail that acquiring technomoral virtue will require that we commit ourselves to a similar practice.

In the chapters that follow, we turn to seven core elements of the practice of moral self-cultivation that emerge from these culturally and historically distinct traditions. Together they offer us a shared frame of reference and a developmental scaffold for a technomoral virtue ethic of global scope. They are:

1. Moral Habituation
2. Relational Understanding
3. Reflective Self-Examination
4. Intentional Self-Direction of Moral Development
5. Perceptual Attention to Moral Salience
6. Prudential Judgment
7. Appropriate Extension of Moral Concern

This framework is never formally articulated in Aristotelian, Confucian, or Buddhist ethics, much less all three. Among them, only Buddhism explicitly formalizes the practice of moral self-cultivation (e.g., the Eightfold Path), and the framework above is not the Eightfold Path. Given sufficient articulation, however, each element of this framework would be recognized by a studious devotee of Aristotelian, Confucian, *or* Buddhist ethics as a conceptual near-relative of some essential feature of his or her own tradition. Put another way, the sketch above compiles a set of *shared conceptual resonances* that bind the three accounts of moral self-cultivation into a single family of ethical thought.

For the purposes of cooperative discourse, this means that if I were to engage in a conversation about the practice of moral self-cultivation with someone well-versed in one of these traditions, bringing any one of these seven concepts into the conversation would help us to illuminate or *make common sense* of the subject of our shared concern, namely, how moral self-cultivation occurs. This framework thus provides a shared frame of reference upon which a pluralistic, globally accessible account of technomoral virtue can begin to be constructed.

3.2.1 The Open Circle of Moral Self-Cultivation

The framework analyzed in the coming chapters does not exhaust the content of any classical account of moral self-cultivation; nor do these traditions perfectly mirror one another in their understanding of the practice. Yet all classical virtue traditions understand the practice of moral self-cultivation as having a temporal dimension, in which certain habits of the practice are a focus of the early stages of moral cultivation while others are more closely associated with mature moral practice. However, in none of these traditions is this developmental process seen as rigidly fixed; the specific manner in which an individual moves through the process is understood to be relative to the particular capacities and moral psychology of that individual. For example, a person who is strongly family-oriented might find it much easier to cultivate relational understanding than to learn to appropriately extend moral concern beyond her kin; on the other hand, for a person whose early family life was profoundly troubled, cultivating a deep and refined understanding of close human relationships may be a lifelong struggle.

In all three traditions, moral habituation must be taken up at a very early stage, since the process of moral self-cultivation involves increasing refinements and enrichments of this habitual practice. All three regard the reliable exercise of prudential judgment as a mark of advanced moral development. Likewise, in all three traditions, the ability to appropriately extend moral concern requires a combination of emotional sensitivity and cognitive discernment that must be acquired in earlier stages of the practice. Still, conceptual and structural differences between the models of self-cultivation will remain. For example, Buddhism's commitment to 'right understanding' (*samyag-drsti*) as the first step on the Eightfold Path may, depending on how it is interpreted, conflict with the Aristotelian view that proper habituation to right action in youth must precede the acquisition of moral understanding. Likewise, Aristotle's view of moral habituation as a corrective to naturally disordered human impulses appears to be in tension with the view of the Confucian philosopher Mengzi, who claimed that moral habituation and practice bring our naturally good impulses to their full and proper expression.[5]

Such differences entail that the family resemblance among classical virtue traditions will never collapse into identity. We must remain attentive to the important divergences in the moral, metaphysical, and/or psychological commitments that inform each view. However, differences in how each classical tradition *orders* the various elements of moral self-cultivation (to the extent that they do explicitly order them) are not as significant as they may appear, because these are elements of a unified practice rather than a series of discrete 'stages' through which the individual passes on their way to acquiring a virtuous character.

While certain parts of moral practice are more commonly associated with early or late phases of self-cultivation, the ultimate aim is for them to be cultivated in an integrated and harmonious way, and expressed *together* in the more or less perfected character of a virtuous person. How does this come about? According to all three traditions, the maturation and refinement of each element of moral self-cultivation continually informs and enriches the others. In the words of the Buddha: "the preceding qualities flow into the succeeding qualities; the succeeding qualities bring the preceding qualities to perfection, for going from the near shore to the far shore."[6]

Not only must the individual elements of moral self-cultivation be jointly pursued, but in order for the practice to enable a good *life*, each element must necessarily become a lifelong habit, an enduring component of one's moral personality. These habits are not used as mere *means* to become virtuous; rather, they become constitutive and enduring elements of virtue itself. None is a ladder to virtue that the cultivated person then kicks away. Indeed, we will see that each of these habits will be implicated not only in the cultivation, but also in the *retention* and gradual *perfection* of the specific technomoral virtues to be highlighted in chapter 6. For the virtues are conceived as endlessly perfectible; barring the possible exception of 'moral saints,' ordinary human beings are never finished with the work of becoming virtuous.[7] Indeed the 'cultivated person' is typically called so only in relation to her earlier, uncultivated self, or in relation to other, less cultivated persons—not in the sense of a task of moral becoming that has been completed and set aside. Thus the temporal order of moral self-cultivation is secondary to its cohesion as a *holistic* practice.[8] We can conceive of each element as part of an open circle that is to be repeatedly retraced, deepened, and expanded. With this in mind, we can now turn to deeper analyses of the seven basic elements of moral self-cultivation. Each analysis uncovers shared conceptual resonances among Aristotelian, Confucian, and Buddhist understandings of these elements, as well as important discordances. Each analysis closes with a brief summary of the common resonances and their implications for the 21st century cultivation of technomoral virtue.

3.3 Moral Habituation

In all virtue traditions—classical, modern, and contemporary—the practice of moral self-cultivation is, at its core, a transformative process of moral *habituation*. It begins with the initial setting down of some basic patterns of moral activity that in turn open up the possibility for more specific, refined, and intentionally directed habits of moral activity to develop. Before we go any further, however, the concept of 'habit' we are employing here must be carefully defined.

'Habit' within the Aristotelian virtue tradition must be understood in terms of the Greek word *hexis*, from which the meaning of the contemporary English term 'habit' departs considerably, as we shall see. Aristotle himself discusses *hexis* in multiple senses, such that its meaning varies with context and in conjunction with closely related terms, such as *ethos*.[9] Seeking equivalents for the term 'habit' in Confucian and Buddhist thought introduces even more semantic complexities. Thus to arrive at a clear idea of what we mean, we must exclude meanings of the term that are incompatible with or irrelevant to moral virtue, in any classical context.

The most important thing to grasp about habits of moral practice is that they are never mindless appetites of the sort we might associate with a smoking or drinking 'habit.' Nor are they involuntary, uncontrollable reflexes like blinking or breathing. Moral habits imply motivating reasons. If I always take a certain route to my favorite restaurant, and my spouse says "Why do you always go *this* way?" I might respond "No reason, I guess it's just a habit"—meaning that I did it without a motivating reason. I could have done just as well had I taken a different route. This cannot be what we mean when we speak of moral habits.

What moral habits *do* share with other sorts of habit is that they arise, at least in part, from the repetition of a pattern of action. An act does not become a habit until I do it repeatedly, and for it to be a habit I must do it more or less consistently when invited by the relevant situation. Thus as with other types of habit a *moral* habit is a practical disposition that emanates from past behavior; however, unlike other sorts of habits:

a) the habituated action is done for motivating reasons;
b) the action is normatively valued as positive, such that there is a (defeasible) expectation that I *should* perform it in the relevant situation(s);
c) the action pattern shapes, and continues to shape, my cognitive and emotional states in a manner conducive to the broader practice of moral self-cultivation.

Aristotle is clear that we first become moral not by studying moral philosophy or listening to moral arguments, but by repeatedly doing moral things until they become habitual: "it is by doing just acts that the just man is produced . . . "[10] An already well-habituated person can enrich their excellence by acquiring a philosophical understanding of ethics, but no poorly habituated and intemperate person could be "made well in soul by such a course of philosophy."[11] With the exception of those rare "gently born" individuals naturally inclined to virtue, he thinks most of us find it exceedingly difficult to be virtuous, and must be gradually acclimated to it.[12]

On Aristotle's view, basic moral habituation helps to remove the most fundamental obstacle to human flourishing—namely, our naturally intemperate responses to pleasure and pain, for "it is on account of pleasure that we do bad things, and on account of pain that we abstain from noble ones."[13] Yet in contrast to the Buddhist conception of moral habituation, the idea is *not* to acquire an appropriate detachment from pleasures and pains. Rather, it is to gradually condition ourselves to "use these well"—that is, to feel the right kinds of pleasure and pain, in the right amounts, at the right times and places, and with regard to the right objects.[14] Thus the goal is to become habituated to responding to pleasure and pain in a way that strikes the appropriate *mean*, relative to the circumstances. For example, we should want to be pained by "things that are terrible and to stand our ground against them," for that is virtuous; but one who is pained by nothing, or pained by everything, would be deficient in character.[15]

How does one know where the virtuous mean lies? Here is where one's access to human models of moral excellence becomes important. The mean is defined by what a truly virtuous person would do in the same situation; so the *imitation* of exemplary persons is a crucial component of early moral habituation.[16] As a first step to becoming good, I should study what is done by truly good people (in Aristotle: the *phronimoi*: cultivated persons with practical wisdom), and consistently try to do the same.[17] In young children, the process will not be wholly voluntary in the beginning: laws and the oversight of parental and other authorities must compel or elicit obedience to virtuous (or at least nonvicious) patterns of action. Even though Aristotle holds that *genuinely* virtuous actions are voluntary, and done purely for the sake of what is noble, he thinks that instruments of social compulsion such as laws and incentives (when intelligently employed by virtuous authorities) actually enable, rather than obstruct, the process of deliberate *self*-cultivation that eventually leads to genuine virtue.[18]

This is because laws and other sources of external moral authority gradually acclimate the young person to noble actions that are naturally painful (or at least not naturally pleasant), such as honoring debts or respecting the status of elders. As Aristotle notes, noble actions become *less* painful as one becomes more accustomed to doing them.[19] Eventually, such actions start to become pleasant to the morally habituated person, and an attraction to nobility for its own sake gradually emerges. Consider the way in which a child is often pained by sharing food with his or her siblings, and has to be induced by rewards or punishments to do so. Yet by the time we are mature adults, we may take great pleasure in baking a cake for a beloved sister or gifting her with a bottle of rare whisky, even when we do not expect to be able to consume any ourselves. At this stage, the natural obstacles to a deliberate undertaking of moral *self*-cultivation have been greatly diminished, as the person has already begun to acquire the essential moral virtue of self-control or temperance (*sôphrosunê*).[20]

Aristotle acknowledges, however, that the majority of people are highly resistant to making this transition to "noble joy and noble hatred," and wonders if "perhaps we must be content if, when all the influences by which we are thought to become good are present, we get some tincture of excellence."[21] Moral habituation, then, is a process that approaches its aim by small degrees, and that can easily stall well short of the ultimate goal of genuine virtue. Moreover, as we will see, for Aristotle habituation to noble acts is a necessary but not a sufficient condition of moral virtue; it must be complemented and enriched by further refinements of moral practice, most notably those that help to cultivate the intellectual virtue of practical wisdom or *phronēsis.*

Confucian moral philosophy contains no direct equivalents of Aristotle's terms *hexis* or *ethos*, nor of the English term 'habit.' However, robust conceptual resonances can be found between the role of habituation in Aristotle's ethics and the Confucian discourse on the concept of *li*, where *li* denotes ritual action and its power to shape moral character. Outside of the context of moral self-cultivation, the term *li* can simply represent socially and/or religiously mandated conduct, such as the elaborate rituals in traditional Chinese cultures that govern the proper conduct of births, marriages, and funerals. When employed by Confucian thinkers, the concept of *li* as ritual practice retains its link to social custom but also takes on a richer set of meanings, many of which resonate with the Aristotelian discourse on habit. For ritual practice not only functions to *express* moral motivations, attitudes, and conceptions of the world, "but it also helps cultivate those attitudes in us."[22] Thus whether we are naturally disinclined to have moral attitudes and motivations, as the Confucian philosopher Xunzi believed, or whether those attitudes and motivations simply need proper nurture to develop from their natural 'seeds,' as his Confucian rival Mengzi claimed, repeated ritual action or *li* can make it easier to cultivate what Aristotle called "noble joys and noble hatreds."

As Kongzi (Confucius) himself said, "By nature people are similar; they diverge as the result of practice."[23] Ritual practice, then, gradually brings human thoughts, feelings and desires into better alignment with the Way (*Dao*), the cosmic pattern with which the actions of a virtuous person harmonize. This function invites a comparison to Aristotelian habituation.[24] However, there are significant differences to be noted. First, and most obviously, the concept of the *Dao* has no application in Aristotle's cosmology. Second, the Aristotelian concept of a moral habit does not carry the implication of sacred tradition that is essential to the Confucian notion of *li*. Moreover, the Aristotelian notion of 'habit' offers little in the way of action-guiding content. At best we can say that a moral habit is a disposition to act in a way that "strikes the appropriate mean for the situation at hand, especially as regarding pleasures and pains." The concept of *li* as 'ritual practice,' on the other hand, is inseparable from a second sense of *li* as the 'rites,' which embody a vast repository of culturally specific, standardized, and highly formal social practices with action-guiding content spelled out in rich detail, often carried out with musical and other aesthetic modes of expression. As comparative scholars have noted, there is simply nothing like this in Aristotle's account of virtue.[25]

Kongzi's disciple Mengzi employs the concept of *li* in a third sense, to represent a specific virtue of 'propriety' or 'respect.' Here we find an unanticipated conceptual resonance. As Lee Yearley has noted in his comparison of Mengzi and Aquinas, there is no virtue comparable to *li* in Aquinas' account, which is largely based upon Aristotle. But Yearley describes the virtue of *li* as a disposition toward yielding or deference, one that corrects a certain deformation in human character, namely, "people's love of mastery and movement to self-aggrandizement."[26] If we remember that *li* in the first sense is a ritual practice of conforming to *li* in the second sense (the rites themselves), then it seems that it is this repeated action of yielding to others, as mandated by the rites, which produces the 'yielding disposition' of *li* as a virtue in the third sense.

The only plausible way to understand this process is to explain it in a manner similar to Aristotle's account: yielding to others is something that in an uncultivated state I find difficult or unpleasant to do, because it opposes the natural pleasure I take in mastery and self-aggrandizement. But the repeated practice of yielding, which I am socially compelled to do by the rites (in a manner culturally distinct from, but comparable to the laws in Aristotle's account), gradually corrects this problem by getting me *in the habit* of yielding to others, a habit that becomes gradually less unpleasant as I become more accustomed to doing it. If I engage in ongoing efforts of self-cultivation, I may even begin to find it *pleasant* to yield to others when morality calls for it, and at that stage my character will embody the full virtue of *li* in the third sense. In this way, ritual conduct not only expresses who one is and what one values, but as Kongzi states, taking up the habit of ritual practice is what first allows me to "overcome myself" as the human I am, enabling me to then consciously choose to become more "humane."[27] This emphasis on ritual practice as *restraining* and *corrective* of the uncultivated state strongly mirrors the Aristotelian account of moral habituation.[28]

Another conceptual resonance with Aristotle is found in Kongzi's 'Doctrine of the Mean,' which, as in Aristotle's account, suggests that a cultivated person strikes an appropriate balance between extreme courses of action.[29] We should recall that the uncultivated person in Aristotle's account lacks the moral discernment to recognize the mean relative to particular circumstances. He must initially look to moral experts (*phronimoi*) who have this discernment and try to imitate their conduct. In this way he becomes more habituated to and comfortable with moderate actions, which enables the gradual emergence of his own perception of the Mean.

This highlights one final Confucian resonance with Aristotle's account of habituation—namely, the role played by the 'humane' or cultivated person (*junzi*), who serves as an exemplar for the uncultivated. In the *Analects*, Kongzi compares the virtue of the *junzi* to the wind, which sets the direction in which the grass (the common people) will bend.[30] He advises that "when you see a person of worth, concentrate upon becoming their equal."[31] The rites cannot be learned solely from sacred texts; they are only fully embodied in their social practice, and observing their practice by a more cultivated person helps me to appreciate not only their deeper meaning, but also the great sensitivity, nobility, discernment, beauty, and harmonious ease displayed in their successful fulfillment.[32] Like the *phronimoi*, then, the *junzi* (or those who approximate that ideal) serve to illustrate not only *what* should be done, but *how* and *why*. The admirable and attractive moral spectacle that such exemplary persons create makes the path of virtue seem pleasant to pursue, sustaining the uncultivated person's will to become more consistent in moral practice.[33]

Thus the concept of *li* in the first sense, as *ritual practice that follows a pattern expressed in the actions of exemplary, cultivated persons in one's community, enabling a self-transformation through which one may himself become cultivated*, has a powerful semantic relation and a similar developmental function to the Aristotelian notion of habit, despite being far richer in action-guiding content and having associations with sacred tradition not found in Aristotle.

The Buddhist framework for moral self-cultivation, as formalized in The Noble Eightfold Path, contains no direct equivalent of *li*. Buddhist religious practices do incorporate many types of ritual—that is, conduct which is understood to be both sacred and rooted in tradition. Yet by rejecting the Hindu inscription of Vedic belief in a fixed social order, the Buddha detached the idea of spiritual action from conventional action. Although individual schools of Buddhism vary so widely in cultural and religious practice that our analysis can only have general applicability, a Buddhist does not primarily cultivate virtue by learning and embodying social conventions. While a Buddhist is not typically required to oppose the patterns of action mandated by custom, law or other social authorities, her path to moral self-cultivation does not naturally begin by following these channels, as it does for both Kongzi and Aristotle.

Yet there *are* clear resonances in Buddhist ethics with the Aristotelian notion of habit and the Confucian concept of *li*. Perhaps the clearest involves the concept of *śīla*, often translated as 'morality, 'virtue' or 'right conduct.' The latter is nearer to the primary meaning of *śīla* as 'that which accords with the precepts of Buddhism,' where these precepts take the form of a set of rules proscribing certain forms of speech (for example, gossip), bodily actions (such as stealing and killing), and livelihood (trading in weapons, meat, or human slaves).[34] *Śīla* has a modest resonance with the Confucian's second sense of *li* as 'rites,' insofar as both offer a fixed body of action-guiding content. Still, *li* are far more numerous, specific, and prescriptive than *śīla*, insofar as they define the appropriate fulfillment of a wide range of traditional social roles and obligations. *Śīla*, on the other hand, offers only a few highly general prohibitions of conduct, which are intended to govern Buddhist practitioners without regard to their social role or status. In this respect *śīla* may function in a manner nearer to that of Aristotle's conception of law, as an instrument that primarily restrains the uncultivated person from becoming habituated to vicious actions that might naturally seem pleasant or otherwise attractive.

As we noted in chapter 2, Buddhist scholars regard the precepts of *śīla* not as constitutive of moral goodness but as defeasible, contextual, and dependent rules which supervene upon a deeper conception of virtue as a cultivated state of enlightened existence. This explains why, as comparative scholars David E. Cooper and Simon P. James note, *śīla* is just one of the 'three trainings' (*trishiksha*) that make up the basic structure of the Noble Eightfold Path.[35] It is joined by *prajñā* (wisdom) and *samādhi* (concentration), each with its own component elements that, together with the three abovementioned components of *śīla*, make up the full eight steps on the path. The first two steps—'right view' and 'right resolve'—together compose *prajñā*, while the last three steps—'right effort', 'right mindfulness,' and 'right attention'—compose *samādhi*. As Cooper and James note, moral implications infuse these other dimensions of Buddhist practice as well.[36] Thus it is at this more encompassing level of the three trainings (*trishiksha*) that the strongest conceptual resonances with moral habituation and ritual practice begin to emerge.

Indeed the concept of habitual practice infuses the entire Eightfold Path, so much so that there is no single concept matching it. The nearest concept capturing the entire action-guiding content of the Path is the *Dharma*, the full body of Buddhist teachings and doctrine, but this is ultimately *too* broad, as we wish to highlight not the teachings and doctrines themselves, but the way in which habitually following their guidance gradually transforms *who the individual is and what she can readily and reliably do*. As in Aristotelian and Confucian ethics, each component of the Path enables a person to train herself to think, believe, act, and feel in ways that appear difficult or unappealing in her uncultivated state, but are performed with increasing ease and joy as her practice of moral self-cultivation continues.[37] As in the other two virtue traditions, the Path is also understood to lead a person along a 'Middle Way' that avoids destructive extremes.[38] Given the scope of the *Dharma*, however, the total habits to be acquired extend well beyond the conventionally approved social behaviors that can be encoded in traditional rites or mandated by law. Habits of meditation, understanding, and other forms of disciplined, self-controlled activity (*yoga*) gradually reshape the affective and cognitive life of the individual until she is the sort of being who can enter into the enlightened state of *nirvāna*, or, if that state is not yet attained, can at least reduce her own suffering and that of other beings. We shall return to these more refined habits of moral practice as we carry out the remaining six stages of our analysis of moral self-cultivation.

There is one final resonance to highlight here: the role of exemplary and/or cultivated persons in Buddhist moral habituation. As with the *phronimoi* and the *junzi*, human examples of moral excellence play an important role in Buddhist ethics. While the Buddha himself, and others who have attained Buddha-nature, are certainly *the* exemplary persons who ultimately constitute the goal of moral self-cultivation, both the ordinary, uncultivated person (*prthagjana*) and the devoted aspirant on the Eightfold Path (the *bodhisattva*) may lack direct access to a living Buddha. The moral exemplars who provide uncultivated persons with *living* models of virtuous conduct are the *Sangha*. These are traditionally members of the monastic community of monks and nuns, although sufficiently enlightened laypersons are also among the *Sangha*. As one of the "Three Jewels" (along with the Buddha and the *Dharma*) enabling moral self-transformation, the *Sangha* are those living persons who, as comparative Buddhist scholar Peter Harvey puts it, "have been permanently changed, to some degree, by insight" into *Dharma*.[39] He goes on to describe the inspiration provided by these socially revered and attractive examples of human flourishing as motivating the uncultivated person to make the initial commitment "to develop virtue, a generous and self-controlled way of life for the benefit of self and others."[40] Here we see once more the role of exemplary persons who motivate the initial desire to transform oneself for the better, who illustrate for the beginner the mean state where virtuous conduct lies, and who encourage the extended commitment to the practice of self-cultivation (in Mahāyāna Buddhism, the *bodhisattva vow*) that a person needs in order to form mature moral habits. It is this habituation that begins to restrain, correct, reshape, or otherwise transform the uncultivated state into one from which a virtuous life appears not only possible, but increasingly attractive.

Where Buddhism appears to depart from Aristotle, and to a lesser extent from Confucianism, is in the view that this initial move to self-transformation must also be motivated by a certain understanding or knowledge concerning the nature of reality, that is, "some acquaintance with the Buddhist outlook and an aspiration to apply it."[41] The reason for the difference is plain: unlike the *phronimoi*, who may manifest civic excellence without pursuing theoretical wisdom, or even the Confucian *junzi*, who are highly learned concerning the rites in relation to *Dao* but require no further metaphysical philosophy, the admirable qualities of the exemplary *Sangha* are simply inseparable from the metaphysical teachings of Buddhism. Unlike the *phronimoi* and the *junzi*, many of the *Sangha* live a monastic life starkly different from that of the masses, and the only explanation for this way of life is that they hold a very different view of reality from everyone else. The motivation to emulate them simply cannot arise without some basic understanding of *why* they have chosen to live as they do. Yet as Harvey and others note, the purpose of this initial wisdom (*prajñā*) is primarily to motivate the immediate taking up of the habitual practice of right conduct (*śīla*) and thereafter concentration (*samādhi*), which begin to transform the individual such that deeper wisdom can be attained. This in turn enables more refined cultivation of the virtues.[42] Thus the initial habituation to moral activity remains, if not the starting point of Buddhist moral self-cultivation, at least very near to it.

Moral Habituation: Shared Resonances and Technomoral Implications

The complex interaction and mutual enrichment of the seven elements of the practice of moral self-cultivation, across all three classical accounts, will be increasingly evident as the analyses in Part II continue. For now, let us summarize the conceptual resonances among our three classical conceptions of moral habituation. Despite their considerable differences, in all three the practice of moral self-cultivation requires:

1) A gradual transition from an uncultivated state to a morally *habituated* one:
2) One typically motivated and guided by moral *exemplars* in the community;
3) One effected by *repeated moral practice* of right (or nearly right) conduct that strikes the appropriate mean relative to the circumstances;
4) A practice that gradually *accustoms* the individual to actions which were previously seen as painful or unattractive;
5) Which eventually leads to greater comfort, ease, pleasure, and even *joy* in performing moral action, enabling the cultivation of the virtue of temperate self-control or discipline;
6) Which in turn enables and strengthens the ongoing *commitment* to moral self-cultivation required for more specific moral habits to be developed and integrated;
7) The full development and *integration* of which constitute the achievement of, or increasing approximation to, a genuinely virtuous character—that is, a cultivated state of moral excellence that promotes a life of flourishing with others.

Does this also describe how we may begin to cultivate the *technomoral* virtues? If so, we will want to learn from the coming chapters what *kinds* of repeated practices can produce the moral habituation that makes virtuous conduct in the technosocial domain more comfortable, attractive, and even enjoyable. We will also want to know how we may identify those moral exemplars who can show us how to engage in technosocial practices *well*, and in a way that strikes the appropriate mean relative to our circumstances.

Given the condition of acute technosocial opacity diagnosed in the Introduction, we will also want to know how our moral habits can be made sufficiently *flexible* and responsive to emerging and unanticipated technosocial developments. We also must ask how moral habituation can engender greater *self-control* and discipline, such that we, and not our technologies, become the authors of our habits and desires. Finally, given our special concerns about collective human wisdom concerning technosocial practices on a global scale, we will want to know how we can habituate ourselves to effective moral deliberation in the global commons, and how we, to use Kongzi's phrase, can 'overcome ourselves' as beings naturally inclined to make selfish, narrowly tribal, provincial, and shortsighted decisions concerning the use of technology.

4

Cultivating the Foundations of Technomoral Virtue

THIS CHAPTER CONTINUES our exploration of the seven habits of moral self-cultivation shared by classical virtue traditions, habits that can also guide and inspire a new and much-needed global practice for the 21st century: *the cultivation of technomoral virtue*. In chapter 3 we saw that the human practice of moral self-cultivation always presupposes a process of general habituation to moral acts, which gradually weakens or removes natural impediments to becoming virtuous. As these obstacles fall away, the deliberate and sustained pursuit of virtue becomes possible, by cultivating the more advanced moral habits of *relational understanding*, *reflective self-examination*, and *intentional self-direction of moral development* upon which virtue is founded.

4.1 Relational Understanding

To see what *relational understanding* is and why it is essential to the practice of moral self-cultivation, it helps to recognize how classical virtue traditions conceive of the human person: namely, as a *relational being*, someone whose identity is formed is through a network of relationships. While some virtue traditions regard one's relationships with other living things, objects, places, or deities as part of one's unique identity, *all* virtue traditions acknowledge the central importance of our formative relationships with other human beings: our family, friends, neighbors, citizens, teachers, leaders, and models. In contrast, both Kantian and utilitarian ethics portray the self as an independent rational agent who can and ought to act *autonomously*, without relying on the external guidance of others. On this view, a child's agency will remain bound by relational ties, but an adult exercising his own reason is an independent being, free to decide entirely for himself how to live. Moreover, on such a view, the ideal moral agent will adopt a universal and impersonal moral perspective—in critical parlance, the 'view from nowhere.'[1] This means they act from the same motivations and principles as would any other ideal moral agent—as opposed to acting from a unique moral situation defined by the agent's own personal obligations and relational ties.

In contrast, most classical virtue traditions view the self as partly or wholly constituted by concrete roles and responsibilities to specific others in one's moral community.[2] The responses of others to my actions continually inform, and often correct, my sense of who I am in the world. Though these responses can often be unfair or misguided, reliable self-knowledge could not be formed or maintained in their absence.[3] Contrary to Kant (and Nietzsche), the purely autonomous agent can never be a fully realized, self-aware human being. Indeed the ideal moral agent is *not* the person who most successfully detaches her deliberations from her own relational context, but rather the one who understands and responds to that context most fully—that is, a person of practical wisdom. Social ties constitute the very field upon which ethics plays out, and ethical norms exist to preserve and enrich the flourishing of dependent human beings in community, not in isolation. In most classical virtue traditions, then, the cultivated person always acts from within her own unique context of important relations, roles, and responsibilities, while seeking perpetually more refined understandings of these relations and the moral obligations and ideals to which they give rise.[4] In this context, 'relational understanding' is not a noun but a verb, not a possession but an active and sustained *achievement*.

Aristotle gives us a glimpse of what relational understanding looks like in his accounts of justice and friendship. In explaining the nature of justice, he states that, because "man is by nature a 'political animal,'" a person without a political community is either "a bad man or above humanity."[5] Indeed, he says, because justice is the full extension of human excellence toward others, not just toward oneself, the political virtue of justice "is not part of excellence but excellence entire . . ."[6] He goes on to distinguish justice, understood as excellence or virtue of character, from individual *acts* of justice. Of course, the person with a just character will be able, far better than most, to discern what is just in a particular case and what is unjust. As it turns out, however, this presupposes a nuanced and accurate understanding of the particular relationships at stake in each case; for what one person owes to another in a particular case is highly dependent on the kind of relational ties that bind them. As mentioned in chapter 1, Aristotle's view is complicated by his belief in the natural appropriateness of unequal relations between elite Greek men and everyone else (e.g., slaves, women, non-elites, and non-Greeks)—assumptions that from a contemporary standpoint we have good reasons to reject. Nevertheless, even his remarks on this topic reveal the way in which no exemplary person could ignore the deep structure of distinct kinds of human relations; rather, such a person would understand such matters in far more depth and with greater nuance and precision than the ordinary person.[7]

In his extensive analysis of friendship in Books Eight and Nine of the *Nicomachean Ethics*, Aristotle continues to highlight the essential connection between the cultivation of virtue and relational understanding. At the start of Book Eight, Aristotle employs the term 'friendship' (*philia*) as a kind of generic catch-all for relationships of affinity, covering kinship (including the parent-child relationship), civic or political friendship, and even the most attenuated form of affinity between strangers who share nothing in common but their humanity.

As his analysis continues, however, it becomes clear that friendship in any robust form is more demanding than *mere* affinity, and requires a reciprocal and mutually recognized wish for the other's good that emerges from *shared activity*.[8] For Aristotle a human relationship is not a formal category but a worldly fact; it is a form of living together (*suzên*) in an active sense. It is this active living together from which emerge the reciprocal obligations, affections, and concerns that make up the field of ethics. After all, he notes, even if (contrary to fact) a human being could be wholly self-sufficient, "the good man will need people to do well by."[9] Thus for Aristotle, the virtuous person will not only enjoy friendship, he will understand what friendship *is*, not just in its essence but in its fine-grained variations and ethical nuances.[10] This is why Aristotle devotes a considerable portion of his discussion of friendship to describing the various hierarchies and differential obligations that mark different types of friendly relations; for example, he insists that a person cannot become virtuous without learning along the way why "it is a more terrible thing to defraud a comrade than a fellow citizen, more terrible not to help a brother than a stranger, and more terrible to wound a father than anyone else."[11]

Of the three main types of friendship Aristotle analyzes (friendships of utility, friendships based on shared pleasure, and complete or 'perfect' friendships), only virtuous persons are capable of enjoying the third and most rewarding type. In this most intimate relationship between two people, the joint pursuit of virtue enables each friend to function as a "second self" for the other: a mirror of noble character that provides the self-knowledge essential to one's ongoing moral refinement, and the reinforcing pleasure of seeing one's own moral successes reflected in the excellent actions of one's friend.[12] Indeed, it is the *sharing* of virtuous activity that defines such friendships. This means that for Aristotle, understanding the ethical duties and concerns generated by human relations requires more than a grasp of the formal classifications of relational ties (brother, neighbor, wife) and their general patterns. It also requires understanding the nature of the activity *actually* shared in any particular relationship. Thus he notes that while the relationship between husband and wife is generally unequal and often based merely on utility or pleasure, "this friendship may be based also on excellence, if the parties are good . . ."[13]

In conclusion, Aristotle holds that virtue presupposes that one habitually seeks to understand the fine-grained structure and implications of particular human relations. Yet it is interesting that in his analysis of these implications, Aristotle regards kin relationships as a species of the more fundamental category of friendship and its various forms; and in general seems to treat familial ties as simply one especially close type of civic bond. Indeed, Aristotle explicitly states that the State is prior to the family and the individual.[14] As we are about to see, this aspect of his view is incommensurable with the Confucian moral tradition, in which ethics begins with and remains, even in its political forms, a direct extension of moral relationships within the family.

Youzi, a disciple of Kongzi, is quoted in the *Analects* as saying:

> The exemplary person (*junzi*) applies himself to the roots. Once the roots are firmly established, the Way will grow. Might we not say that filial piety (*xiao*) and respect for elders constitute the root of virtue (*ren*)?[15]

This quotation introduces two core Confucian virtues: *xiao* and *ren*. Neither has a direct conceptual parallel in Aristotle, or any other classical virtue tradition, but these distinctive virtue concepts help to explain why Confucians also believe that the development of a virtuous character presupposes the active exercise of what I have called 'relational understanding.'

The concept of *xiao* is generally translated as 'filial piety': it refers to a reliable disposition toward appropriate respect and loving treatment of one's own family, especially one's parents. It is noteworthy that there is a special virtue-concept associated with this behavior, especially since there are other Confucian virtue-concepts that might be thought to do the same moral work, such as respectful yielding or propriety (*li*) and benevolence (*ren*). In the Confucian worldview, however, the family really is the 'root' of moral virtue, and not simply because families have direct influence over the moral education of children. Family is central in Confucian moral philosophy because all ethical action, even between unrelated adults, is seen as an extension or modification of natural family love and respect. Thus a Confucian cannot become morally cultivated without first acquiring a nuanced and fine-grained understanding of family relations and the various obligations, affections, and cares they engender. Indeed, study of the rites (*li*) functions in large part to educate persons about the proper expression of *xiao*.

The concept of *ren* is the subject of much debate among comparative Confucian scholars; it is most commonly translated as either 'benevolence' or 'humanity,' or sometimes simply as 'human excellence.' It is a linguistic composite of the Chinese characters for 'person' and 'two,' and has associations with earlier terms for 'manliness' and 'nobility,' but in the moral context of Confucianism it appears to have two special meanings: first, as a specific virtue of benevolence toward others, and second, as a complete or perfected state of general human virtue (*de*).[16] The use of a single term to capture both meanings is revealing. For the perfection of human virtue in classical virtue ethics just *is*, for reasons noted at the start of this chapter, a perfection of the person as a relational being. To know *how to be excellent* is ultimately to know *how to treat well* those others to whom one is related by kinship, civic, or other ties. As Kongzi sums up the '*ren*-minded' *junzi*: "He cultivates himself in order to bring comfort to others."[17] All other Confucian virtues, including wisdom (*zhi*), courage (*yong*), honesty (*cheng xin*), propriety (*li*), appropriateness (*yi*), and empathy (*shu*) are ultimately aspects of this good treatment of others (*ren*).[18]

This brings us back to the family, where we first learn how to treat others well and in which our moral motivations to do so have their deepest roots. Indeed this is what Youzi means in telling us that the *junzi* or cultivated person 'applies himself to the roots.' Thus *xiao* or filial piety turns out not to be something different from human excellence (*ren*), but its grounding achievement. Moreover, it is not easy to treat one's family well, for the most demanding and enduring moral obligations (and frustrations) arise in precisely this context.[19] Thus it is not as if *xiao* is an elementary virtue that is transcended in later stages; the practice of moral self-cultivation requires constant efforts to preserve and refine it. Nor is it the task entire; for the moral lessons of family virtue must be extended into the political community.[20] If the Way (*Dao*) is to be followed, *ren* must eventually encompass *all* meaningful human relationships, even interactions between strangers.[21] Thus working to understand the nature of human relationships, starting with those of the family and extending into the political realm, becomes a core habit of the person seeking to cultivate herself. As Kongzi states, while the substance of *ren* is to "love others," the substance of wisdom (*zhi*) is to *know* others—for only through relational understanding can I know how properly to love them.[22] As he advises one seeking to cultivate himself: "Do not be concerned about whether or not others know you; be concerned about whether or not you know others."[23]

As did Aristotle, Confucians assert a relational hierarchy of graded moral obligations and affections, such that a morally cultivated person is *not* the one who treats everyone the same, but the one who best understands how *particular* human relationships generate weaker, stronger, or qualitatively different moral requirements.[24] Kongzi's most celebrated disciple, Mengzi, devoted much effort to defending this view against the claims of the rival Chinese philosopher Mozi, who argued for a moral policy of indiscriminate and equal love of all.[25] In defending the importance of understanding and respecting relational distinctions in moral life, something he held to be rooted in human nature, Mengzi describes Mozi's advocacy of indiscriminate love as an extreme position, one no less in violation of the virtuous Mean than is radical egoism.

Yet Mengzi *also* criticizes any form of discriminating love that would ignore the special circumstances of each case. As did Aristotle and Kongzi, Mengzi routinely calls for the use of prudential judgment in applying fixed moral principles such as 'honor one's elders above all.' As Mengzi notes in a discussion with one of his disciples, the cultivated person has the discernment to suspend or modify customs of relational hierarchy (for example, to honor a younger brother before an uncle, or a village elder before an elder brother) when the occasion demands it.[26] Blind, unthinking conformity to the conventional social order does not bring relational understanding—indeed it 'cripples' the Way by singling out one thing "to the neglect of a hundred others."[27] Genuine virtue is never rigid; it is an active, discerning and contextually sensitive appreciation of human relationships, and the particular claims they make upon us.

Thus the pursuit of a nuanced and flexible understanding of human relationships, especially their distinctions and gradations within the State or family, is a core habit of the practice of moral self-cultivation in both Aristotelian and Confucian virtue ethics. When we look for this element in Buddhism, however, we might seem to encounter a problem. While the pursuit of understanding or 'right view' is the first step on the Eightfold Path, in Buddhism this 'right view' holds that all beings are causally interconnected (*pratītya-samutpāda*, or 'dependent co-arising') and that all sentient creatures ultimately embody the same Buddha nature. This *appears* to undermine the moral importance of understanding particular relational distinctions. The ethical doctrine of Buddhism is, in fact, far less wedded than Aristotelian and Confucian ethics to the conventional social order and its attendant roles. Indeed the Buddhist virtue of equanimity (*upeksā*) seems to require the cultivated person to practice ethical "neutrality" and to extend loving kindness (*maitrī*) and compassion (*karunā*) to *all* creatures.[28] Is this a serious obstacle to the claim that all three virtue traditions share a family resemblance, with respect to the importance of understanding the specific moral claims that issue from our relationships?

While these differences are real and profound, they do *not* render the habit of relational understanding unnecessary for a Buddhist who aspires to become cultivated and eventually enlightened. Quite the opposite: a Buddhist must pursue a correct understanding of human relationships with even *greater* vigor and dedication, precisely because on this account, the correct view of human relationships *contradicts* the one reflected in conventional family and political hierarchies. The *Lankāvatāra Sutra* of Mahāyāna Buddhism states that:

> in this long course of transmigration here, there is not one living being that . . . has not been your mother, or father, or brother, or sister, or son, or daughter . . . let people cherish the thought of kinship with [living beings], and, thinking that all beings are [to be loved as if they were] an only child, let them refrain from eating meat.[29]

The relational understanding presented in this passage is part of the transformation in belief and affection that a Buddhist must undergo in order to become cultivated. Thus it turns out that Buddhists too see the habitual pursuit of relational understanding as essential to the cultivation of moral virtues such as compassion and equanimity. Moreover, the passage does not ask the aspiring practitioner to abandon her preexisting understanding of the moral importance of human relations such as kinship. To the contrary, the passage asks the practitioner to dwell cognitively and affectively on the full moral weight of those relationships, to *cherish* rather than abandon that thought of filial love, and *then* to combine it with an important new fact (the metaphysical kinship of all beings) that allows the prior motivation of familial love to be correctly extended.

The same pattern appears in the *Jātaka* story no. 68, where two elderly strangers ask the Buddha to extend to them the kindness ordinarily reserved for elderly parents, for as they explain, they have been already *been* his parents, and his grandparents, hundreds of times over.[30] In chapter 5 we will have cause to discuss the important element of 'extension' of moral concern in the practice of moral self-cultivation. But for now, simply notice that the conceptual link between moral self-cultivation and refined relational understanding is not, as we first feared, *weaker* in the Buddhist virtue tradition than in other traditions; it may even be stronger.

Furthermore, the interest in *particular* human relations and the distinctions among them is in no way made irrelevant to ethical life by this move; they are simply placed in a larger context. For one thing, Buddhists strongly value social cohesion and equitable flourishing, and must therefore be highly attentive, rather than indifferent, to the structure and harmonious functioning of particular social and political arrangements.[31] Also, proper Buddhist equanimity cannot be attained without a lengthy process of moral habituation that must *begin* with an understanding of how to act ethically in particular relationships.

For example, part of the initial 'right view' to be accepted by the Buddhist aspirant is the following assertion: " 'there is mother and father': it is good to respect parents, who establish one in this world."[32] While it offers a limited perspective, it is not false, nor does it contradict the lesson of the *Jātaka* story. It also conveys more useful ethical guidance to the person first seeking to improve her character than would be the case if she were told right off the bat that her parents in metaphysical fact had no unique claim on her respect or gratitude. Indeed Buddhist practice is exquisitely sensitive to the gradual character of personal transformation; the early aspirant cannot absorb or usefully apply the same form of truth as can a learned member of the *Sangha*.[33] Thus one always starts from what one already knows: here, the natural importance of family relations; once one has learned to do well with that, one is gradually able to understand more profound relational truths, which in turn enables even more refined practice and learning.[34]

As in Aristotelian and Confucian ethics, the Buddhist habit of relational understanding will ideally be exercised in conjunction with another habit of moral self-cultivation, discussed in chapter 5: prudential judgment in responding to the morally salient features of particular situations. Prudential judgment is part of all Buddhist practice, but most explicitly articulated in the Mahāyāna tradition, which reserves for sufficiently cultivated and adept practitioners the practice of 'skillful means' (*upāya*)—that is, the ability in unusual or challenging circumstances to suspend or modify the precepts in unconventional ways, where made necessary in the interests of compassion and the enlightenment of others.[35] Thus striving for relational understanding does not mean adhering to rigid relational scripts, but rather aiming to exercise superior relational *judgment*.

Relational Understanding: Shared Resonances and Technomoral Implications

We may now summarize the conceptual resonances found among the three classical conceptions of relational understanding. In all three, notwithstanding significant differences, the practice of moral self-cultivation requires:

1) The habitual pursuit of an increasingly nuanced and accurate yet holistic understanding of how one is bound to other members of one's moral community by friendship, kinship, political, metaphysical, or other morally salient ties;
2) A lifelong effort to progressively improve one's understanding of these relationships and the distinctive moral obligations, affections, and concerns they generate—in turn fostering character virtues such as justice, empathy, care, and civic friendship;
3) The use of this understanding in increasingly prudent judgments that respond ably to the *particular* moral claims that these relationships make upon us in actual circumstances.

In what way might this part of moral self-cultivation assist us in developing *technomoral* virtues? How can a deeper and more nuanced understanding of our relations to others improve our conduct and motivations in the contemporary technosocial arena? Would our practices become any more *just, caring, civil,* or *compassionate*, any more likely to promote human flourishing, if more of us had a deep understanding of how our technomoral choices affect others across the globe, and how our fates increasingly depend on the technomoral choices of other global actors? Could this sort of understanding foster the *moral perspective* needed to see our lives as given meaning by a larger moral whole? Can this perspective be sought not as an academic exercise, but as part of a sincere and habitual effort to understand the pull that our relationships have, or ought to have, on our natural moral sentiments and sense of obligation to others? It will not surprise the reader to hear that I believe the answer to all these questions to be "Yes."

In Part III we shall see concrete indications of how today's emerging technologies, from social media and digital surveillance tools to social robots, can both challenge *and* enable the contemporary moral pursuit of relational understanding. For now, let us turn to the third foundational habit of the practice of moral self-cultivation.

4.2 Reflective Self-Examination

This component of moral self-cultivation has well-known roots in classical Greek philosophy. After all, it was Socrates who stated to his Athenian jury that "the unexamined life is not worth living for a human being."[36] By this Socrates meant that a good life presupposes a lifelong habit of reflective self-examination, in which one turns a critical eye upon one's actions and dispositions, and measures them against the moral norms and virtues to which one aspires. Socrates warns his fellow citizens that their desperate efforts to be happy by means of caring for their social reputations, bodies, and personal property are doomed to fail because in devoting all their energies to these, they neglect the one form of self-care that actually has the power to deliver a good life: the lifelong improvement of the soul.

While Aristotle's *Nicomachean Ethics* identifies key points of disagreement with Socrates (for example, on the topic of weakness of will or *akrasia*), it does not directly address the topic of the examined life. On some level, Aristotle seems to assume that his readers already accept the Socratic mandate of reflective self-examination. This is perhaps a not unreasonable assumption, given that the lectures from which the *Nicomachean Ethics* are composed were addressed to committed students of philosophy rather than the broader Athenian populace.

Still, a close reading of the *Ethics* does tell us something about *how* Aristotle saw the activity of self-examination contributing to the cultivation of moral virtue.

For instance, in his discussion of the difficulty of striking the appropriately virtuous mean that characterizes right actions, Aristotle notes that the difficulty is magnified by the fact that where the mean lies in any given case, and how best to strike it, is relative not *only* to particular circumstances and to general human flaws, but also to the distinctive weaknesses of the individual seeking to act rightly. As he points out, even though *in the abstract* the mean lies 'in between' two vicious extremes, such that we would seem to want to try equally hard to avoid either extreme, in practice we must make a more vigorous effort to pull away from that one extreme that embodies *our* own weaknesses of character.[37] Thus most of us, when facing a moral situation in which fear is a relevant factor, must try harder to avoid cowardice than rashness if we are to get the right result. For example, consider the massive emissions software fraud that Volkswagen perpetrated on its customers and environmental regulators for years before it was uncovered in 2015. What kept the software engineers involved in the fraud from blowing the whistle? It wasn't likely a lack of moral knowledge; it would have been obvious to anyone that the fraud was unlawful, malicious, even contrary to the long-term interests of company shareholders. Yet for most, doing the right thing would have required pulling hard against a strong natural fear of betraying a powerful entity from whom one receives a livelihood. A virtuous person would have recognized this weakness in themselves, and the need to muster the moral will to counteract it.

But this is not true of everyone; *some* people are naturally prone to being rash, and more likely to do wrong by needlessly blundering into danger than by trying to avoid it. A few VW engineers might have had little or no fear of their employer's wrath, but found themselves impulsively drawn to the risky high-wire act of perpetuating such a massive con. It is thus of *critical importance* to my practice of moral self-cultivation that I know which kind of person I am. For if I am among the minority who are naturally rash, but I rely in matters of courage on the general advice of moral experts aimed at the timid majority, I am certain to fail. Reflective self-examination not only of past actions, but also of my values, beliefs, and priorities can help me to identify those vices to which *I* have in the past tended to succumb, and hence aid me in devising more effective strategies for self-improvement.

The importance of the habit of reflective self-examination is also in the background of Aristotle's discussion of moral incontinence in Book VII of the *Nicomachean Ethics*, where he distinguishes this from mindless self-indulgence. In the realm of pleasures and pains the truly virtuous state is called *temperance* (*sôphrosunê*); this is the state in which one is attracted only by pleasures that are truly good. But this is a very rare excellence; most modestly cultivated persons are, at best, *continent*—that is, we still have desires that we know it would be bad to indulge, but we exercise self-control and typically refrain from doing so. In the contrary condition of *incontinence*, I know that I desire what is not good, but I lack the self-control to refrain from pursuing such things.

A different condition of vice would be *self-indulgence*, in which I do not know that my desires are bad, and I therefore do not even try to restrain them. Now notice that the difference between incontinence and self-indulgence is one of *knowledge*; in the case of incontinence I have engaged in sufficient self-examination to at least know which desires of mine I *ought* to control, while in the case of self-indulgence I could not begin to cultivate myself even if I wanted to, for I do not know which of my moral habits need correcting. Aristotle states that the latter condition, though it makes one helpless to improve, is the result of a failure of moral habituation that is itself blameworthy, for "they are themselves by their slack lives responsible for becoming men of that kind. . ."[38]

For the practice of moral self-cultivation to succeed, then, I must at an early stage of the process get in the habit of actively 'taking care'—that is, engaging regularly in the sort of reflective self-examination that Socrates describes as the highest form of self-care.[39] Interestingly, however, the function of reflective self-examination in Aristotle is not exclusively critical. In his discussion of friendships among virtuous persons, he notes that one of the great joys of such friendships is the way in which, as I contemplate the virtues of my beloved friend, I am also able to see a reflection of my own excellence. This is because through living together (*suzên*), our virtues have largely developed in tandem by performing noble activities together, whether these be acts of public or private generosity, heroic military courage, wise legislation, or philosophical discussions of the Good itself.[40] Whether or not this is a realistic vision of contemporary friendships, especially those in which joint activity often happens in online or virtual spaces, it suggests that, for Aristotle at least, reflective self-examination allows me to do more than learn from my failures—it also allows me to recognize and find joy in my moral achievements. Even further, he suggests that self-examination ought to produce *only* moral pride in the exceptionally cultivated and "great-souled" person (the *megalopsuchos*), as moral shame is desirable only in those who have yet to learn to avoid shameful actions.[41]

When we turn to the Confucian tradition, it is far less clear whether the cultivated person will allow himself much latitude to contemplate his own virtue; after all, Kongzi himself repeatedly denies his own achievement of a cultivated state.[42] Yet he does demonstrate a willingness to chart his own moral progress, as when he describes how he gradually moved from the moment when he first dedicated himself to self-improvement at age fifteen, to his acquisition of a temperate character more than five decades later, when at last "I could follow my heart's desires without overstepping the boundaries of what is right."[43]

Still, the majority of Kongzi's extensive remarks on reflective self-examination emphasize its critical power—that is, the importance of being able to identify one's moral shortcomings.[44] As he says, "the real fault is to have faults and not to amend them."[45] Perhaps the best interpretation of Kongzi's attitude toward the habit of reflective self-examination is that it should seek the mean between excessive self-criticism and excessive self-praise, but, as with Aristotle's admonition to pull harder against the extreme to which we are more naturally subject, we must restrain our natural tendency for self-praise and make far greater efforts to be humble. After all, Kongzi laments, "I have yet to meet someone who is able to perceive his own faults and then take himself to task inwardly."[46] Given his repeated encouragement of this very practice, it is clear that Kongzi does not mean to suggest that reflective self-examination is humanly impossible to practice; only that to do this sincerely, consistently, and accurately is an ideal of which virtually all humans fall somewhere short. Only a *junzi* ('exemplary person') "does not grieve that people do not recognize his merits; he grieves at his own incapacities."[47] This interpretation also explains why Kongzi repeatedly encourages his disciples to see virtue in others, since our excessive tendency toward self-aggrandizement happens to be paired with an excessive tendency to find fault with others.[48]

Reflective self-examination, then, is an uncommon and difficult practice for most. As Mengzi says, even though 'watching over one's character' is a fundamental moral duty akin in importance to caring for one's parents, "the multitude can be said never to understand what they practice, to notice what they repeatedly do, or to be aware of the path they follow all their lives."[49] Later we will explore how some of our current technological fixations seem to exacerbate this age-old problem. But for Mengzi, the problem is not that we are constitutionally *unable* to consistently examine ourselves; for if that were true, morality would be an unattainable goal for any of us. Rather, the real obstacle is our overall lack of interest in being good—that is, in our own moral cultivation.[50] Once we acquire that interest, nothing stands in the way of taking up a habitual, lifelong practice of reflective self-examination. One of Kongzi's earliest disciples, Master Zeng claims that:

> Every day I examine myself on three counts: in my dealings with others, have I failed to be dutiful? In my interactions with friends and associates, have I in any way failed to be trustworthy? Finally, have I in any way failed to repeatedly put into practice what I teach?[51]

This passage should also remind us that the practice of reflective self-examination is far less likely to deliver reliable or useful self-knowledge if practiced at a level of high generality ('Am I courageous?' 'Am I loyal?' 'Sure I am!') than if it is practiced at the level of concrete moral choices that engage specific relational obligations ('Was I brave enough in speaking out against that injustice?' 'Did I behave like a loyal son when my father's reputation was slandered by my friends?'). Thus the habit of reflective self-examination cannot be separated from the habit of relational understanding, since what I need to examine is whether I have adequately attended to those to whom I am connected in any number of different ways. It should be increasingly apparent from our analysis that the core elements of moral self-cultivation mutually implicate and reinforce one another—not as stages of a sequential process but as interlinking and interdependent parts of a continuous circle.

Is there in Buddhist ethics a conceptual resonance with the practice of reflective self-examination? Given the extensive Buddhist literature on 'mindfulness,' a practice of attending closely to one's own changing mental and physical states, it seems obvious that there is. However, the resonance is not a simple correspondence, for several reasons.

First, since Buddhists generally deny the substantial reality of the personal self (*ātman*), one might think that a Buddhist cannot commit to a practice of reflective *self*-examination. After all, without a concept of the self, what is there to examine? Fortunately, this obstacle evaporates on deeper inspection. While it is true that Buddhists do not accept the reality of a substantive self, they fully acknowledge the *apparent* self—the self as an appearance that others encounter and interact with in the world. And since Buddhist ethics is a practical concern with human suffering in this world, it in no way ignores or denies the ethical significance of the (apparent) personal self.[52] Moreover, since each of us has his or her own unique history of incarnations and karmic interactions with others, the practice of moral self-cultivation must be tailored to the needs of that unique history, even if the goal of enlightenment toward which we strive is the same.[53]

With the obstacle of the 'no-Self' doctrine (*anātman*) cleared away, it is evident that Buddhism *does* recognize habits of reflective self-examination as an essential part of the practice of moral self-cultivation. As in Aristotelian and Confucian ethics, the Buddhist practitioner is explicitly encouraged to examine the moral quality of her interactions with others; to identify her own particular and general human faults and to learn from them; to take responsibility for correcting her faults so that her actions may find the virtuous mean relative to her circumstances; and to examine and take appropriate joy in her moral successes.

In truth, the habit of reflective self-examination plays a more expansive role in Buddhism than in the other two virtue traditions. Whereas the aspiring Aristotelian *phronimos* will exercise the habit of regularly examining his actions and desires to see if they are in conformity with accepted standards of virtue, and the Confucian *junzi* will regularly examine his interactions with others to see whether they express the cardinal virtues of ritual respect (*li*), appropriateness (*yi*). and benevolence (*ren*), the Buddhist field of self-examination is far more encompassing.

We can see this from a quick review of The Noble Eightfold Path. The concept of reflective self-examination resonates most closely with the third of the 'three trainings' that compose the path: *samādhi* or 'concentration.' As noted in chapter 3, this section of the path is composed of three steps: 'right effort,' 'right mindfulness,' and 'right attention.' Right effort represents the conscious striving to avoid or overcome unhealthy states of mind and replace them with healthy ones; right mindfulness represents a habit of discerning awareness of the mental and physical states I experience, as well as their interrelations; while right attention represents a habit of calming and clearing the mind with meditative practice. Reflective self-examination, in the sense we have developed thus far, seems to resonate quite closely with the first two habits of *samādhi*: right effort and right mindfulness. Together, these habits allow me to monitor and improve upon my actions and character. It must be noted that many, perhaps even most, lay Buddhists do not engage in dedicated habits of meditative and reflective practice; and Buddhism does not rule out the ability of rare individuals to become enlightened without such methods. Still, for the vast majority of devoted *bodhisattvas* and *Sangha*, such habits are core elements of self-cultivation.

These habits have as their objects not moral conduct alone, but *every component* of the Eightfold Path. I must make an effort to remain aware not only of whether my speech, actions, and livelihood are conforming with ethical precepts (*śīla*), but also to maintain awareness of the two aspects that constitute wisdom or *prajñā*: namely, whether the views and beliefs that occupy my mind are true ('right view'), and whether my attitudes, desires, and emotions are appropriate ('right resolve'). Beyond this, these habits of right effort and right mindfulness are directly reflexive upon the domain of *samādhi* itself: I must make an effort to be mindful of whether I am striving to be (and continuing to succeed at being) mindful! Now before one dismisses this task as impossible, or at least well beyond the ability of nearly every living person, it should be emphasized that the practice of *samādhi* is assumed to develop gradually and very slowly; an ordinary, relatively uncultivated person who tried to leap into a state of total mindfulness would fail instantly.[54] The point is simply that the vast majority of persons cannot progress very far along the path of moral self-cultivation without undertaking and gradually building the reflective practice of *samādhi*.

While Buddhism's account of reflective self-examination extends well beyond the explicitly ethical realm, we should remember that both Aristotle and Kongzi required virtuous actions to be accompanied by appropriate moral feelings, thoughts, and beliefs. Thus the conceptual resonance among the three remains quite strong, since in all virtue traditions the practice of self-examination must be directed not only at what I *do,* but also what I *feel, think,* and *believe.*

Reflective Self-Examination: Shared Resonances and Technomoral Implications

This leads us naturally into the fourth core habit in the practice of moral self-cultivation, namely, the intentional self-direction of moral development. But first, let us summarize the shared conceptual resonances we found in the third:

1) In order to make reliable progress toward a morally cultivated state, one must develop a habit of reflective self-examination;
2) This habit must evolve into a lifelong practice that aims to discern how well one's actions, feelings, thoughts, and beliefs conform with the moral self to which one aspires;
3) It should identify and pay close attention to those particular weaknesses and faults to which one is generally subject as a human being, as well as those which are distinctive of one's individual character; it should foster the virtues of humility and honesty.
4) The practice should engender a sense of responsibility for correcting the faults identified so that one's actions may more reliably strike the virtuous mean;
5) The practice should also engender appropriate joy in the ongoing experience of moral self-improvement and increasingly virtuous living.

Could more widespread habits of reflective self-examination *today* help human beings take more personal *and* collective responsibility for the global consequences of their technomoral vices? Could it engender more honesty and humility with respect to the current limits of our moral vision? Might this habit gradually refine our moral judgment of the best courses of technosocial action, and allow us to experience greater joy in living with the technologies we create? Only by understanding how this habit fits into a complete practice of moral self-cultivation will we be able to answer these questions, so let us continue our study of that practice.

4.3 Intentional Self-Direction of Moral Development

Given the mutually implicating and enriching structure of the elements of moral self-cultivation, each subsequent habit will have been anticipated in some way by our explication of the preceding ones. Indeed, the fact that moral self-cultivation must be intentionally self-directed may seem quite trivial now. Yet its core dimensions have yet to be made explicit.

In the Aristotelian model of moral development, which explicitly addresses the moral habits of young children, intentional self-direction is not possible at the beginning of the process. For it presupposes four things that children necessarily lack in Aristotle's view: the mature exercise of deliberative reason, reliable knowledge of general moral principles, accurate perception of the proper ends of a human life, and habituated self-control of the emotions and appetites. These deficiencies can only be corrected over time, by a combination of proper parental influence, wise laws, guidance from moral exemplars, accumulated experience and understanding of human relationships, and maturation of the rational faculties. Even those rare children who possess naturally virtuous dispositions and need less in the way of moral habituation face the remaining obstacles of limited practical experience and immature cognitive powers. Thus even the most naturally precocious youth are incapable of directing their own moral cultivation.

Gradually, however, young people become able to exercise the faculties of both practical and theoretical reason, and given sufficient moral habituation and experience, a young person becomes increasingly capable of forming both the independent will to cultivate himself and the ability to direct his own moral progress as a form of self-care (with the continued support of virtuous friends and exemplars). At some point, he may finally embody the prerequisites for genuine moral virtue: namely, that he acts well, "from a firm and unchangeable character," by choice and with knowledge [of the goodness of the acts], choosing those virtuous acts *for the sake of virtue itself*.[55] Once these achievements are joined with that part of practical wisdom described in Book VI of Aristotle's *Ethics* as perception of the proper ends of a human life, one is fully capable of directing one's own moral improvement.[56] Unfortunately, Aristotle thinks that the majority of adults fail to reach this stage of moral maturity, mostly due to a lack of the initial moral habituation needed to develop self-control and temperance.[57]

Those who do reach this final stage have an intrinsic motivation to pursue their own moral improvement. For the exemplary person or *phronimos*, this motivation takes shape in the virtue of *magnanimity* (*megalopsuchia*): a form of moral ambition or 'greatness of spirit' that finds pleasure in one's truly superior moral achievement.[58] Although this virtue is distasteful to many modern readers, it is important to remember that while the *phronimos* is motivated by the pleasure he takes in noble actions, this does not reduce his virtue to mere hedonism. For the pleasure is taken *in* the noble activity itself, not *from* the activity. It is the act and not the pleasure that is the desirable thing.[59] Nor does this pleasure in his own moral superiority make him an egotist: Aristotle points out that while the *phronimos* in seeking the good life "assigns to himself the things that are noblest and best," the pursuit of the good life is not the zero-sum game it is commonly thought to be, for "if *all* were to strive towards what is noble and strain every nerve to do the noblest deeds, everything would be as it should for the common good, *and every one would secure for himself the goods that are greatest*, since excellence is the greatest of goods" (emphasis added).[60] If being a noble and morally cultivated person is what you value most, Aristotle has good news for you—that's an unlimited resource we don't have to fight over.

Such moral striving marks my transition to awareness of my *responsibility* for the state of my moral being and the total quality of my life. It represents a commitment to a project of lifelong self-perfection, since as Aristotle notes, the full moral measure of a life cannot be taken until that life is complete.[61] This commitment not just to moral acts but to a *moral life* presupposes both courage and ambition, as well as an imaginative broadening of perspective to one's being in the world, taken as a moral whole.

The importance of intentional self-direction of moral development is equally explicit in Confucianism. Kongzi famously claimed that he would only continue to teach the kind of student who, after having Kongzi lift up one corner of a problem for him, returned with the other three.[62] Otherwise, the student would never learn how to solve moral problems for himself, he would forever rely on the Master to give him the answers. Moral cultivation requires our own personal effort to develop morally, and following the example and instruction of others will only get you so far. Even Mengzi and Xunzi, who disagreed sharply over the question of whether the basic moral orientation of human nature is good or bad, agreed upon one central aspect of moral self-cultivation: it requires a lifelong, cumulative effort of self-improving activity, a habit that requires a far greater and more sustained commitment than merely *willing* oneself to be good (what Mengzi dismisses as an attempt to become good by an illicit "seizure of righteousness").[63]

Like Aristotle, Mengzi identifies a passive element in moral development, such that certain environmental conditions are ordinarily needed in order to bring a person to a point in their natural development where they have the motivation, habituation, and preliminary understanding needed to actively direct their own moral cultivation. These conditions include exposure to the Way (*Dao*) via the rites and appropriate social models of virtue, the experience of familial love and concern, and protection from abject poverty and violence.[64] Yet even where the environmental conditions are suboptimal, the individual is encouraged to actively take up the effort to improve herself: "When the Way prevails everywhere, use it to pursue your personal cultivation. When the Way does not prevail, use your personal cultivation to pursue the Way."[65]

The active component of moral self-cultivation is described by Mengzi as a form of attention (*si*), in which one maintains a focused awareness not simply on potential impediments to moral improvement (such as inappropriate desires), but also on the ultimate goal of that process: the full maturation of the 'sprouts' of virtue that he claims are naturally present in all of us.[66] We may say, then, that for Mengzi, *si* combines two mental functions. The first is the mental habit of reflective self-examination discussed above, through which I discern the extent to which my actions, thoughts, and feelings align with the moral ideal (for Confucians, the *Dao*). The second component of *si* is the goal-oriented activity of steering these actions, thoughts, and feelings in the morally desirable direction. This activity requires one to begin to grasp the meaning of one's own existence in relation to a moral whole—that is, to cultivate a perspective on one's own life in relation to the *Dao* itself.

It is important to note that, for Mengzi, the intentional self-direction of moral development must respect the natural integrity of the process, and cannot be achieved by overly self-conscious or contrived efforts. He likens the latter to the disastrous attempts by a farmer to help his crops grow faster by constantly tugging at their sprouts.[67] The Confucian tradition describes fully virtuous actions as performed with harmonious ease. Yet this is compatible with lifelong moral striving. As Kongzi notes, "the task of self-cultivation might be compared to the task of building up a mountain: if I stop even one basketful of earth short of completion, then I have stopped completely. It might also be compared to the task of leveling ground: if I have only dumped one basketful of earth, at least I am moving forward."[68] Among classical Confucians Xunzi emphasizes this striving the most, since he holds that moral self-improvement requires violating and wholly remaking (with the guidance of *li*) one's 'ugly' inborn nature. He describes this as requiring a lifelong campaign of unwavering focus, one that progresses not by moral leaps but by many small steps. Yet even *he* claims that the result is joyful ease in living well:

> When he has truly learned to love what is right ... his mind will feel keener delight than in the possession of the world. When he has reached this stage, he cannot be subverted by power or the love of profit; he cannot be swayed by the masses; he cannot be moved by the world. He follows this one thing in life; he follows it in death. This is what is called constancy of virtue. He who has such constancy of virtue can order himself, and having

> ordered himself, he can then respond to others. He who can order himself and respond to others—this is what is called the complete man.[69]

One may rightly question when one can realistically expect to reach the stage Xunzi describes here, and what this stage actually entails; is the complete man *perfected*? It seems not, since Xunzi suggests that moral learning ends only in death.[70] Even Kongzi states that the *junzi*'s burden of becoming humane (*ren*) is heavy, and ends "only with death."[71] If this is right, then does the *junzi*, who can 'order himself' with apparent ease, still make an intentional *effort* to be good? The tension here can be reduced if we recall that moral self-cultivation happens by degrees. Since we are never fully perfected, even the cultivated person must still make an effort to order herself. But not every effort is a brutal, desperate struggle; compare the elegant, joyful effort expended by an accomplished pianist or gymnast with that of an amateur. The process of moral cultivation is a self-overcoming, but the self I have to overcome at the start is *far* more opposed in her thoughts, feelings, and desires to my ideal self than is the self that I am trying to overcome after having finally 'learned to love what is right.' Thus the complete human is not perfected; rather, he is complete in that the whole of his *actual* being—his beliefs, attitudes, values, desires, feelings, and even physical movements—is at last engaged in the effort of becoming his ideal self. This is what Mengzi means when he says that "A *junzi* steeps himself in the Way because he wishes to find it in himself. When he finds it in himself, he will be at ease in it; when he is at ease in it, he can draw deeply upon it; when he can draw deeply upon it, he finds its source wherever he turns."[72]

We may say, then, that the intentional self-direction of moral development is a long, uphill, and endless climb, but that the path's incline becomes less steep, the footing more secure, the trail better marked, and the scenery more lovely, familiar, and welcoming the further I go.

Buddhist ethics offers similar ideas about the intentional self-direction of moral development. While these ideas resonate with the Aristotelian and Confucian traditions, they do not correspond perfectly. For one thing, Buddhist metaphysics entails that persons do, if only rarely, accomplish total self-perfection (that is, the achievement of *nirvāna* or the Buddha-nature) in their own lifetimes. However, we have said that for our purposes we will be concerned only with the 'ordinary' paths of moral self-cultivation consistent with engaged family and political life, as opposed to the extraordinary lives of monastic or philosophical contemplation that many Buddhists and Aristotelians respectively regard as highest.[73]

According to the philosopher Asanga, co-founder of the Yogacara school of the Mahāyāna tradition in Buddhism, we can only pursue intentional self-direction of moral development after having received correct guidance from the habituating power of the moral precepts of *śīla,* along with the initial 'right views' of the self and its relation to others and to the world. Yet once this foundation is set, one must *activate* this wisdom with one's own proper resolve and sustained mental efforts to cultivate oneself (i.e., "to have a quite purified intention, to make correction after failure, and to avoid failure by generating respect and remaining mindful after that.")[74]

If we examine this quote, we find that it resonates with the idea of reflective self-examination as a habit of self-care that entails minding and correcting one's moral failures. Moreover, a 'purified intention' aims to perfect oneself not for the sake of social approval or expedience, but for the sake of moral nobility itself. Such 'right resolve' is in fact the second step on the Noble Eightfold Path.[75] Among our three virtue traditions, the likeness in moral resolve or intention is somewhat stronger between Buddhist (especially Mahāyāna) and Confucian ethics, since both emphasize the motivating power not *only* of the noble quality of the virtuous path, but also its power to bring comfort to others.[76] While Aristotle regards virtuous living as incomplete without friends to benefit and do good by, it is primarily the other two virtue traditions that emphasize the reduction of others' suffering as a central part of the cultivated person's moral resolve.

Yet in Buddhist ethics, the *mere* resolve or will to perfect oneself is insufficient. We must also actively 'remain mindful,' in the sense intended by the 'right effort' that is the Eightfold Path's sixth step. More specifically, this means that we must make an enduring effort to actually oversee and steer our moral progress in the right direction. How this is done resonates in some strong respects with the Aristotelian and Confucian virtue traditions, while departing in others. In their analysis of Buddhist self-cultivation, David E. Cooper and Simon P. James use the term 'self-mastery' to characterize this process. As they note, it is not that one is somehow trying to enslave or degrade his true self (after all, there is ultimately no 'self' to master). Rather, through gradually liberating one's mind from unhealthy and base thoughts, desires, and feelings, a person gains mastery "over what, so to speak, renders him less than a person and obstructs his development into a rational, self-directing being, no longer at the beck and call of whim, caprice, and 'craving.'"[77]

Cooper and James note here another strong resonance with the Aristotelian (and, it turns out, Confucian) tradition insofar as the long-term goal of Buddhist self-cultivation is to eventually move beyond the merely continent condition that comes from submitting my desires to conscious restraint, attaining instead a state of temperate desire (in Aristotle, *sôphrosunê*).[78] Such discipline is characterized not by white-knuckled conformity to virtue but a "relaxed" and "graceful" dwelling in it—what in the Confucian tradition is referred to as "harmonious ease." Yet there remain ways in which the Buddhist tradition stands apart from Aristotelian and Confucian accounts of self-directed moral improvement. First, as Cooper and James note, the very advanced Buddhist practitioner is often encouraged to cultivate himself in greater solitude than Aristotle would have thought necessary or ideal.[79] The tension with the Confucian tradition here is even stronger, since the complete virtue of *ren* is *only* acquired with others.[80]

Moreover, Buddhism advocates "guarding the sense doors": turning the mind away from or recontexualizing ordinary sense-perceptions so that one's understanding of reality as fleeting and inconstant is not obstructed by illusions of sensory power and permanence.[81] Nothing like this effort is discernible in the Aristotelian or Confucian moral traditions; whatever part it could play in Aristotle would belong to theoretical rather than practical self-cultivation, and it could find no conceptual home in the moral worldview of Confucianism, which largely takes the world of sensory appearance as it is given.[82] Still, at a more abstract level, 'guarding the sense-doors' does resonate with the other virtue traditions, as it conveys the idea that the self-direction of moral development must be guided by a truer, more complete moral *perspective* on reality—a view of one's life and its meaning as part of a larger whole.

This resonance can be further explicated if we return to the earlier quotation from Asanga. He refers to "generating respect," and gives this a causal role in avoiding moral failure. Peter Harvey analyzes this remark as related to the concept of self-respect or *hri*, a state of moral conscientiousness "which causes one to seek to avoid any action which one feels is not worthy of oneself and lowers one's moral integrity." Along with *apatrapya*, a sense of moral shame for the potentially harmful consequences of my actions, *hri* is seen by Buddhists as "the immediate cause of virtue" in the world.[83] The concept of *hri* resonates strongly with the emphasis in Aristotelian and Confucian ethics on discerning *what sort of future self is actually worthy of my becoming in this world*— that is, acquiring a proper perception of what sort of life and activity is fitting for the kind of worldly [human, sentient] being that I am. As we saw in chapter 3, all three traditions hold that I am incapable of reliably directing my own moral improvement until I have acquired at least a glimpse, and eventually a steady view, of the proper ends of a human life. Moreover, this cannot be a purely *intellectual* grasp of what it is good for me to become; this ideal of the self living nobly must engage my affections, as something to be *loved* and *enjoyed* for its own sake. This love and joy in the good becomes the moral motivation to pursue my own self-perfection, and to courageously persevere in the considerable and lifelong efforts that I must dedicate to its ongoing achievement.

Intentional Self-Direction of Moral Development: Shared Resonances and Technomoral Implications

In this habit of moral practice, we can identify the following conceptual resonances:

1) It presupposes attainment of a certain degree of moral habituation, relational experience and understanding, and a habit of reflective self-examination. Certain material and social deprivations or misfortunes can inhibit the attainment of these conditions.
2) The activation of intentional self-direction as a form of self-care implies a genuine moral desire or resolve to attain moral goodness; such resolve must arise not from mere social expedience or passive submission to external authority, but from a sincere, internal desire to become morally cultivated for its own sake.
3) Moral intention is insufficient without some accompanying insight into the type of being that is *worthy of my becoming in this world*; this achievement of a holistic moral perspective upon my existence must eventually be joined with a temperate and magnanimous character that *enjoys* existing and acting nobly.
4) The active exercise of this intention entails a lifelong effort to attend to, or be mindful of, my moral development and courageously direct it along the path I have chosen;
5) The enduring exercise of this intention may not lead one to moral perfection, but does lead one by degrees to a more graceful, relaxed, harmonious, and joyful way of life.

What does all this mean to us in the technosocial context? First, we should ask how emerging technological practices and institutions can foster the basic material and social conditions needed to enable the above-described capacity. We should also ask more broadly how modern technosocial life affects our resolve to cultivate ourselves, or our ongoing motivation to continue with the effort. For example, how might certain emerging technologies, in specific use contexs, inhibit our ability to remain attentive to our moral development? How might other emerging technologies, or other use contexts, be used to *strengthen* our moral attention? Are modern technosocial culture and its dominant liberal values compatible with the idea that only some types of life or activity are actually *worthy* of human attainment and enjoyment? Finally, is the goal of living in a more graceful, relaxed, harmonious, and joyful way consistent with modern technosocial habits? If not, how should we address these conflicts?

Building upon the early foundation of *moral habituation* described in chapter 3, chapter 4 has articulated three more cornerstones of the practice of moral self-cultivation: *relational understanding, reflective self-examination*, and *intentional self-direction of moral development.* Together, these four habits can be understood as providing the foundation upon which moral character is built. Yet as Aristotelians, Confucians, and Buddhists all recognize, genuine moral wisdom, of the sort found in persons of *exemplary* moral character, requires still further efforts.

In chapter 5, we will review the final three habits essential to the shared practice of moral self-cultivation, which together transform the foundations of morally virtuous character into a robust and mature moral wisdom. This is a critical part of our inquiry, for it is the global cultivation of *technomoral wisdom*, above and beyond mere technical knowledge or even moral discipline, without which 21st century humans facing an increasingly unsettled, risky, and opaque future may have little chance of learning how to live well—or perhaps even how to continue living at all.

5

Completing the Circle with Technomoral Wisdom

THE FINAL THREE habits of moral self-cultivation that we will explore (moral attention, prudential judgment, and extension of moral concern), along with the foundational habits analyzed in chapters 3 and 4, enable what is called moral or practical *wisdom*. Practical wisdom is often classified as an intellectual virtue because it involves cognition and judgment; yet it operates within the moral realm, uniting cognitive, perceptual, affective, and motor capacities in refined and fluid expressions of moral excellence that respond appropriately and intelligently to the ethical calls of particular situations. While the acquisition of practical wisdom does not *conclude* the effort of moral self-cultivation, it represents its summation as a *complete practice*.

5.1 Moral Attention

The importance of moral attention is easy to miss from a cursory reading of Aristotle's *Nicomachean Ethics*, yet it is deeply embedded in his account of practical wisdom (*phronēsis*). In Book VI of the *Ethics*, he describes practical wisdom as involving a form of moral perception. That is, a person of practical wisdom can almost always discern and attend to those features of a particular situation that are most salient for the purposes of ethical judgment. Turning to geometry for an analogy, Aristotle explains that, just as a geometrically wise person will 'see' that what is *mathematically* significant about three intersecting lines is that they form a triangle, a practically wise person (a *phronimos*) will reliably 'see' and attend to the *morally* significant facts of concrete situations.[1] For example, we can expect that a *phronimos* would notice that the elderly widower whom he passes on the street every day appears increasingly gaunt and in tatters, and in need of someone's—at this moment, *his*—active concern and generosity. Almost all of us have some capacity to perceive and attend to moral salience in the world. How many of us would fail to see, for example, that what is significant about the sight of a starving coyote stalking an unattended toddler is that it calls for the observer's courageous and immediate intervention? What distinguishes the well-cultivated person or *phronimos* from those who are morally ordinary is how reliable and accurate his faculty of moral attention has become, as a result of habitual moral practice and considerable experience. The practically wise will notice and respond to morally significant phenomena to which the rest of us will often be oblivious or indifferent.[2]

This perceptual capacity of *moral attention*, what John McDowell calls a "reliable sensitivity" to the morally salient features of the given situation in which I am called to act, does not by itself constitute the encompassing virtue of practical wisdom.[3] A person of exemplary virtue exercises her moral attention jointly with other habits of practical wisdom, especially those analyzed later in this section: prudential judgment and appropriate extension of moral concern. Of course practical wisdom also assumes the foundational elements of moral character, such as proper habituation, relational understanding, self-examination, and moral intention. The perceptual element of moral attention thus serves as a bridge between my understanding of the *general* aim of moral action (to promote the goods proper to a human life); my knowledge of the *particular* facts of my circumstances; and my judgment of the *best course of action* toward that goal that is available to me in these circumstances. It is only by attending to the salient features of *this* moral situation confronting me that I am able to discern how to act in a manner that is prudent and appropriate to my immediate context, while also effective in promoting the ultimate ends of a good life.[4]

By cultivating a knack for perceiving the moral contours of any situation around us, we are better armed for the unexpected and often confounding moral dilemmas that life inevitably brings. The world does not give us advance notice of every kind of moral scenario it might throw at us, and there is no definitive moral playbook that tells us in advance how to read or respond to any situation. As Martha Nussbaum explains, only the skill of attending to moral salience can ready us "for the actual flow of life, and for the necessary resourcefulness in confronting its surprises."[5] What better skill to cultivate for an opaque future in which the only predictable thing is more surprises? Moral attention, as a type of sensitivity to changes in one's moral environment, is also far better suited than abstract principles or fixed rules to accommodate the many contextual subtleties of actual human relationships, something we have said the virtuous person will understand well. Such a sensitivity promotes justice and moral care by allowing for the different ways in which the 'right and the good' can be expressed by particular people, in particular relationships.[6] Finally, when we consider the important role that empathy plays in the perception of human relations and their moral salience, we recognize that Aristotelian virtue ethics avoids the all-too-common trap of treating our emotions as either irrelevant to, or hived off from, the intellectual virtue of moral wisdom.[7]

In Confucian ethics we find strong resonances with the concept of moral attention. Already in chapter 4 we encountered a term employed by Mengzi, *si*, often translated as 'attention.' There *si* seemed to play two key roles in moral self-cultivation: attending to the *self* in reflective examination, and attending to the *goal* or aim of one's moral development (for Confucians, a life lived according to the *Dao*/Way.) But moral perception in Aristotle does something different. Its primary focus is neither the self, nor the intended goal of the self's cultivation, but rather the features of the particular situation that call *to* the self to respond in a moral way. Is there any indication in the Confucian literature that *si* plays a similar role in moral life?

We may find one in Arthur Waley's analysis of Kongzi's remark in *Analects* 2:15 that "if one learns without thinking (*si*), one will be confused. If one thinks without learning, one will be in danger." Kongzi is drawing a distinction between thinking and the learning of ritual, not unlike Aristotle's distinction between knowing moral particulars and knowing general moral principles. Though he translates *si* as 'thinking' rather than 'attention,' Waley explains that in this context, *si* should be understood as "a fixing of the attention . . . on an impression recently imbibed from without and destined to be immediately re-exteriorized in action."[8] One way of unpacking this is that *si* is the cultivated person's *attention to the morally salient features of his or her present circumstances* (that "impression recently imbibed from without"). These features call upon the virtuous person to form, with the help of prudential judgment, an appropriate moral response (one "destined to be immediately re-exteriorized in action"). Thanks to *si*, this response will be context-sensitive but still informed by ritual 'learning' (*li*). If this interpretation is roughly correct, then there is a strong conceptual parallel in Kongzi with the attentive moral perception central to Aristotle's account.

This interpretation gains further support from repeated references in the classical Confucian canon to cases in which a cultivated person or *junzi* was able to perceive a morally salient feature of a given situation, one that prompted him to modify the standard application of ritual learning (*li*) in a manner that a) responded appropriately and skillfully to a situational novelty; b) could not have been specified in advance by the ritual itself; c) was seen as perplexing or improper by less cultivated persons who clung rigidly to the letter of the ritual teachings; and d) was evoked by the cultivated person's affective response to the salient fact, seen against the background of a larger moral whole (the *Dao*). Each of these factors parallels Nussbaum's analysis of the advantages of Aristotle's perceptual account. The cases in which they appear also show how the habit of attending to moral salience fosters appropriate flexibility in one's responses and a richer moral perspective that takes in the total meaning of the situation.

While it would consume too many pages to survey the relevant cases in detail, they include Kongzi's observations that in a certain case declining a salary would not be principled but wasteful (*Analects* 6:5); that truthful speech desperately needed by one man is wasted on another (15:8; 16:6); that compassionate respect for a blind person is more important than honoring one's social station (15:42); and that withdrawing from politics to protect one's moral purity is essential at times but foolhardy at others (18:8). In the last case, he notes that he holds "no preconceived notions of what is permissible and what is not"; an *astonishing* remark from someone who spent his life championing the importance of full adherence to the rites (*li*). This suggests that for Kongzi, the action-guiding content of ritual learning becomes fully manifest only *after* one has attended closely to the morally salient facts of the case at hand. Similar cases are emphasized by Mengzi: from the man who must perceive that ritual propriety does not bar him from grabbing his drowning sister-in-law; to the man who rightly conceals his marriage from his abusive parents; to Mengzi's resolute defense of his own gossip-inducing decision to spend more on his mother's funeral than on his father's.[9] In each of these cases, attention to morally salient facts enables a prudentially wise response but does not itself constitute or determine the appropriate action. The action itself issues from a skillful judgment of what should be done *in light of this moral salience*, a form of judgment that we shall take up in the next section.

Whether it is best captured by the concept of *si* or by the related Confucian term *zhi*,variously translated as 'moral wisdom' or 'intelligent awareness,' moral attention in Confucianism has both cognitive and affective components, as in Aristotle's ethics.[10] One must grasp *why* a particular fact is morally salient while also having that salience evoke the appropriate moral emotions. In the Confucian tradition this sensitivity may be associated less closely than in Aristotle with immediate, quasi-perceptual moral 'seeing,' since the moral agent aspiring to be cultivated is often encouraged to reflect upon these salient similarities or differences for an extended period of time.[11] Yet there is debate on this point: Stephen Angle has argued for a view of Confucian moral attention as the intuitive and unreflective exercise of a morally perceptive faculty, while P.J. Ivanhoe argues that the sensitivity in question is more akin to a connoisseur's deliberately cultivated sense of taste.[12] This may simply be a difference between an ordinary person and a highly cultivated one; the *junzi* may directly perceive the moral salience that a less cultivated person requires considerable training and reflection to discern.

To what extent do we find in Buddhism conceptual resonances with the concept of moral attention? As we might expect given the already-noted importance of mindfulness in Buddhism, cultivating moral attention is perhaps even *more* central to Buddhist ethics than to other traditions. However, it is difficult to find a single concept to which it may be assigned. The Buddhist concept of the mind and its functions (*citta*) is notoriously complex. Several of its associated terms might be properly translated in ways that implicate attention to moral salience, but seeking an exact equivalence to the concept is probably a fool's errand.[13] For one thing, the cultivated Buddhist directs her primary efforts of concentration inward—not in order to turn *away* from the moral or spiritual salience of the world, but rather to purify the mind of illusions and preoccupations in such a way that this salience can finally break through.[14] This habit of moral attention in Buddhism enables the cultivation of a virtuous perspective in which particular moral saliences are properly related and scaled with respect to a larger moral whole. While the result of an attentive mind in all three virtue traditions is a more ethically sensitive being, each tradition describes the mental activity that gets one to that state somewhat differently, with the Buddhist account standing furthest apart.

Moreover, Buddhism's emphasis on the reduction of suffering means that the cultivated Buddhist's moral attention has a more specific aim. Whereas a cultivated person in the other two traditions attends to everything from situational calls for filial obedience to the political need to respond appropriately to an expensive gift, the kind of moral attention most valued in Buddhism is a "heightened sensitivity to the suffering of these beings" that fosters empathy or compassion (*karunā*).[15] The ability to perceive and attend to the presence of suffering is a capacity without which one could never hope to become an enlightened Buddhist. If I do not *see* the suffering of others I cannot work to reduce it, no matter how much I might wish in the abstract to be compassionate and caring. Still, for the cultivated Buddhist, compassion will not be a blind, unthinking emotional response, nor a detached and merely cognitive appreciation of a duty to relieve suffering; rather it will synthesize moral feeling with that 'right view' and understanding that puts suffering into its proper perspective.[16]

A concept closely related to *karunā* (compassion) is *muditā* ('empathetic joy'). Here the idea is to be able to attend to others' happiness and to respond with appropriate moral co-feeling (consistent with a cultivated understanding of such emotions as transient phenomena.) Since *karunā* and *muditā* require moral attention be paid to *all* beings, including strangers and animals, their scope of application is far broader than in Aristotelian and Confucian ethics, where the moral calls to which I am supposed to attend arise almost exclusively from local human relationships. This difference is of deep metaphysical significance to Buddhists, given the principle of dependent co-arising (*pratītya-samutpāda*) that holds all beings to be causally interconnected. It is also of great significance for those moral saints on a supramundane (*lokottara*) path to moral enlightenment. Still, for the ordinary moral aspirant, attending more fully and generously to all whom I encounter in daily living may be a suitable short-term goal.

While all three accounts regard moral attention as an essential capacity of the cultivated person, none fully articulate *how* this perceptual capacity might develop and be habitually refined—that is, how it can best be *cultivated*. Aristotle states only that it is a 'natural' capacity, insofar as we are cognitively and affectively 'fitted' by nature to develop it, but notes that it only *does* develop with sufficient practical experience.[17] Nor do Confucians spell out precisely how the *junzi* comes by his superior moral sensibility. How much experience is required to develop it, and more importantly, what *kinds* of practical experience best enable its mature and habitual exercise? Does it require a deliberate effort to cultivate and fully refine? Or does it develop passively, given sufficient time and experience? How do we *learn* to pay moral attention? These questions receive some answers in Buddhism, in the form of meditative and moral practices that heighten one's attention to the suffering and joys of others, but in general this aspect of moral self-cultivation remains undertheorized in classical virtue ethics.

Moral Attention: Shared Resonances and Technomoral Implications

Here are the conceptual resonances we found among the three accounts:

1) The practically wise person displays, in addition to the foundational habits of moral development, a heightened attention or sensitivity to the morally salient features of particular situations and circumstances in her environment;
2) Moral attention integrates a cognitive grasp of *why* certain facts of the situation are morally salient with an appropriate emotional response to that salience. The habit of paying moral attention directly fosters virtue by making us appropriately sensitive to the specific features of our environments that call for morally excellent action.
3) The deliverances of moral attention inform our practical judgment of how to concretely respond to a given situation; they also enrich and enlarge the moral perspective that enables us to successfully contextualize and adapt the rules, conventions, or normative principles that would generally apply. Moral attention thus allows the cultivated person to respond more quickly, flexibly, and appropriately to novel, fluid, or unpredictable situations.

In our emerging global condition of acute technosocial opacity, the value of moral attention is clear. If this habit helps humans respond better to changing, unpredictable, or unanticipated moral environments, then we need it to be as widely cultivated among us as possible. But how can attention to moral salience be more reliably and consistently cultivated in technosocial life? Which emerging technological practices can be used to aid its development, and which will tend to impede it? What *sorts* of moral attention should 21st century humans most want to cultivate? Greater attention to the joys and suffering of all creatures, as called for in Buddhist practice? Closer attention to the demands of justice and merit, as Aristotle would suggest? More attention to human needs for familial respect and benevolence, as the Confucians would emphasize? Or are there *other* dimensions of moral salience even more relevant to contemporary technosocial life? We'll take up these questions in Part III.

5.2 Prudential Judgment

The term 'prudential judgment' is sometimes used interchangeably with 'practical wisdom.' In our account, however, while prudential judgment presupposes the foundational development of good moral character, it is just *one* of the habits that refine this character into practical wisdom, the exemplary virtue of the morally cultivated person. Thus far we have said that the person engaged in moral self-cultivation must begin from 1) a state of preliminary habituation to moral acts, and must go on to acquire; 2) a habit of seeking relational understanding; 3) a reflective habit of self-examination; 4) the habit of intentionally steering one's activity and character toward the aims of a humanly good and noble life; and 5) an attentive environmental sensitivity to moral salience. We have also said that cultivation of each of these five habits enriches and further perfects the others.

Yet these five elements would be wholly incapable of producing reliably ethical action without the habit of prudential judgment, defined as *the cultivated ability to deliberate and choose well, in particular situations, among the most appropriate and effective means available for achieving a noble or good end.* As technical as that may sound, most will recognize this ability in someone we know. Some people have 'good moral sense'—they often seem to judge just how to make a bad situation better, and they rarely if ever make things worse. They are not overly distressed by unexpected or novel problems to which existing moral rules and conventions do not easily apply, nor do they fall victim to 'analysis paralysis' when those rules or conventions appear to conflict. They reliably display self-control and flexibility, and maintain appropriate moral perspective. They rarely if ever overreact or 'go overboard'; they seem to know how to strike the appropriate mean relative to the situation, which means that they are *also* capable of a mounting a firm and vigorous response when it is called for.

We have seen references to this habit already in this chapter, but let us now bring it into focus. For Aristotle, prudential judgment is the very core of *phronēsis* or practical wisdom, even though *phronēsis* also incorporates and presupposes other elements. Of practical wisdom generally, he says that the person who has it aims, deliberates, and calculates well concerning how what is "best for man" can be attained by action in the present.[18] Prudential judgment, the core skill we are speaking of now, is the deliberative and calculative part. It *presupposes* the moral discipline and understanding needed to have the right 'aim' at what is 'best for man' overall. It also assumes a correct perception of the salient moral goods, attainable in my present situation, which may promote that aim. Yet it remains necessary to *judge and choose a course of action*—the specific means by which those goods can be most effectively attained.

Aristotle describes this prudential judgment as involving reasoning rather than simple intuition or guesswork, but it is the kind of reasoning that is both correct and *expedient*—that is, effective within the time allowed for a decision.[19] Thus a person who always calculates the right course of action, but not before the window of opportunity to put that decision into practice has closed, is not prudent. Deliberative prudence is a habit of judgment which is 'discriminating of the equitable' relative to the situation— that is, it strikes the correct mean between excessive and deficient response. It is also appropriately *flexible* judgment, insofar as it recognizes when particular circumstances drive the mean or 'equitable' course of action in a direction that requires, for example, forgiveness rather than legalistic standing on principle.[20] Deliberative skill *alone* is not practical wisdom, since the latter integrates the full range of moral habits that we have been analyzing. Skill in choosing the best means to a perceived good is meaningless unless I am habituated to aim at the correct good in the first place, and it isn't virtuous unless I am motivated to choose that good for its own sake. A person who skillfully selects means to attain ends that are not *actually* good ends for a human being to seek, or who aims at the right ends but for the wrong reasons, is merely clever, not wise.[21]

When we turn to the Confucian moral tradition, we find that this habit of prudential judgment is also recognized, though not explicitly theorized to the extent we find in Aristotle. It is embedded within a Confucian concept often translated as moral wisdom: *zhi*. Bryan Van Norden identifies *zhi* as having four basic components:

> (1) the disposition to properly evaluate the characters of others and oneself; (2) skill at means-ends deliberation: the ability to deliberate well about the best means to achieve given ends, and to determine the likely

> consequences of various courses of action; (3) an appreciation of and commitment to virtuous behavior; and (4) intellectual understanding [as of ritual].[22]

We can see from this that (1), (3), and (4) have already been discussed in the context of other habits of the practice of moral self-cultivation, and they also resonate closely with other components of practical wisdom identified by Aristotle. It is (2), skill at means-ends deliberation, which closely approximates the specific part of practical wisdom that Aristotle calls "deliberative excellence" or prudential judgment. This skill is highlighted frequently in the Confucian tradition by means of parables and historical cases, including those briefly summarized in our earlier discussion of moral attention, and in this remark by Mengzi:

> Wisdom is like skill, shall I say, while sageness is like strength. It is like shooting from beyond a hundred paces: it is due to your strength that the arrow reaches the target, but it is not due to your strength that it hits the mark.[23]

Here we see the archery metaphor so often used in classical virtue traditions to represent the combination of right aim and skillful execution of moral action. The metaphor is especially helpful insofar as it helps us to see how all the ingredients of practical wisdom are integrated: the knowledge or learning of the basic principles behind the art; the habituated desire to practice it; the love of doing it well; the understanding of what 'doing it well' amounts to (Where is the target?/ What is genuinely good for a human being?); the knowledge of what my particular strengths or weaknesses are in this area (Do I tend to pull left or right?/ Do I tend to excessive anger or fear?); the perception of the salient features of the particular situation (Where is the wind blowing from today?/Is my typical policy of being direct and honest unusually dangerous in this case?); and, finally, the *skillfulness in putting together an action* that will, given all these other considerations, most likely yield a successful result.

In addition to wisdom (*zhi*), comparative Confucian scholar Jiyuan Yu regards deliberative skill as strongly associated with the virtue of *yi* or "appropriateness." Yu describes the power of *yi* as ensuring that the moral course of action is not obscured from my view by my benevolent love for others (*ren*) *nor* by my respectful adherence to ritual propriety (*li*).[24] It finds a "fitting" balance between these basic moral attitudes, which is another way of saying that it discerns the virtuous or 'equitable' mean for the situation. As noted in the previous section, this also depends upon the cultivated person's ability to see when the salient moral features of the situation invite or necessitate a modification, reinterpretation, or suspension of accepted moral norms. The case in Mengzi of the drowning sister-in-law is perhaps the best known example. Although ritual propriety (*li*) prohibits a man from touching his brother's wife, to stand on ritual principle and refuse to pull her out of the water not only fails to appropriately balance the moral attitudes of love and ritual respect, it violates the very moral meaning of the ritual! The ritual principle in question exists to ensure that a man respects the sanctity of his brother's relationship with his wife; but how does he respect it by allowing it to be destroyed by her untimely death?

This reminds us that prudential judgment *can* be exercised without the full possession of practical wisdom; one does not need to be an especially cultivated person to recognize what is called for in the kind of case Mengzi is discussing. That said, a person who is not highly cultivated will very often err in the other direction—that is, they will excuse themselves too readily from important moral duties for reasons of mere expedience or comfort, falsely chalking it up to 'extenuating circumstances' that make such an exception permissible. It was this sort of tendency, in fact, that led Immanuel Kant to insist that affective desires and emotions (our 'inclinations') cannot reliably guide or inform our moral deliberations.

Kant had a point, though he pushed it too far. Practical wisdom, which merges cognitive and affective modes of moral sensibility with finely cultivated habits of moral reasoning and judgment, offers more reliable and useful action-guidance in a greater variety of circumstances than Kant's abstract categorical imperative. But while *some* people are practically wise, not very many are, even though very many people would falsely label themselves as such, or are so labeled by others. Thus the trick is to be able to distinguish those with genuine practical wisdom from those who possess what comparative Confucian scholar Lee Yearley calls "semblances of virtue."[25] A practically wise person, then, is one who in addition to the deliberative skill of prudential judgment also possesses the other cognitive and affective elements of a highly cultivated moral nature—elements that allow prudential judgment to be exercised reliably and *well*.

Explicit reference to this exemplary power can be found in the Mahāyāna Buddhist doctrine of 'skillful means,' or *upāya kaushalya*. This doctrine has a far broader scope of meaning than the concept of prudential judgment, but it does include something like a (limited) license to use skilled and careful judgment to modify or even violate established moral precepts (*śīla*), though only when demanded by the broader interests of morality itself.[26] This license is highly restricted to practitioners of considerable moral and spiritual cultivation, for precisely the reason mentioned above: only a person of genuinely virtuous character and deep moral wisdom can be trusted to modify, reinterpret, or suspend such precepts. As comparative Buddhist scholar Damien Keown notes, while the doctrine of skillful means preceded the emergence of the Mahāyāna tradition, it originally allowed only relatively minor departures from accepted rules of conduct. Only in certain Mahāyāna texts did the view emerge that a sufficiently advanced *bodhisattva* or spiritual adept could, in extraordinary circumstances, commit occasional violations of even the most important moral precepts (for example, prohibitions of lying, stealing, and killing).

However, these texts do not imply that the highly cultivated person has somehow transcended or earned her freedom from the duties of morality. Indeed, as Keown notes, the very same text that asserts that "even what is forbidden is allowable" for an advanced *bodhisattva* also claims that such a person will "strive *unfailingly* not to violate the precepts" (emphasis added).[27] The point is that the employment of skillful means or *upāya* is *itself* a form of striving to meet the ultimate demands of morality, namely, to seek above all things "the welfare of others with compassion."[28] Only when keeping to the letter of the precepts would gravely endanger others' welfare am I licensed to violate them, and only if I am sufficiently cultivated to genuinely know *what* I am doing, *why* I am doing it, and *how* to do it well. Even then the cultivated person must take responsibility for the profound consequences of this violation. Consider the story of the Buddha who, when faced with a 'lose-lose' moral situation as a *bodhisattva* in a former life, chose in the interests of compassion to kill a robber bent on murder, even though he knew that the karmic cost to him would be a "hundred thousand aeons" of being reborn in hell, while the would-be robber/murderer would find himself reborn in heaven.[29] Prudential judgment is not therefore a license to shed moral constraints; it is a way of making sure that the ultimate aims of morality remain controlling.

Prudential Judgment: Shared Resonances and Technomoral Implications

Let us summarize the family resemblance among the three classical virtue traditions concerning prudential judgment:

1) The practically wise person must express (in addition to the other habits of a morally cultivated person) a well-honed capacity for prudential judgment.
2) Prudential judgment can be defined as *the cultivated habit of deliberating and choosing well, in particular situations, among the most appropriate and effective means available for achieving a noble or good end.*
3) The prudential judgment of the morally cultivated person enables her to deliberate and act well even in cases involving unanticipated, novel, or rapidly evolving circumstances for which established rules or scripts of conventional moral behavior do not prepare us.

4) In exceptional circumstances, a sufficiently cultivated person may employ prudential judgment to modify, reinterpret, or suspend accepted moral norms, rules, and conventions, but only when the more encompassing interests of morality require it.

Along with the habit of moral attention that informs it, prudential judgment is essential to coping with the circumstantial novelty, rapid change, and uncertainty about the future that humans increasingly face in the 21st century, a condition I have called acute technosocial opacity. As noted in chapter 1, there are no fixed rules, systems, or principles of ethics existing today that can offer sufficient action-guidance for the long-term ethical development of emerging technologies such as artificially intelligent robots, or the ethical use of new media or pervasive digital surveillance systems, or the ethical reengineering of our biology. Each of these emerging domains, not to mention future technomoral domains that we cannot even anticipate today, presents challenges far too complex, dynamic, and context-sensitive to be resolved *simply* by applying standing moral rules or principles intended to govern all possible cases.

That said, important questions are left unresolved by our examination of prudential judgment. First and most broadly, how can this kind of deliberation, along with the other dimensions of practical wisdom, be more widely cultivated in the human family? Most importantly, how can this deliberative excellence be employed not just in the realm of private judgments but in *collective* moral deliberations regarding the global development and use of emerging technologies? Finally, how can we keep our deliberative skill from being exercised outside of practical wisdom, in ways that serve the interests of personal, national, or corporate expedience rather than the broader aims of human flourishing in our technosocial environment? We'll take on these and other related questions in Part III.

5.3 Appropriate Extension of Moral Concern

This final habit of moral self-cultivation is perhaps the simplest to articulate, yet the most difficult to put into practice. By 'concern,' we mean not mere worry but an actively caring aim, an authentic decision to *do well* for others. By 'appropriate extension,' we mean the ability to expand one's basic attitude(s) of moral concern (whether in the form of fairness, love, respect, or compassion) to the right beings, at the right time, to the right degree, and in the right manner. In keeping with the holistic account we have given, the development of this ability does not mark the cessation of moral cultivation, but continues to inform and enrich the entire process, improving our habits; our relational and self-understanding; and our moral intention, attention, and prudential judgment, which then individually and jointly aid us in further improving our extension of moral concern. Still, the ability to extend moral concern in a virtuous manner marks a certain kind of completion of the moral self; not a *perfection* of the self, in the sense of having no more work to do, but a completion in the sense that every active moral potential of a person has been directed toward the project of self-cultivation. In this sense, the completion of the moral self may be thought of as the closing of an endless circle that now runs with increasingly less resistance, since no active element of one's intellect, affect, or desire naturally impedes it.

While the extension of moral concern is a central and explicit theme in both Confucian and Buddhist thought, its role in Aristotle's ethics is more obscure. Nevertheless, it does play an integral part in the core Aristotelian virtue of justice, and in the activity that for Aristotle makes justice redundant and, among practical activities, is most closely associated with the good life: complete friendship. On Aristotle's view, complete friendship is the most intimate and enduring form of friendship. Unlike friendships based solely on mutual pleasure or utility, this form of friendship can only emerge between virtuous persons who pursue *together* the noblest activities in life. A friend of this sort becomes a 'second self,' someone in whom my own life's worth is reflected, and whose well-being matters to me as much as my own.

The connection between justice and complete friendship in Aristotle's account is highly complex, but the basic conceptual link is *reciprocity*.[30] For Aristotle, justice entails that people get from one another the good that they deserve. What people deserve from one another is relative to the circumstances between them, and to the individuals in question; perfect reciprocity is not perfect equality.[31] For example, justice between an apprentice and a mentor entails reciprocity—each owes something to the other—but it does not prescribe perfect equality. The mentor deserves a form of respectful deference that would be inappropriate for the apprentice to receive, and the apprentice deserves a form of nurturing concern that the mentor would be wrong to expect in return. Laws and other moral norms exist in order to compel and/or reward reciprocation with one another in appropriate ways. When Aristotle says that "complete friends have no need of justice," he simply means that in this type of relationship the appropriate sharing of the good need not be compelled, nor rewarded from without; it is already guaranteed by the noble character of the individuals and the moral care they spontaneously extend to one another.[32]

This spontaneous care, which he identifies as "love" (*philia*), is something that Aristotle took great pains to explain; he characterizes it as emanating from a kind of self-love, but one that is neither blameworthy nor selfish. Much has been written about this aspect of Aristotle's account, and it suggests a strong resonance with the concept of 'moral extension.' Love, well-wishing, and pain or joy for my friend in virtue seem to well up from within as spontaneously as do these emotions for myself.[33] If I do not want to describe my love for another as a weaker, less spontaneous, and derivative form of affection, then one way of avoiding that is to conceive it as essentially the *same* original love—just extended outward and broadened in scope to include two persons rather than one. This fits well with Aristotle's description of complete friendship as an empathetic bond between a virtuous person and his "second self."[34] It also explains why he regards parental love as extending outward from the original recognition of my child as a literal "part of myself."[35] Whether or not we are comfortable with these accounts of friendship or familial love, they suggest that Aristotle indeed validates the extension of moral concern.

For Aristotle, then, the highly cultivated person will readily extend appropriate moral concern to all those that deserve it, that is, he will be just. His justice will also involve *care*, for "it is fitting to go unasked and readily to the aid of those in adversity."[36] Furthermore, he will be just and caring in the way of a temperate man—that is, he will find it easy to genuinely desire justice for its own sake and not for personal expedience or reward. However, he will not typically extend to others more than they actually deserve, for this would be inequitable, excessive, and therefore *unjust*. That is to say, his judgments with regard to extending concern will be prudent and honest, informed by attention to the morally salient features of the situation calling for concern.

For Aristotle, limitless moral concern for another does not extend further out than the realm of virtuous or complete friendship, which he regards as a selective, rare, and nearly exclusive bond. A virtuous person will, if he is sufficiently lucky, have friends who are a "second self" to him; but he will not be able to have more than a few; perhaps no more than one.[37] While citizens who are not friends of this sort will not be bound by mutual love and limitless empathy, *if* their State is a just one they will still extend to one another that broader moral concern for the welfare of their fellows that is manifest in the virtue of civic friendship, or *philia politikē*.

The Confucian account of 'moral extension' reveals important conceptual resonances but also striking differences. Unlike Aristotle's account, where the general moral concern to be extended is one of benevolent justice between friends (i.e., an equitable fairness in giving to others what they deserve), in the Confucian tradition the moral attitudes of concern to be extended are those of familial love and respectful yielding. Hence in Confucian ethics, family care is the affective model for all other human bonds, including political relationships between fellow citizens and between ruler and subject. Still, while the moral quality of the attitude remains that of filial care (*xiao*), its intensity must be appropriately modulated when extended outside the biological family (so one should extend the same caring *attitude*

to one's elderly neighbor that one naturally has for an elderly uncle or grandfather, but it should not be of the same *degree*). The Confucian model, then, does not seem to allow for the *unbounded* mutual care between unrelated persons that we find in Aristotle's account of virtue friendship. This is because in the Confucian model, the ideal moral attitudes to be extended do not originate in relationships among equals, but between child and parent, or older and younger siblings.

The doctrine of appropriate extension of moral concern is most explicitly articulated in the teachings of Mengzi, who repeatedly cites it as the key to the moral perfection of virtue and the mark of a highly cultivated person. The most notable discussions of extension occur in the context of Mengzi's visits with King Xuan of Qi.[38] The king opens the discussion to his personal shortcomings as a ruler, such as his incontinent weaknesses for wealth and women, and his tendency to overtax his subjects and expose them to unnecessary military campaigns. The discussion between Mengzi and the king is not really about moral knowledge, for the king is fully aware that he should be a more caring and compassionate ruler. The discussion is about why he finds it so difficult to actually extend this moral concern to his subjects.

Mengzi points out that the king is not incapable of empathy or care, for he was once famously moved to spare a terrified ox that was headed to slaughter, replacing it with a lamb. As Mengzi puts it to the king, "all you have to do is take this very heart here" [the king's heart, the one moved by the suffering ox] and "apply it to what is over there" [the suffering people].[39] Nor should this be difficult, since if he can be overtaken by spontaneous compassion for the suffering of a mere animal, he can certainly be moved by the suffering of his own people. The problem is not an inability to have the right kind of moral experience, but rather the king's refusal to allow himself that experience with his people (since allowing it would mean being moved to curtail his own indulgences and ambitions). Although the king claims to see the point, Mengzi ultimately fails to move him to change his ways, and much of the secondary literature on this story is concerned with whether this reveals a flaw or an inconsistency in Mengzi's account of moral motivation.[40]

For our purposes, however, what matters is the success of this story in highlighting how a very simple norm—extend your existing moral attitude of concern from its natural or current target(s) to all those who deserve it—is nevertheless one of the most challenging moral habits to practice. Of course one general source of difficulty is determining who deserves our moral concern, and how much of it. As with Aristotle, classical Confucian thinkers like Mengzi did *not* think that 'appropriate' extension entailed caring for all others to the same degree, or in the same manner. Since some people have a far stronger moral claim on our concern than others, one has to apply one's relational understanding, moral attention, and prudential judgment to the situation to inform the judgment of who in particular deserves our moral concern, and how much. This is especially challenging in novel, highly complex, or changing social situations.

But King Xuan is not struggling with *this* kind of case. Even in the absence of prudential judgment and moral attention, his basic ritual learning regarding political relationships is sufficient to tell him that his own subjects deserve much better care than he is giving them. The king also has sufficient awareness of his own moral failings to be interested in discussing their causes with Mengzi, so a lack of honesty or humility isn't the chief obstacle. Self-control is clearly a problem area for him (hence the wealth and women), but is this the real issue? As Mengzi notes, the king could cherish wealth and still share it with the people if he was simply willing to extend to his people his own care; for then providing comforts for them would seem appealing in the same way that providing comforts for himself and his own family appeals to him.[41] He just seems unwilling, or unable, to be moved by empathy to take them into his care.

Mengzi repeatedly addresses this tension between the immense moral power of becoming humane (*ren*: a power which "subdues inhumanity as water subdues fire,") and our individual and collective failures to make the necessary effort ("Nowadays those who practice humanity do so as if with one cup of water they could save a whole wagonload of fuel on fire.").[42] As contemporary Confucian scholar Tu Wei-Ming notes in his commentary on Mengzi's metaphor, our only hope for counteracting the "dehumanizing tendencies in society" today that all too easily convince us that our species is hopelessly evil (or at best, hopelessly weak to resist evil), is to throw our individual and collective energies into the effort to expand our moral concern: "If we really want to save a whole wagonload of fuel on fire (or our world from being incinerated), we have no other long-term recourse but to educate ourselves to increase our supply of humanity."[43]

Perhaps it is fitting that the appropriate extension of moral concern, perhaps the most powerful moral tool we can hope to command, is the most challenging to cultivate among the seven moral habits of self-cultivation that we have considered. In the coming chapters we will have cause to return to this difficulty as a special challenge for 21st century life, one that has a complex and evolving relationship with emerging technologies. Before we can consider this, however, we should see if Buddhist ethics can shed further light on the habit of moral extension and how it can be more effectively and reliably cultivated. After all, the duty of moral extension is arguably the *overarching* ethical principle of Buddhism. As we will recall from our discussion of 'skillful means' (*upāya kaushalya*), the only possible justification for a *bodhisattva's* skillful suspension of the precepts of morality (*śīla*) would be that this course of action was the only way for her to extend generosity (*dāna*), loving-kindness (*maitri*), compassion (*karunā*), sympathetic joy (*muditā*), and equanimity (*upeksā*) toward *all* creatures.

Yet this reveals a fundamental difference between the ideal practice of extension in Buddhism and its ideal practice in Aristotelian and Confucian ethics. Buddhism does require the extension of moral concern to be 'appropriate'—for example, compassion and loving-kindness must always be expressed in a mindful, morally attentive, and prudent way. To express compassion by refusing to give a lifesaving injection to a child terrified of needles would not be a virtuous act. 'Appropriate' will also still be relative to the particular situation; I have to be flexible enough to adjust my compassionate treatment of others to *what they need* relative to their particular stage of moral and spiritual development, their distinctive karmic history, and their present material circumstances. But for the most highly cultivated person, an 'appropriate' extension of moral concern no longer entails a discriminating calculation of graded love or degrees of concern based on who the subject of my compassion is in her worldly relation to me. All other things being equal, a Buddhist will *not* be more concerned with extending her compassion to her cousin than to a new neighbor or stranger on the road, nor will her compassion for her cousin be more intense. Indeed, a key story of the Buddha in a former life has him giving away his own wife and anguished children to wandering brahmins—just for the sake of the future awakenings of other suffering beings, for whom he had endless compassion.[44]

We saw earlier that the Buddhist principle of dependent co-arising (*pratītya-samutpāda*) refers to the causal interconnectedness of all beings. To embed this principle concretely in ethical life, Buddhist practice encourages you to remember that all beings have at one time been in a familial or other important relation to you; so again, the idea here is not to deny or devalue the natural bonds of concern that cement human relationships, but to *extend* and multiply those caring bonds among the entire human family, or even to *all* beings, past, present, and future.[45] This goal would have appeared clearly absurd to Mengzi or Aristotle, for whom the natural human affinities into which we are born reflect a stable pattern or moral order, the integrity of which must not be destroyed. Let us set aside the question of who is right. Still we must acknowledge that a failure to extend sufficient concern to those impacted by our own technosocial habits has brought our planet to the brink of a sixth mass extinction; permitted untold environmental, political, and economic injustices against the least empowered among us; and endangered the futures of billions not yet living—humans and nonhumans alike. This jarring reality brings us back to the problem that challenged Mengzi and the king: *how can we make it easier for ourselves to put the proper extension of moral concern into practice?*

Buddhism offers a novel strategy for moral development that helps us face this challenge. As comparative scholars Cooper and James note in their analysis of Buddhist virtue and its applications to environmental ethics, we can cultivate the ability to appropriately extend compassion and care through meditations specifically designed with this aim. In the Mahāyāna Buddhist canon, one is called "Exchanging Self and Others."[46] It presupposes that one has first meditated on the idea that "all beings are alike in wanting happiness and wanting to avoid suffering." It follows from this that it is unreasonable to consider only one's own happiness, but how does one convert this intellectual insight into practical motivation? The practice of Exchanging Self and Others involves an exercise in which one actively imagines oneself as another, and then takes a perspective upon one's own real self from that vantage. Examples of this practice include "identifying with another being and from that perspective engendering jealousy towards oneself . . . "[47]

The point of such exercises is that it is not enough for a virtuous person to *intellectually* grasp her moral duty to extend compassion, or even to understand that it would be irrational not to do so. We must also find ways to *feel* compassion, which is an experience that goes beyond the intellect. As Mengzi found in his interactions with the king, this experience cannot be summoned simply by the knowledge that you should have it.[48] To this knowledge, Buddhism adds the habitual practice of *perspective-taking*, an act that recruits moral attention (hence the emphasis on 'viewing' from the 'eyes' of another being), but broadens the scope of ordinary empathic response with the help of *moral imagination*.[49]

Still, even the most dedicated meditative practice will fall short of an easy fix. At best, we can *gradually* strengthen and expand that basic capacity for empathic concern that belongs to any person who is not morally vicious. If I am filled with compassion for my mother or my child when they are visibly in pain (rather than being moved *only* by my own motivation to restore quiet, or to look like a dutiful daughter or parent) then perhaps, if I desire, I can habituate this basic capacity to become more easily activated in other situations that call for it, through the repeated mental and emotional exercise of moral imagination and perspective-taking. More will be said in the next chapter about the potential relationship between this practice and many of the technomoral virtues, including *honesty, humility, justice, courage, empathy, care, civility, flexibility,* and *moral perspective*. For now, however, let us summarize the results of this section, and its implications for technomoral wisdom.

Appropriate Extension of Moral Concern: Shared Resonances and Technomoral Implications

We found a close but imperfect family resemblance among classical virtue traditions with respect to the appropriate extension of moral concern:

1) A cultivated person will be capable of extending natural moral attitudes of caring concern (such as love, fairness, benevolence, respect, or compassion) beyond their initial scope, as called for by the general and situational demands of morality.

2) A fully cultivated person with practical wisdom will be able to extend her moral concern *appropriately*—that is, to the right beings, at the right time, to the right degree, and in the right manner.
3) The proper extension of moral concern presupposes the effective integration of the other habits of the practice of moral self-cultivation, especially moral attention and prudential judgment. Perhaps for this reason, it is the most demanding requirement of the practice. In fact, the ability to extend one's moral concern reliably well is arguably a necessary and sufficient condition for practical wisdom: the cultivated state of the exemplary person with complete (though not *perfected*) virtue.

Having seen how important moral extension is for the cultivation of the moral self, we must ask how our present capacity for this extension is impacted by emerging technosocial developments. Are certain kinds of emerging technology expanding or narrowing the scope of our moral concern, or do they exert no influence on this aspect of moral practice? Might some emerging technologies change *how* we express our moral concern for others? Might they allow some forms of care, compassion, and civility to be expressed more easily than others? Could some emerging technologies exacerbate moral tribalism, neglect, or incivility, shrinking the circle of our moral concern for others? Might other technologies have the opposite effect, encouraging the exercise of moral imagination and perspective-taking that enrich our capacities for moral extension? How can we drive more resources into development of the latter, rather than the former? What is the potential risk to human flourishing if we can't, or won't?

Glaring lapses in collective and individual practical judgment have led to widespread and growing environmental degradation and resource depletion, global economic and climate instability, and an increasingly chaotic and violent geopolitics, all of which point to the fragility of human flourishing in our present moral condition. Against this backdrop, the chances of the human family being able to live well with emerging technologies such as social media, artificial intelligence, social robots, biomedical enhancement, and ubiquitous surveillance—*as well as* any number of disruptive future technologies whose emergence we cannot possibly hope to anticipate—seem to depend upon our ability to encourage the wider, even global, cultivation of the *technomoral virtues*. In chapters 3, 4, and 5, we examined seven habits of moral self-cultivation that diverse virtue traditions have held to facilitate the cultivation of moral wisdom and human flourishing. In the next and final chapter of Part II, we will look at twelve *specific* virtues that will best characterize 21st century people of *technomoral wisdom*—those with an exemplary ability to cope, and even flourish, in the uncertain and chaotic technosocial environment that is rapidly becoming the global norm.

6

Technomoral Wisdom for an Uncertain Future

21ST CENTURY VIRTUES

SO FAR IN Part II we have seen that among three classical virtue traditions, rooted in diverse cultures and embodying very different visions of human excellence and the good life, there is a shared family resemblance that goes well beyond a thin commitment to the notion of moral character. This family resemblance can be articulated as a framework for the practice of moral self-cultivation, with seven core elements: *moral habituation, relational understanding, reflective self-examination, intentional self-direction of moral development, moral attention, prudential judgment,* and the *appropriate extension of moral concern.* As these seven elements of moral practice become well-developed and integrated, they jointly enable the attainment of practical wisdom. Concrete expressions of practical wisdom will always reflect particular cultural and historical views of human flourishing. Yet the *path* to becoming practically wise remains marked by these seven habits of moral practice in each of the three moral traditions examined in Part II.

If cultivating these same habits is a viable path to a new kind of practical wisdom—*technomoral wisdom*—then we *can* meet the cultural challenge for a global virtue ethics articulated in chapter 2, and with more resources than the thin tissue of abstract concepts previously recognized as shared by virtue ethical traditions. These resources are embedded in a framework of moral practice that provides any human person seeking to become cultivated with robust action-guidance. Without ignoring the profound cultural differences among virtue traditions, this action-guidance focuses upon the *roots* of virtue rather than its ultimate deliverances. This framework does not tell us *what* it is for a human to live well, to flourish in a particular culture or community. Rather, it specifies the *how*: a set of behavioral, cognitive, perceptual, and affective habits needed to cultivate oneself in *any* moral world. So we still have to ask: "Who qualifies as an exemplary human being, and what does a 'life lived well' look like, in *our* present technosocial context?"

While the *how* of becoming virtuous provides a great deal of general guidance for living well, the *what* of virtue must reflect the specific setting of the moral practice. For the first time since our protohuman ancestors diverged out of Africa, we each find ourselves today situated not only in our familiar cultural and regional settings, but also in a new setting: a globally networked and increasingly interdependent human community marked by rapid, complex, and ever more unpredictable technosocial and planetary change. Even those still systematically excluded from meaningful participation in this new global information society are subject to its economic, political, and environmental effects. Having explored *how* a human becomes a person of moral character, we must now ask *what particular sort* of human character we require to flourish in *this* newly shared community. We need an account of technomoral virtue.

6.1 A Taxonomy of Technomoral Virtues

For better or for worse, global technosocial practices and their effects increasingly bind the fate of the human family together, in all our multiplicity and difference. Whatever virtues of character are most likely to increase our chances of flourishing together in these conditions we shall call the *technomoral virtues*. As noted in the Introduction, these will not be character traits unlike any seen before. Any plausible virtue ethic must be rooted in a more or less stable range of human moral capacities, and in the absence of some radical alteration to our basic psychology, there is no reason to think that humans will suddenly acquire a wholly new repertoire of moral responses to the world.[1] Indeed this underlies our assumption that classical accounts of our moral capacities still have much to offer us.

Yet 21st century virtues, even those bearing old names such as wisdom and courage, must be cultivated with a *new and explicit adaptation* to our emerging global technomoral environment. As Martha Nussbaum has argued, while the core meaning of a virtue such as wisdom, courage, or justice is fixed by reference to some enduring domain of human experience (e.g., knowledge, or hardship and risk, or the distribution of goods), the concrete or 'thick' meaning of each virtue is determined by the distinctive shape of that domain in our present cultural context, and what specific dispositions enable us to flourish there.[2] *Our* present moral context, the one to which our virtues must be more effectively adapted, is one of increasingly rapid, transformative, global, unpredictable, and interdependent technosocial change. Of course the virtues needed to flourish in these conditions cannot simply replace local and settled conceptions of human excellence, for it is unreasonable to think that we can live well globally without also flourishing in our own families, states, and cultures. There will inevitably be tensions between those expressions of technomoral virtue that aid our pursuit of the global human goods highlighted in chapter 2, and those that promote more locally and culturally circumscribed visions of the good life. Still, given our increasing technosocial interdependence, both global and local visions of human excellence and flourishing *must* begin to be better integrated into our lives, and our character.

As with all taxonomies of virtue, ours remains subject to open-ended elaboration and revision, for two reasons.[3] First, as technosocial conditions change over time, our virtues will have to evolve with them. Secondly, even if it were possible to give an exhaustive account of every single character trait likely to promote human flourishing in our present condition, limitations of space would preclude our giving it here. We must be content with pointing out the virtues plausibly *most* crucial to such flourishing; especially those under pressure from our contemporary technosocial practices. The twelve technomoral virtues that shall be our focus are:

1. Honesty
2. Self-Control
3. Humility
4. Justice
5. Courage
6. Empathy
7. Care
8. Civility
9. Flexibility
10. Perspective
11. Magnanimity
12. Technomoral Wisdom

Let us now examine them in depth.

6.2 Honesty: Respecting Truth

[Related Virtues: Trust, Reliability, Integrity]

Honesty is among the most culturally universal virtues. Aristotle identifies honesty (*aletheia*) as a cardinal virtue. The Confucian term (*cheng*), often translated as 'honesty,' also incorporates broader meanings of integrity or "self-completion" as the highest good to be attained by moral practice.[4] In Buddhism, honesty is embodied in the joint exercise of 'right view' (*samyag-drsti*) and 'right conduct' (*śīla*), which includes the precept of truthful speech. The reasons for the importance of this virtue are fairly obvious. Flourishing in interactions with other people, which is the primary task of ethics and all social life, would be virtually impossible without a general expectation of honesty, as Kant famously noted. While there is tremendous cultural variance among human conceptions of *when* honesty is demanded, to *whom*, and what *degree* or *form* of candor suffices for it, virtually every human community values the distinction between the honest and the dishonest person.

Why then describe honesty as a *technomoral* virtue? After all, if honesty has been a moral ideal in almost all societies, regardless of their technosocial complexity, then one might not expect it to have any special significance for our inquiry. Yet the virtue of honesty is powerfully challenged by the technosocial developments of the last century, and those to come. This becomes clear as soon as we grasp the conceptual links between *honesty, information*, and *communication.* Honesty is about the appropriate and morally expert communication of information. Yet this practice has been and continues to be radically and globally transformed by ICTs, to use the conventional shorthand for information and communication technologies.

ICTs, especially those founded on digital computing, have revolutionized how we communicate and how we expect information to be communicated to us. The exploding literature on the problem of trust in online environments—including e-commerce, e-government, and social media—is just one aspect of this worldwide phenomenon.[5] The new communication habits enabled and fostered by ICTs are shaping how we define the truth, when and how often we tell it, when we expect to be told it, where and in whom we expect to find it, how we package it, how we verify it, and what we do with it and are willing to have done with it. ICTs have also created a fierce global debate about privacy norms, which are placed in a profound tension with new digital ideals of 'transparency,' 'sharing,' and 'open community.' From the international controversy about the United States NSA's spying and exposures by Wikileaks, Anonymous, and Edward Snowden; to the EU's 'right to be forgotten'; to debates about online anonymity and self-censorship on social networks; to the fallout from the Ashley Madison hack and data dumps, human beings around the globe *have never been less certain what honesty as a virtue looks like.*

Technomoral honesty is not just a personal concern but a social and political one. ICTs have transformed the nature and reliability of evidence in scientific, political, and media contexts, along with public perceptions of whose information, if anyone's, warrants *trust.* Indeed public trust in information supplied by traditional media outlets, academic researchers, governmental bodies, and religious leaders has fallen precipitously in recent years as new media practices and ICTs widen the gap between the availability of information and traditional notions of expertise and authority. In the United States, for example, public trust of mass media hit an all-time low in 2015.[6] Meanwhile, the Internet continues to function well as a space for loudly contesting the informational claims of others, but far less well as a medium for building reliable public consensus.

Yet we cannot surrender the moral ideal of honesty. Arguably, the future flourishing of humanity depends *more than it ever has* on our ability to obtain, verify, and share reliable information concerning problems such as global climate change; infectious disease and public health; threats to air and water supplies; depletion of energy resources; and ongoing existential threats from nuclear, biological, and other weapons as well as cosmic perils such as near-Earth objects. Thus as traditional norms, standards, and habits of appropriate disclosure crack under the weight of radical changes in information and communication practices, it is essential that we consider what norms, standards, and habits will replace them. Human flourishing in social environments has never been able to endure without established norms of honesty, and there is no reason to think that this has ceased being the case. So what does the virtue of honesty look like in an increasingly digitized information environment? What does a person who handles information in a 'morally expert' way look like now? What will she look like tomorrow? What *is* the 21st century technomoral virtue of honesty?

Let us define the technomoral virtue of honesty as *an exemplary respect for truth, along with the practical expertise to express that respect appropriately in technosocial contexts*. Recall that merely telling the truth, even *reliably* so, is never sufficient for the virtue of honesty, which requires that we tell the truth not only to the right people, at the right times and places, and to the right degree, but also knowingly, and *for the right reasons*. As philosopher Harry Frankfurt's *On Bullshit* so concisely explains, honesty is not the same as mere true speech, which can be issued for any number of amoral or vicious purposes, including deliberate obfuscation.[7] How respect for truth gets cashed out in particular technosocial contexts will be a central question for us in chapters 7 and 8, when we examine the ethical implications of new social media and surveillance technologies. But recall that knowing how to express a virtue is largely mediated by our experience of those who are already doing it. So who today do we regard as exemplifying honesty in the information society? Edward Snowden and Glenn Greenwald? Rachel Maddow? David Brooks? Bernie Sanders? Donald Trump? John Oliver? "Curator of the Internet" George Takei? Mark Zuckerberg? The 'Impact Team' hackers who dumped the Ashley Madison data? If these are not our best models of technomoral honesty, then we had better ask ourselves who is. Even better than asking ourselves would be asking those outside our local circles of trust—initiating a global intercultural dialogue about the role of technomoral honesty in living well.

6.3 Self-Control: Becoming the Author of Our Desires

[Related Virtues: Temperance, Discipline, Moderation, Patience]

This virtue is the ability to reliably and deliberately align one's desires with the good. There is a long-standing ambiguity in this notion, which is why Aristotle in *Nicomachean Ethics* Book VII carefully distinguishes temperance (*sôphrosunê* or right desire) from mere continence (*enkrateia* or willful restraint of wrong desire). Likewise, we have seen that Confucian and Buddhist moral traditions also emphasize self-discipline and the cultivation of right desire, giving the latter higher esteem than the suppression of wrong desire.

Yet we saw in chapters 3 and 4 that the willful restraint of wrong desire—aided by moral education, law, or human exemplars—is the first step on the path to developing right desire. So for our purposes, let us define the virtue of self-control as an umbrella notion that captures both the self-restraint of moral continence and the deliberate cultivation of right desire that yields genuine temperance. That said, why should we consider self-control to be a distinctly *technomoral* excellence of character? Self-control is a requirement for *any* person of virtue, even a monk living on a remote mountaintop entirely cut off from modern technology.

Consider, however, how emerging technologies increase the number and variety of potential objects of our desire. Compared with past eras, ICTs in combination with global transportation systems grant us access to a vastly expanded range of available goods, more aggressively advertise to us their selection and enjoyment by others, and increase the speed with which we can attain, consume, and replace them. New media advertising techniques fueled by the power of 'Big Data' subject us to a constant flow of solicitations custom-tailored to inflame our desire, to strengthen our existing consumption habits and expand them to new types of goods, and to convince us that our wants and needs are one and the same.

The ethical implications of these phenomena are significant on individual, local, and global scales. Our desires and consumption habits reflect the physical and emotional health of our persons and our societies. They shape the activities that bind our family and community lives; the kinds and amounts of natural resources that are extracted, used, priced, and distributed; and the type and amount of environmental waste that is produced by those activities. Not only material goods but increasingly, *virtual* goods, relationships, and experiences fill the ever-expanding catalog of things we are invited to desire and pursue. Online app and game developers encourage us to spend collective billions of human hours growing virtual crops in Farmville, massacring pigs with Angry Birds, or solving new puzzles in the Candy Kingdom. Advanced techniques of software design psychology magnify the addictive (the preferred software nomenclature is 'sticky') qualities of apps, driving users to make more and more in-app purchases, share our monetizable information or contacts, or just keep playing into the wee hours of the night to reach whatever surprises await us on the next game level.

The problem of how to evaluate the worth of these virtual goods in relation to others will only be exacerbated by the increasing sophistication of digital environments and artificial intelligence. Life online already challenges our self-control on multiple levels, causing many of us to resort to software lockout tools such as Freedom to keep us out of the digital cookie jar for a fixed period of time. Such tools allow us to complete an important task or spend time with people we highly value without being seduced away by the apparently irresistible distractions of Facebook, Gmail, Minecraft, and Instagram. Others simply surrender to the juggernaut of technosocial distraction. But what does all of this say about our state of personal discipline and self-control?

Let us define the technomoral virtue of self-control as *an exemplary ability in technosocial contexts to choose, and ideally to desire for their own sakes, those goods and experiences that most contribute to contemporary and future human flourishing*. As we noted in chapter 1, flourishing is not *mere* self-satisfaction or pleasure, states of mind easily attained by a sociopath or a heroin addict with a steady supply. Flourishing means actively *doing well* as a human being; it involves facts about how our lives are going socially, politically, physically, intellectually, and emotionally. This invites the question: how do we determine which goods and experiences help us to live well? Consider the much-discussed digital fragmentation of cultures. Gone are the days in which one could safely assume that one's local peers had read the same books, seen the same movies or news shows, engaged in the same leisure activities, or visited the same places. Even if we look only toward people we admire as intelligent, discerning, and culturally literate, we will be hard-pressed to find much uniformity in their consumption habits. As a result, cultural narratives about what desires individuals should try to cultivate in order to facilitate not only a good life for themselves, but a good *shared* life in community, are rendered increasingly incoherent—as are the institutions of cultural education, art, and ritual that traditionally steered individuals onto the path of right desire.

Of course, these developments have positive aspects, not least among them a great proliferation and global exchange of diverse ideas and images of the good life. There are very, very many things in life worth wanting, and global information networks bring ever more attractive options into our view. Yet it is impossible to choose them all. So the question remains: how do persons develop the virtue of self-control in this new technosocial environment? A virtuous person in any time or place must be able to discern and eventually *become the author of right desires*—that is, self-conscious and authentic desires for goods that promote her flourishing with others, as opposed to those that diminish her life. As global information technologies dilute or blunt the effect of those cultural mechanisms that previously assisted us in making these distinctions, what new mechanisms for shaping human desire are taking their place? In short, how will 21st century humans acquire the individual and collective virtue of self-control, and what are the short and long-term global consequences if we cannot? Our discussions of new media in chapter 7, surveillance technologies in chapter 8, and biomedical enhancement in chapter 10 will engage this question.

6.4 Humility: Knowing What We Do Not Know

[Related Virtues: Modesty, Reverence, Wonder]

Compared with honesty and self-control, the moral importance of humility is less universally acknowledged. While humility was often recognized in ancient Greece as a check on destructive *hubris*, for Aristotle the virtuous mean was nothing like humility, but rather the justified pride of the man who correctly recognizes himself as 'great of soul' (*megalopsuchia*). For Aristotle, to abase oneself is dishonest if one is actually a good and noble man, and if one is *not* a good and noble man, one's primary aim should be to improve one's character, not to lower one's self-regard.

The influence of Christian perspectives restored humility to prominence in the medieval virtue-ethical tradition; in the 13th century, St. Thomas Aquinas described humility as "keeping oneself within one's own bounds, not reaching out to things above one, but submitting to one's superior."[8] Since then humility's reputation as a virtue has fluctuated; it was celebrated by Kant, famously rejected by Nietzsche, and in the late 20th century came to be increasingly displaced by psychological discourse on the importance of "high self-esteem."[9] In the East, regard for humility as an important character virtue has been more stable. Our prior analyses showed that for Confucians, humility issues from the habit of reflective self-examination and expresses itself in the yielding disposition that constitutes the cardinal virtue of ritual propriety or respect (*li*). We saw that for Buddhists humility is part of the understanding of *anātman* (not-self) that corrects one's distorted sense of personal importance and ego-attachment, both of which obstruct the cultivation of right view (*samyag-drsti*) and right action (*śīla*).

What role should humility play in a 21st century account of technomoral virtue? Arguably, 21st century life demands greater humility than we presently enjoy. Insofar as the global political, cultural, economic, and environmental consequences of emerging technologies are proving themselves to be increasingly complex and difficult for us to predict, increasingly systemic and far-reaching in scope, *and* increasingly challenging for us to mitigate or undo, we can no longer afford the modern illusion that our technosocial innovations are conducive to human *mastery*.

For example, the 20th century confidence that modern medical technologies would soon grant humans permanent mastery over disease has been progressively undermined by the rapid rise of antibiotic-resistant bacteria; the spike in the emergence and global reach of formerly unknown pathogens; and the persistent spread of 'lifestyle' diseases associated with technosocial changes in our diet, air quality, physical activity, and exposure to chemical compounds. Rising infant mortality and stagnant or declining projected lifespans in technologically advanced nations like the United States are a signal that far from ensuring our mastery of human health, our present technosocial practices may be pushing this goal further from our grasp.

Similarly, our attempts to master the new technosocial frontier of the Internet have proven equally weak: cyberbullying, cybercrime, cyberespionage, 'hacktivism,' and even the potential for cyberwarfare are expanding rapidly, and not even the most powerful nations and institutions are presently able to prevent their spread.[10] The intentional conduct of 'bad actors' is not the only source of unpredictable ICT phenomena; consider the aggregate effects and complex interactions of otherwise benign electronic agents, such as the high frequency trading algorithms that contributed to the "Flash Crash" of May 2010 and the Knight Capital fiasco in August 2012. Limited understandings of human-technology relations also produce massive design failures such as Microsoft's notorious AI 'teen' chatbot Tay, which began spewing neo-Nazi propaganda on Twitter within a day of her 2016 release. Given that new algorithmic techniques for machine 'deep learning' resist thorough human inspection or prediction, it should be clear that the digital domain—to which human beings are increasingly linking their material, political, cultural, and economic resources— is no more the subject of human 'mastery' than are the organic domains of human health, ecology, or climate. We must come to terms with the challenge of *acute technosocial opacity*. Among character traits likely to promote human flourishing in this condition, humility is high on the list.

Let us define technomoral humility as *a recognition of the real limits of our technosocial knowledge and ability; reverence and wonder at the universe's retained power to surprise and confound us; and renunciation of the blind faith that new technologies inevitably lead to human mastery and control of our environment.* Not only is this faith undermined by present conditions, it prevents us from honestly confronting the hard choices we must make about technosocial *risk*. For we cannot evade present risks to human flourishing simply by *forgoing* technosocial innovation. As Nick Bostrom notes, that path exposes us to equally profound, even existential risks.[11] In truth, risk to the future of humankind cannot be eliminated. A humble appreciation of this radical exposure to risk is essential to the ability to make prudential judgments about how best to proceed in our present state of technosocial opacity: judgments not only of *which* technosocial risks to take, but also how to plan for inevitable technosocial *failures,* and especially how to preserve the resources, resilience and flexibility needed to cope in midstream with the unforeseen consequences of our failures *and* successes.

Technomoral humility, like all virtues, is a mean between excess and a deficiency. The deficiency is blind *techno-optimism*, which uncritically assumes that any technosocial innovation is always and essentially good and justified; that unanticipated negative consequences can always be mitigated or reversed by 'technofixes' (more and better technology); and that the future of human flourishing is guaranteed by nothing more than the sheer force of our creative will to innovate. The other extreme, *techno-pessimism,* is equally blind and uncritical: it assumes that new technological developments generally lead to less 'natural' or even 'inhuman' ways of life (ignoring the central role of technique in human evolution), and that the risks to which they expose us are rarely justified by the potential gains. This attitude sells short our creative potential, our adaptability, and our capacity for prudential judgment. Humility, then, is the intermediate state: a reasoned, critical but *hopeful* assessment of our abilities and creative powers, combined with a healthy respect for the unfathomed complexities of the environments in which our powers operate and to which we will always be vulnerable. The importance of technomoral humility will be a central topic of discussion in chapters 9 and 10, when we examine emerging developments in robotics and biomedical enhancement, respectively.

6.5 Justice: Upholding Rightness

[Related Virtues: Responsibility, Fairness, Reciprocity, Beneficence]

The virtue of justice is perhaps the broadest and most varied in its interpretation; in chapters 4 and 5 we discussed the important role of justice (*dikaiosunē*) in Aristotle's ethics, where it represents ethical excellence writ large and, more narrowly, the appropriate extension to others of what is deserved. We contrasted this with a Confucian account in which just treatment of others, especially in the political realm, is subsumed under a broader conception of humane benevolence (*ren*). In classical Buddhism, the topic of political justice is neither as central nor as well developed as one finds in Aristotle or Confucian literatures.[12] Still, a universal sense of justice may be considered equivalent with the *Dharma* itself: the complete body of Buddhist learning that mandates unconditional concern for the welfare and dignity of all creatures.

Our present concern is the kind of justice most critical to flourishing in the global technosocial environment. Let us call this *technomoral* justice. This is divisible into two interrelated but distinguishable character traits. The first is a *reliable disposition to seek a fair and equitable distribution of the benefits and risks of emerging technologies*. The second is a *characteristic concern for how emerging technologies impact the basic rights, dignity, or welfare of individuals and groups*. Technomoral justice entails reliably upholding rightness (values of non-harm and beneficence), along with fairness (moral desert), and responsibility (accountability for the consequences of one's actions).

The absence of technomoral justice is an increasingly destabilizing social phenomenon. Consider the rising tensions in San Francisco, Silicon Valley, and other communities worldwide where the ballooning wealth of high-tech investors and workers has had dramatic effects on the standard of living enjoyed by long-term area residents and workers. In 2013, local activists in the Mission district of San Francisco, protesting rising rents and evictions by landlords looking to capitalize on cash-rich tech renters, repeatedly surrounded and blocked the corporate luxury shuttles that transport tech employees of Google, Apple, and other Silicon Valley companies to and from their work campuses. Local and national media wrote about the growing perception of an insular tech community indifferent to the well-being of their fellow citizens. Adding fuel to the fire were a series of objectionable public statements by tech entrepreneurs such as startup AngelHack CEO Greg Gopman. He shared his disgust at the homeless "degenerates" and "trash" of San Francisco who fail to observe the self-segregating ideal of cities where the "lower part of society keep to themselves. They sell small trinkets, beg coyly, stay quiet, and generally stay out of your way. They realize it's a privilege to be in the civilized part of town and view themselves as guests."[13] Such sentiments are, if nothing else, suggestive of a vicious deficiency of *technomoral justice* as defined above.[14]

Other emerging deficiencies of technomoral justice include the average technology consumer's detachment from the profound environmental, economic, and political harms created by the mining and disposal of rare earth elements (REE's) used in electronic devices, and countless other harmful externalities generated by global technosocial development yet imposed on the most disenfranchised residents of the planet. Socially destabilizing asymmetries of power are magnified by large-scale data mining, pervasive digital surveillance and algorithmic profiling, and robotic and drone warfare. Still we must not ignore the potential for emerging technologies to expose, mitigate, or remedy injustice. For example, digital surveillance tools are used by oppressive regimes and powers *and* against them. Deciding how such tools can be developed and used more fairly and responsibly, on both an individual and a collective human level, is the domain of technomoral justice. In chapters 7, 8, and 9 we will take a closer look at the effects of new social media, digital surveillance, and military robotics on this domain.

6.6 Courage: Intelligent Fear and Hope

[Related Virtues: Hope, Perseverance, Fortitude]

The moral excellence of courage is much discussed in virtue ethics, though its meaning varies widely by cultural and historical setting. In Book III of the *Nicomachean Ethics*, Aristotle defines courage (*andreia*) as the mean between cowardice and rashness—the disposition to fear "the right things and with the right motive, in the right way and at the right time."[15] The model of physical and martial courage found in soldiers and athletes was central to Aristotle's understanding of this virtue. Yet later discussions of courage in the Western philosophical tradition, such as that of Aquinas, placed greater emphasis on *spiritual* or *moral* courage. Such courage is a reliable tendency to fear grave wrongdoing and a compromised character more than one fears the other dangers or injuries that one might invite by acting rightly, and in general to intelligently balance measured and justified fears with measured and justified confidence and hope.

Unlike self-control, which guides one to pursue only those things that are actually good for oneself to obtain, courage necessitates risk and sacrifice. In all forms of courage, the courageous agent is willing to endure some injury, forgo some legitimate good, or otherwise incur a real loss in order to do what is necessary and right.[16] This willingness distinguishes genuine courage from its pale facsimiles: acts of seeming bravery by persons blinded by passion or ignorance of the danger, overconfident in their ability to evade it, or motivated by some greater personal advantage they expect to gain, such as social acclaim.[17] What is distinctive about the virtue of spiritual and moral courage is that it must manifest itself in an *enduring* orientation to one's world, not just in rare moments of mortal danger such as those presented in battle. Aquinas emphasizes this 'steadfastness' of courage (*fortitudo*) as the backbone of all other virtuous dispositions; to live rightly is to maintain the fortitude to make sacrifices, take risks, or incur injuries (even death) in the name of the good, when that good is not trivial but of considerable spiritual or moral gravity.[18] Lee Yearley's analysis of Aquinas's account of courage is particularly helpful here; he notes that it has both a positive and a negative aspect. Courage demands positive *perseverance* in one's active and hopeful pursuit of the good, and *forbearance* and *patience* in enduring the pains and losses inevitably associated with this lifelong pursuit.[19] The virtue of spiritual and moral courage is thus a constant renewal of the choice to live well rather than badly, whatever else this may cost us.

As Yearley notes, courage also plays a central role in Confucian ethics. Kongzi describes courage (*yong*) as a virtue not in isolation, but only when supportive of moral rightness (*yi*); yet it seems that the virtuous person cannot do without it, since Kongzi more than once identifies it as part of the threefold essence or Way (*Dao*) of the cultivated person (*junzi*): "The wise are not confused, the Good do not worry, and the courageous do not fear."[20] Historical commentaries on these passages suggest that this was interpreted to mean that the *junzi* is marked not by narrowly martial courage but by moral courage.[21] As Yearley notes, Mengzi's discussions of courage diverge still further from the martial conception, even more so than Aquinas, who is still influenced by the heroic Aristotelian model.[22] In Mengzi, martial bravery is a mere "semblance" of courage, while the only genuine article is moral courage, the tendency to fear no evil more than one fears failing to live rightly: "Life I desire, righteousness too I desire; if I cannot get to have both, rather than life I choose righteousness. . . . That is why there are troubles I do not avoid. . . . Not only men of worth have a mind which thinks like this, all men have it; it is simply that the worthy are able to avoid relinquishing their hold on it."[23]

Mengzi is rebutting the objection that moral courage is psychologically unrealistic. While he paints a picture of moral self-sacrifice that few of us are confident we could mirror, his point that 'all men' have this original inclination is significant. As Yearley notes, Mengzi sees this as the root of natural self-respect, the unwillingness to allow oneself to be debased. In the rest of the passage, Mengzi notes that a starving beggar will refuse food given with abuse. Why? Because his self-respect and dignity, already endangered by his low social status, in the end are more precious to him than mere survival. He can endure more hunger, even death, but he will not endure his own further debasement. How does this relate to moral courage?

In Mengzi's view, the majority of human beings are numb to the threat to our own dignity posed by the ethically compromised ways in which we allow ourselves to live. We all share the beggar's natural inclination to preserve his own dignity, but due to the relative comfort of our lives this inclination is no longer actively engaged, it is not part of our daily awareness. This failure to habitually care for our self-respect and moral dignity results in a lack of moral courage; when presented with a choice between giving up some material or social comfort to which we are accustomed, and surrendering even more of our moral respectability, our will to endure the former to save the latter is lacking. I expect that many readers will find Mengzi's point no less resonant in our contemporary world.

The Buddhist virtue tradition also grants an important role to moral courage. Of the 'Six Perfections' of moral character in the Mahāyāna canon, one is often translated as courage (*vīrya*) or vigor, but others in the list are also associated with moral courage as we have defined it, especially forbearance (*ksānti*) and strength (*bala*). Buddhist courage also critically depends on the interaction of two related virtues discussed in chapter 4: *hri*, the disposition to be ashamed of doing evil, and *apatrapya*, the fear of doing evil. The proper habituation and intelligent application of *hri* and *apatrapya* produce a disposition of moral courage, insofar as such a person will reliably and correctly fear living in a morally debased manner more than she fears having to endure the worldly consequences of living rightly.

What role does moral courage have to play in 21st century technosocial life?[24] Certainly the rise of robotic warfare and biomedical engineering confronts us with many new moral *and* material fears. It has never been more important that humans be able to judge rightly *what* we should most fear, what we can and must hope for, how *much* to fear or hope, how best to *act* on our fears and hopes, and how much confidence we should have in our efforts to manage the risks presented by our technosocial choices. This is even more true given what we have said earlier in this chapter about humanity's increasing self-exposure to existential danger. No longer must we simply weigh the relative risks of our own material extinction or moral debasement, or even that of our near kin or nation; the material and moral fate of the entire human family is now implicated by many of our individual and collective technosocial choices.

Thus it is highly unlikely (barring some astonishing run of blind luck) that our species will continue to flourish on this planet for very long unless communities have at least *some* success in encouraging their members to individually and collectively cultivate the virtue of technomoral courage. By this I mean *a reliable disposition toward intelligent fear and hope with respect to the moral and material dangers and opportunities presented by emerging technologies*. As with ordinary moral courage, technomoral courage presupposes the tendency to give proper priority to the preservation of our moral well-being and dignity over the preservation of material comfort and ease, or in some cases, even our physical safety. However, there is a critical difference between the classical forms of moral courage and *technomoral* courage. Someone with moral courage in the classical sense could safely assume that life in general would endure, even if his or her own did not. The choice to risk one's own life rather than accept profound moral debasement was therefore of a different nature than many of the choices 21st century humans face, as we collectively confront choices of technosocial policy having planetary implications.

For example, imagine that you are confronted with the choice of supporting or resisting a global geoengineering initiative that has a fair chance of dramatically reducing the existential risk to the biosphere posed by rapid global climate change. Assume that putting this plan into action would necessarily involve committing various grave injustices: the forced displacement of certain indigenous peoples from their lands; the imposing of taxes and other material externalities on communities who were not themselves responsible for the reckless degradation of the environment that the plan aims to mitigate; the destruction of certain ecosystems in order to improve the conditions for human settlement, and so on. Your support of this plan would implicate you in all of these injustices and thus debase your moral character; but assume for the sake of argument that in resisting the plan you would be accepting a significantly increased risk to future human survival. If no better options existed, which would you judge to be the more morally courageous course of action?

Contemplating such challenging scenarios should make clear why it is essential that we begin seeking *now* a better understanding of what counts as courage in our contemporary technomoral situation, and determining how this virtue of intelligent hope and fear can be more widely cultivated among us. In Part III, discussion of the importance of cultivating technomoral courage will be central to chapter 9 on robotics and chapter 10 on biomedical human enhancement.

6.7 Empathy: Compassionate Concern for Others

[Related Virtues: Compassion, Benevolence, Sympathy, Charity]

The virtue of empathy has a strong pedigree in Confucian and Buddhist virtue traditions, and we will see that seeds of this virtue are planted in Aristotle's ethics. Yet it is not always classified as a virtue by contemporary thinkers, and is defined in so many ways that it invites especially careful discussion.[25] Before exploring its classical and technomoral significance it will be helpful to first establish a conventional distinction between empathy and sympathy.

Although these terms are used in a variety of contexts, and translate a range of closely related concepts in other languages, for the most part "sympathy" refers to a form of benevolent concern *for* another's suffering. "Empathy," on the other hand, is often used to describe a form of co-feeling, or feeling *with* another, synonymous with compassion. Empathy may include sharing feelings of pain, fear, shame, excitement, joy, or other passions. Though we can speak of purely cognitive empathy (i.e., the correct belief that another is experiencing a certain emotion), empathy is often associated with bodily passivity, an experience of being physically *affected* by another's emotion. It is an intimate, personalized form of relation; I can have sympathy for the misfortunes of a group of anonymous individuals who are more or less interchangeable in my mind, but I typically can only empathize with a specific person or set of persons. Even a stranger with whom I empathize will appear in my experience not as an abstraction, but as a concrete, nonfungible individual with specific emotional experiences.

Empathy and sympathy often interact; once I find myself empathizing with the specific frustrations and suffering of a wounded veteran in my community who is struggling to adjust to returning home, I may be more inclined to sympathize with the millions of other veterans facing similar challenges. Likewise, a well-cultivated and informed general sympathy toward victims of cyberbullying may make me more likely to empathize with a student in my child's school suffering severe depression and anxiety as a result of online harassment—even if my own child is one of the bullies. Empathy and sympathy can also stimulate and be stimulated by caring activity; thus they are intimately connected with the virtue of care described in the next section. Yet one form of empathy qualifies as a moral virtue in its own right. This is empathy understood not as simple shared affect, nor as mere understanding of another's mental state, but as a *cultivated disposition* reliably uniting these affective and cognitive aspects in active concern for others. Following psychologist C. Daniel Batson, we can classify this "empathic concern" as "a motivational state with the ultimate goal of increasing another's welfare."[26] Defined as a technomoral virtue, it is a *cultivated openness to being morally moved to caring action by the emotions of other members of our technosocial world.*

There is empirical evidence for the claim that empathic concern can be actively cultivated as a disposition; as with any virtue, empathy can be expressed more or less wisely.[27] It makes a great difference to my character whether I am moved by empathic concern at the right time and places, by the right people, and with the right intensity. For example, if my brother is a malicious hacker facing harsh punishment for intentionally defrauding millions of vulnerable senior citizens, my own moral excellence in this situation requires my mental and emotional energies to be adequately attuned to and moved by the suffering of his victims, and not inappropriately consumed by concern for my brother's troubles. On the other hand, to be wholly indifferent to my brother's present and future suffering would also be vicious. The appropriate cultivation of empathy is a particular challenge in our present technomoral condition and deserves special emphasis. We will say more about this shortly. But given that Western ethicists often neglect the virtue of empathy, it is worth saying a bit more about its classical roots in virtue traditions, and how we might defend our characterization of it as a technomoral virtue.

Empathy occupies a critical role in the life of virtue Aristotle describes,[28] yet he did not classify it as one of the virtues, because he lacks a concept of empathy as a cultivated state or disposition. That said, not only does his account recognize empathy as a capacity for 'feeling with' others, he makes it a constitutive element of friendships of virtue. In empathy, a virtuous person "grieves and rejoices with (*sunalgounta kai sunchaironta*)" his friend, in a manner that contributes to his experience of that person as a "second self."[29] Still, Aristotle sees empathy as a spontaneous passion erupting naturally rather than by choice.[30] Virtues, on the other hand, are states of character arising from deliberative choice and amenable to rational direction, adjustment, and cultivation of expression. In contrast, he describes empathy as a natural feeling affecting not only rational beings but other animals.[31]

Why then does Aristotle make empathy a distinctive mark of *virtue friendship*, noting that this passion is impossible among those lacking in virtue who, due to the absence of a soul worthy of love, cannot even "rejoice or grieve with themselves," much less with a friend?[32] How does it happen that for Aristotle human mothers and even birds can empathize, but a bad man cannot? Aristotle's strong association between empathy and virtue is undeniable, yet puzzling. Note that for Aristotle mere mutual goodwill, in the absence of empathic feeling, is insufficient for friendships of virtue.[33] He says that with those for whom we have only goodwill, we wish them well but "we would not do anything with them . . . nor take trouble for them."[34] Empathy here seems to be a *motivating* factor behind our decision to engage in caring activity with or for another. But is Aristotle right to assume that empathy is never cultivated by choice or directed by reason?

This seems implausible if we consult our own experiences. For *most* humans, empathy is a quivering flame that is always near to being extinguished by apathy or cynicism, or our desire to shield ourselves from suffering. To allow the full experience of empathy is to open oneself up to being moved by the joys but also the *pains* of others, and many of us choose, often regrettably, to turn away from this experience—either by avoiding circumstances that might bring it on, or by altering our thoughts ('putting up a wall') to create emotional distance, perhaps while maintaining a façade of caring behavior. Even after opening ourselves to empathy, we must continually choose whether to allow ourselves to *remain* moved by the other's situation, or to withdraw emotionally and close ourselves off. The latter can be psychologically necessary, even morally justified—thus discerning deliberation is called for in the expression of empathy, as with all virtues. Still, with repetition of the choice to remain open, we may become more accustomed to or even welcoming of the emotional weight.

Empathy as compassionate concern, then, must be exquisitely and quite consciously cultivated if it is to endure and thrive rather than wither on the vine. Alasdair MacIntyre makes a similar point with respect to the virtue Aquinas characterized as pity (*misericordia*). MacIntyre reminds us that "our affections and sympathies are generally, if not always, to a significant degree in our control, at least in the longer run."[35] According to MacIntyre, not only is *misericordia* a virtue, it is one of the "virtues of acknowledged dependence" most crucial to our flourishing as human animals, one that Aristotle mistakenly devalues from his illusory standpoint of masculine invulnerability and privilege.[36]

Had Aristotle realized this, he would have recognized empathy as a virtue essential to a good life. For without appropriate empathy, we could not participate in the highest forms of human fellowship, as even he admits. We must make a distinction that Aristotle did not: between empathy as a natural, uncultivated, and unchosen passion, and the deliberate enrichment of this basic capacity into a moral virtue of empathic concern, one that presupposes the natural capacity to be affectively moved by another but requires intelligent choice and habituation to be properly directed and reliably expressed. This distinction has the further advantage of resolving the earlier puzzle: that is, how a bird can empathize (in the first, uncultivated sense), while a bad person cannot (in the second, morally cultivated sense).[37] In short, Aristotle would likely have recognized empathy as a virtue had he enjoyed a better view of the integration and co-development of moral emotion and moral reason in human beings, as well as the essential nature of human vulnerability in grounding ethical human relations.

In Asian virtue traditions the virtue of empathy is more straightforward. In the Confucian moral tradition, the virtue of *shu* is generally translated as "empathy," "empathic reciprocity," "sympathetic concern," or "understanding," where 'understanding' means not detached intellection but the ability to place oneself in the situation of another, to intuitively grasp not only what they are going through but what it demands as a moral response.[38] The Confucian concept of the 'heart-mind' (*xin*) does not artificially divide empathic feeling from moral intelligence, and Confucians readily conceive of empathy as a cultivated virtue.[39] In the *Analects*, *shu* is described as "one word that can serve as a guide for one's entire life," tempering the fixed requirements of duty or loyalty (*zhong*).[40] It thus plays an essential role in practical wisdom, by motivating the discerning flexibility in the application of moral conventions that is essential to the *junzi*'s exemplary character.

Shu plays a critical role in Mengzi's account of virtue as well; he tells us that "in seeking to be humane (*ren*) . . . there is nothing that comes closer to it than working hard at *shu*."[41] Here we see a parallel with our distinction between empathy as a natural, spontaneous passion and empathy as a deliberately cultivated state. For while empathy is something that we must 'work hard' to acquire as a virtue, Mengzi famously claimed that any normal human being will experience compassionate horror at the sight of a child about to fall down a well, and that "all human beings have a heart/mind (*xin*) that cannot bear to see the sufferings of others."[42] This "commiserating heart/mind" is the natural "seed" or "sprout" of empathic response from which the mature virtue of *shu* may be cultivated, given proper habituation and effort.[43]

The concept of empathy also resonates strongly with two Buddhist virtue concepts we encountered previously: compassion (*karunā*) and sympathetic joy (*muditā*). Their pairing addresses an element of empathy too often neglected in the West—namely, the moral importance of being moved by the joys of others as well as their pains. As distinct from loving-kindness (*maitri*), which is a universalized benevolence without the implication of emotional attachment to a specific other, *karunā* and *muditā* imply the emotional passivity of being *affected* by another's state of being. They therefore convey the emotional concreteness required by our concept. However, from a Buddhist standpoint, the more cultivated one is, the more inclined one is to extend *karunā* and *muditā* without discrimination. While it is still concrete beings who elicit the *bodhisattva*'s empathic concern, for her the distinctions of kinship and political affiliation are increasingly irrelevant to her moral affections.[44]

Buddhist teachings are explicit that empathic concern is a virtue that must be actively cultivated through moral practice. As Buddhist scholar Dale Wright notes:

> For the most part, compassion is something we *learn* to feel . . . we cannot feel compassion simply by deciding to feel it, or by telling ourselves that it is our responsibility to feel it. We do, however, have the capacity to develop compassion by cultivating our thoughts and emotions in ways that enable it. This is the function of the practice of giving. Making generosity of character [*dāna*] an explicit aim of self-cultivation, we sculpt our thoughts, emotions, and dispositions in the direction of a particular form of human excellence.[45]

Contrary to Mengzi, this suggests that whatever innate compassion or empathy we have is extremely limited in its moral power. Yet even here, natural empathy's appropriate expansion is seen as the primary goal of moral practice. Over time, such practice gradually habituates me to outward-facing concerns and exposes me more intimately to others' sufferings and joys. For a Buddhist, these gradually "soften" and eventually erase formerly hard boundaries between self/other, kin/stranger, citizen/alien, human/animal, and other dualistic gradations of empathic concern.[46] Yet Confucians still hold these distinctions as morally significant. Thus we see that there is great cultural variance concerning empathy's appropriate expression, but a strong case for it as a cultivated moral virtue.

Is it however a *technomoral* virtue? In scholarly and media circles it is common to lament the apparent paradox of worsening social isolation, political apathy, and intercultural and religious hatreds in a world that, thanks to information technology, has never been more connected.[47] Whether the paradox is real or just a media projection, and if real, whether and to what extent it implicates technology, are serious questions for future empirical study. Most would agree, however, that the global information commons has certainly not *solved* the problem of moral indifference to the suffering or happiness of others. Such indifference perpetuates the greatest obstacles to human flourishing: war; oppression; sexual, religious and racial hatred and violence; abuse of women, children, and the elderly; and widespread poverty, famine, disease, and environmental degradation.

Not only has digital culture not alleviated the world's empathy deficit, but consider the controversial claim by some researchers that 'digital natives' raised on information technology and new media are increasingly deficient in empathic concern and/or prone to pathological narcissism.[48] Such reports are embraced by many who feel increasing alarm at a society of people constantly absorbed in their screens, declining to make eye contact or respond to requests for assistance or companionship, even from friends and family members. Were further empirical research to bear out some version of the 'digital narcissism hypothesis,' it would certainly bode poorly for the future of human flourishing in the absence of some effective intervention. Of course, even if the hypothesis is correct, it asserts only a correlation. It does not identify the *cause* of this supposed trend. Suspicions that digital culture itself may be partly to blame are understandable, but at this point far too broad and untethered to reliable data to be of much use.

Assuming for now that digital culture does *not* make us into narcissists, can it *facilitate* the wider cultivation of empathy? After all, the information society enables us to be virtual firsthand witnesses to human tragedies *and* joys in a manner never before possible. That said, many lament new media's allowance of emotionally low-investment acts of 'click-concern' and 'hashtag activism,' such as 'liking' or retweeting stories, as socially acceptable responses to events that would normally call for more robust expressions of empathic concern and care. Others worry that digital culture encourages a form of 'empathy burnout,' where endless social media stories and video feeds of suffering and tragedy, as well as the stream of joyous life events shared by our peers, eventually lead to emotional overload, numbness, or resigned indifference.

Fortunately, we do not need to know if digital culture contributes to declining human empathy in order to take action. We *already* know that future human flourishing is unlikely without a great expansion of empathic concern—for we have said that the global empathy deficit is a major obstacle to human flourishing *today*, and has been for millennia. The moral need for more human empathy is not a novelty of the digital age. But the need is *growing*, not just because some new habits of the information society might deplete or inhibit empathy, but because the increasingly networked and interdependent nature of the human family entails that we shall find ourselves exposed to ever more circumstances that seem to call for it. Consider the global soul-searching following the 2014 outbreak of Ebola about which victims of the disease we ought to have been moved to take into our care, and where, by whom, how soon, and how well; the same questions arose in the Syrian refugee crisis of 2015. Determining who we ought to feel empathy for, when, and to what degree, is a moral problem that is getting ever harder, not easier, to get right. If human flourishing on an increasingly networked and interdependent planet is going to be virtually impossible without the broader cultivation of empathy as a virtue, *then we had better start working a lot harder at it*. Chapters 7, 8, and 9 take a closer look at how we might begin to do this in light of the unique challenges for technomoral empathy posed by new media, pervasive digital surveillance, and advanced robotics.

6.8 Care: Loving Service to Others

[Related Virtues: Generosity, Love, Service, Charity]

Now we come to a virtue with no perfect equivalent in classical traditions, though it has close relatives in Christian virtues such as love and charity. Another near analogue is the cardinal Confucian virtue of *ren*, not in its broader sense as 'complete virtue' but in its narrower meaning of 'humane benevolence': a reliable tendency to actively foster the good of others to whom one is bound by familial or political ties. In Buddhism the virtue closest to what is meant by 'care' is *dāna*, usually translated as 'generosity.' As one of the Six Perfections of character in the Mahāyāna tradition, *dāna* expresses itself in habits of material and spiritual giving, which simultaneously foster liberation from personal attachments and deep concern for the welfare of others.[49] The virtue of care corresponds nearly to all, but exactly to none of these classical concepts, although the Confucian virtue of *ren* may be nearest, due to its foundation in the moral responsibilities of familial love.[50] The concept of care as a virtue is also associated with the late-20th-century emergence of *care ethics*, an approach rooted in feminist critiques of traditional rights-based theories of ethics.[51]

We will define technomoral care provisionally as *a skillful, attentive, responsible, and emotionally responsive disposition to personally meet the needs of those with whom we share our technosocial environment*. Care is closely related to the virtue of empathy as compassionate concern, yet these virtues are not the same. Someone may be morally moved by empathy to alleviate another's suffering but lack the moral skill, knowledge, or resources to offer effective care—indeed this is a key reason why charitable organizations and caring professions exist, in order that we may enable others to care for those for whom we empathize or sympathize but are ill-positioned to care ourselves. Conversely, a virtuous person's exercise of moral care need not, in every case or context, be motivated by immediate empathic concern. Caring virtue can be exercised toward beings physiologically incapable of suffering or joy, or in ways that are routine rather than provoked by immediate compassion.

However, there is a sense in which virtuous care, even when not attended by immediate experiences of empathic concern, must still be motivated by a general feeling for the importance of loving service to others. Skillful caring practices and dispositions are often thought to grow out of the labors of maternal, paternal, and filial care; hence the view of many care ethicists that the common experience of women as primary family caregivers is an especially relevant source of ethical understanding. There is much debate about whether an ethics of care should be subsumed under a more encompassing framework of virtue ethics, or whether it represents a theoretically independent way of thinking about ethics.[52] This book takes the former view; care may well be a cardinal virtue, but it is not clear that all other virtues can be satisfactorily defined as extensions of a caring disposition.

What special relevance does the virtue of care have to contemporary techno-social life? Why have we chosen it as a *technomoral virtue*? Consider how systems of social and economic privilege have long allowed individuals to divest themselves of the responsibility for caring practices by delegating these responsibilities to hired substitutes or, increasingly, by using technology to meet needs that previously could only be met by the active labor of human caregivers. On care ethicist Joan Tronto's view, human beings who enjoy such privilege risk becoming less and less capable of competent care, less emotionally comfortable with close proximity to vulnerability and weakness, less attentive and responsive to need, and less responsible for themselves and others.[53] Insofar as the practice of skillful caring is a formative experience in cultivating the moral self, this degradation of care is potentially devastating to our collective flourishing.

Is the use of technology to relieve us of labors and burdens a *bad* thing? Few would object to the modern use of cranes to lift and position massive objects, mechanical filters to clean wastewater, or to the invention of the cotton gin. Future technological innovations will continue to reduce human exposure to the 'three D's': work that is 'dull, dirty, and/or dangerous.' Yet the work of human caring can itself be dull, dirty, and even dangerous; from the rocking of a crying child to the changing of an aging parent's diaper to the efforts of a medic or aid worker in a war zone. Technology promises relief of these labors too; many middle-class parents today already own a mechanical rocking cradle or swing to soothe their infant, and the predicted rise of 'carebots' may soon allow Dad's or Junior's diaper to be discreetly changed by a robot. Likewise, military research agencies such as the United States' DARPA have funded research into robotic field medics, and NGO's are looking to use cheap drone aircraft to safely deliver food and medicine to war-torn villages.

It is hardly necessary to point out that less exhausted and stressed parents, fewer wounded medics, and fewer aid-workers taken hostage by militant warlords would all be good things. However, if we recall the view of care ethicists that we become moral selves largely by teaching ourselves to actively respond to and meet the needs of others, we see the attendant moral cost of a trend toward expanding technological surrogates for human caring. Without intimate and repeated exposure to our mutual dependence, vulnerability, weakness, concern, and gratitude for one another, it is unclear how we can cultivate our moral selves.[54] Even Aristotle, whose moral point of view is famously deficient in its acknowledgment of 'feminine' virtues and human vulnerability, notes that moral excellence and flourishing requires the opportunity to meet the needs of others, not in distant or mediated acts of assistance but in close relations such as those of friendship.[55]

Yet an enriched understanding of 21st century care must intelligently incorporate the assistance of technology rather than dismiss it out of hand; will a caring person of exemplary practical wisdom *refuse* the opportunity to deliver lifesaving medicine or food by drone simply because it is not given with the touch of a hand? Is such a delivery that much less intimate in its care than having boxes of food and medicine shoved from the back of an aid truck as it moves on to the next village? In each of the chapters that constitute Part III of this book, we will take great care not to reflexively regard emerging technologies as the enemy of caring virtue, but instead to consider how emerging technologies can promote human flourishing by being successfully integrated *within* our caring practices. That said, the danger remains that our current cultural trajectory will bypass this integration, and instead seek only the most technologically expedient routes for meeting others' needs without concern for our moral need to cultivate ourselves as fully caring persons. This delicate relationship between emerging technologies and technomoral care will be the central focus of chapter 9's treatment of social robotics.

6.9 Civility: Making Common Cause

[Related Virtues: Respect, Tolerance, Engagement, Friendship]

Civility has close associations with both classical and contemporary virtue concepts.[56] Yet the challenges of 21st century life demand a new conception of technomoral civility. Provisionally, let us define this as *a sincere disposition to live well with one's fellow citizens of a globally networked information society: to collectively and wisely deliberate about matters of local, national, and global policy and political action; to communicate, entertain, and defend our distinct conceptions of the good life; and to work cooperatively toward those goods of technosocial life that we seek and expect to share with others.* It is a disposition to 'make common cause' with all those with whom our fates are now technosocially intertwined.

This virtue is much more demanding than what the narrow use of the English term 'civility' implies, namely self-restrained and polite engagement. Technomoral civility is a far more robust form of cosmopolitan civic-mindedness, a reliably and intelligently expressed disposition to value communal ethical life in a global technosocial context and to act accordingly. A habit of mere politeness does not entail this virtue, for a person who wishes that she did *not* have to suffer the political opinions of others, or invite their input, or share the goods of technosocial life with them, may remain politely 'civil' if behaving rudely or obstinately toward them is likely to frustrate her own aims. Genuine technomoral civility would obviate such a strategy: I would remain polite precisely because my sincere wish *is* to construct and share the good life with others in my world. And while the virtue of civility often involves politeness, this is a defeasible relation. In some cases, the aims of technomoral civility may necessitate behavior that is conventionally impolite, but essential to the vitality of deeper civic connections.

Civility resonates strongly with the Confucian concept of *ren* in its political context—that is, the extension beyond the biological family of humane benevolence toward others in one's political community. As we noted in chapter 5, for Confucians civic virtue retains the emotional tones of love, respect, and care that mark familial bonds, although political affections must be appropriately tapered so as not to obscure the special priority of family life. In Buddhist ethics civic virtue is less explicit, insofar as it is folded into broader moral obligations to others. As we noted in chapter 5, the universality of Buddhist benevolence aims to foster a fundamental transformation and enlargement of one's sense of who *counts* as fellows of one's moral community. Thus on the one hand, Buddhism would seem to license the movement toward a more cosmopolitan, global conception of civility such as the one we have articulated. Yet our conception would not go far enough for a Buddhist. While technomoral empathy and care can and should be extended well beyond civic and even human circles, *civility* is necessarily restricted in scope to those with whom we can deliberate about, share conceptions of, and engage in cooperative activity to together attain the good life.

In this respect our concept of civility remains close to Aristotle's. Aristotle emphasizes the importance of civic friendship (*philia politikē*) in the *Nicomachean Ethics* and in the *Politics*, noting that it "seems to hold states together, and lawgivers seem to care for it more than justice."[57] Much has been written about the relationship between civic friendship and other forms of friendship in Aristotle; less has been written about how Aristotle's notion of civic friendship might apply to contemporary political life. A likely reason is the assumption that Aristotle's notion of civic friendship has limited application to modern liberal societies; after all, his notion is rooted in a classical conception of political flourishing in which civic friendship can be predicated on the close intimacies of a shared life (*suzên*) in near proximity with citizens with whom we have much in common.[58] How could this notion possibly carry over to large liberal states predicated on tolerance of cultural diversity and individual pursuits, much less to a globally networked planet?

Among those who have explored this question is Sibyl Schwarzenbach, who argues that civic friendship is in fact *essential* to the functioning of liberal states. She asserts that much of the work of civic friendship that holds liberal states together is the politically invisible work of women who habitually reproduce relationships of civic affection through practices of care.[59] Moreover, she points out an important and often sidelined dimension of Aristotle's concept of *philia politikē*: namely, that it involves not only citizens' concern for one another's welfare, but more specifically a concern for one another's *character*. That is, "citizens of the best *polis* care about what 'kind of persons' their fellow citizens are."[60] They do this not *merely* as a matter of instrumental advantage (i.e., because vicious neighbors pose a threat), but because they see intrinsic value in sharing life with 'good people.'[61]

Schwarzenbach also notes the importance for civic friendship of the mutual trust and goodwill secured by social justice or fairness, which supports a public perception that all are "in this together."[62] On Schwarzenbach's view, this has clear implications even for large liberal states. She notes that nothing in Aristotle's account requires that I actually *know* or interact with all, or even most, of my civic friends. What it *does* require of each citizen is that he be "informed" about the nature, history, and standard of living of citizens in other parts of the state, and be actively "concerned" about their welfare.[63] A citizen shows this concern by educating himself about their local hardships (malnutrition, poverty, floods, tornadoes, disease); being supportive of collective assistance (not begrudging the effective use of his tax dollars for their aid); and being willing to perform direct public service as appropriate. Cultivating such sentiments and habits would be an essential part of the virtue of technomoral civility.

One might question how such a virtue can be compatible with the pluralistic norms of toleration *and* individualism that define modern liberal societies. How can these cohere with a civic virtue that, as we have said above, entails a strong interest in the moral character of one's fellow citizens? *Whose* moral character?

Which cultural norms of human excellence shall prevail? Schwarzenbach's answer is that in a tolerant liberal society, civic virtue no longer entails a special interest in the *private* morality of others, but rather a narrower interest in their *public* moral character. That is, a civil person will have a strong interest in supporting and maintaining her fellow citizens' capacities for excellence in public deliberation and action. For this only a shared legal system and constitution, effective and widespread civic education, and shared public norms of mutual goodwill, concern, and toleration would appear to be essential.[64]

When we think on a global scale, many of these commonalities are lacking or infeasible, but is it unreasonable to think that *some* minimal set of norms could emerge for global technosocial discourse and action? Don't we already have widely shared ideas of who are the *worst* civic actors on this stage—those who are least capable, cooperative and expressive of a desire to enjoy the goods of technosocial life *together*? If we have sufficient exemplary cases to generate a roughly shared concept of global civic *vice*, why think it impossible to cultivate a rough sense of global civic *virtue*—a state of public moral character expressed by those international actors who are most admirable, reasonable and reliable in their cooperative technosocial endeavors? Let us not forget that while humans are not among the most cooperatively inclined creatures, we are far from the least. From the International Space Station to the Large Hadron Collider to the near-eradication of polio, we are clearly animals capable of effective, extended, and complex international cooperation on a massive scale when we are sufficiently motivated. We demonstrably do not lack the ability to make common cause with others, even on a global scale. We do lack the consistent will.

Unfortunately, contemporary technosocial life is not reliably conducive to the formation of civic will, nor to the social, economic, and political justice that it presupposes. As Schwarzenbach notes, contemporary liberal states do not reliably and explicitly support those acts of *public care* that would constitute the expression of civility among a people. Instead, she argues, the norms of social and political life are presently still oriented primarily toward production of goods and services rather than reproduction of supportive civic ties and friendships. They are also, she claims, still reflective of a distorted view of caring practices that sees these as exclusively personal and domestic, rather than public and political.[65] She suggests that this was in fact *not* the view of Aristotle, who repeatedly spoke in the *Ethics* and the *Politics* of care (*epimeleia*) and of public or *common care* in particular (*koinon epimelein*).[66]

Indeed many if not all of the technomoral virtues we have named fit together as aspects of a potentially *public technomoral character* conducive to flourishing in a globally networked society. The cultivation of this character is arguably an essential precondition of the successful pursuit of the technomoral goods of global human security, community, understanding, wisdom, and justice highlighted in chapter 2. Unfortunately, our present condition is one of widening political and economic inequality in low- and high-tech societies alike, as well as a disturbing coarsening and fragmentation of civic discourse at all levels. The handwringing of political and media experts has done little to mitigate civility's decline, as we can see on Twitter, political talk shows, blogs, and news sites (many of which have famously had to disable their online comments sections as a result.) A global atmosphere of civic insecurity and distrust has been amplified by recent revelations about massive data collection, cyberespionage, and cybersabotage by governments, corporations, and other non-state actors.

At the same time, emerging technologies have a clear *potential* to foster the rekindling of civility and the skills of making common cause. Social media are powerful tools for stimulating our interest in the character, history, and welfare of distant others. They can help us to educate ourselves about hardships others in our local or global community have faced. For example, public education regarding the hardships faced by lesbian, bisexual, gay, and transgender persons has been greatly furthered by social media campaigns such as the "It Gets Better" project, as well as many smaller-scale uses of technology to convey the stories of LGBT people. These are often credited with contributing to a surprisingly rapid boost in public support for LGBT rights in the United States and other countries. Even surveillance technologies can foster civility by provoking public outrage and joint action against forms of pervasive and profound injustice that these technologies make newly visible. In recent years such tools have been used to motivate a broad range of civic actions against the surreptitious abuse of racial minorities, women, the homeless, children, the elderly, the disabled, and other vulnerable populations.

ICTs and new media also offer newly expedient means of rendering global assistance to those whose needs have become known; consider the surges in online donations to victims of earthquakes, typhoons, and floods, or the phenomenon of 'crowdfunding' via sites like Kiva and others that finance small local projects around the world for social benefit. For better *or* for worse, social media are also used as means of collectively *policing* emerging norms of global public character; consider the 2015 Internet uproar over the illegal killing of a lion in Zimbabwe by a Minnesota dentist; or the 2013 'Tweet heard 'round the world' of Justine Sacco, the American public relations executive whose Twitter joke about African AIDS victims just prior to getting on a plane to Johannesburg ignited a gleeful Internet countdown to her inevitable firing and global humiliation, facilitated by the hashtag #HasJustineLandedYet. Such practices are all too often infected with the vices of vigilantism, including malice and reckless disregard, and have a long way to go to become morally mature. Yet for all their immaturity they are *gestures* of collective desire for a more civil, just, and caring global society.

Chapters 7 and 8 will further examine the complexity of emerging interactions between new social media, surveillance technologies, and the cultivation of civility in technomoral life.

6.10 Flexibility: Skillful Adaptation to Change

[Related Virtues: Patience, Forbearance, Tolerance, Equanimity, Mercy]

'Flexibility' is a funny name for a virtue. When we think of someone who is 'morally flexible,' we usually mean someone wholly lacking in virtue! Yet the reader will recall that flexibility was an important characteristic of the morally expert person in Aristotelian, Confucian, and Buddhist virtue traditions, and will not be wholly surprised to see it appear as a *technomoral* virtue. In the technomoral context, let us provisionally define it as a *reliable and skillful disposition to modulate action, belief, and feeling as called for by novel, unpredictable, frustrating, or unstable technosocial conditions.* This virtue is critical to our ability to cope with acute technosocial opacity.

Flexibility is closely related to other moral virtues, including *patience* or *forbearance*. Aristotle defines patience (*praotes*) as a "slowness to anger"; more broadly, a disposition to moderately forbear frustration, disappointment, injury, or insult (though *extreme* forbearance would be deficient, since there are things which will make a just and virtuous man rightfully angry.)[67] Another trait related to flexibility is *tolerance* of the shortcomings of others. As we saw in chapter 4, Kongzi was described as wholly free of the vice of rigidity or intolerance (*gu*).[68] As he says in explaining his own willingness to be employed by an imperfect minister, the *junzi* is "so hard that grinding will not wear him down, so pure that dyeing will not stain him black."[69] The virtuous person can afford to be flexible and tolerant when particular circumstances warrant it, without becoming morally compromised himself. Thus flexibility as a virtue is not *opposed* to integrity or moral uprightness—it is enabled by them.

Like Aristotle, however, Kongzi thought this virtue had limits; as shown by his firm rejection of the idea of repaying injury with kindness. He asks, "With what, then, would one repay kindness?" Instead, he suggests, one should repay injury with "uprightness," and repay kindness with kindness.[70] Kongzi rejects any form of flexibility that *indulges* the vicious rather than leading them toward rightness, or that indicates an unwillingness to make moral distinctions. Thus for both Aristotle and Kongzi, flexibility is a virtue that must remain attuned to and upholding of the *correct moral standard*, even if the manner in which this is done must be patient, sensitive to circumstances, and generous in spirit.

Buddhist virtue ethics has its own virtue of flexible tolerance, patience, or forbearance: *ksānti*, one of the Six Perfections of character in the Mahāyāna tradition. *Ksānti* reflects a disposition to patiently endure suffering, including injuries to body or ego caused by others, and to endure without fear or anger those new, truer perceptions of reality to which Buddhist practice exposes one.[71] The Mahāyāna poet Śāntideva says that "no spiritual practice is equal to tolerance (*ksānti*)."[72] This is because the experience of anger is intrinsically anathema to mental clarity and enlightened perception; in this way, the exemplary Buddhist practitioner must be far less inclined to it than the Aristotelian *phronimos*, in whom *praotes* makes room for appropriate, intelligent, and measured anger. Even the Confucian *junzi* will probably be less flexible and patient than the Buddhist sage, though he will eschew the volatility of anger for something more like sternness.

Ksānti adds an important new dimension of flexibility, beyond mere tolerance for imperfect people. *Ksānti* requires that one also be flexible enough to accommodate new and uncomfortable *truths* or *realities*. One must not only be able to bend *for* others when appropriate, but also to bend *to* newly revealed or emerging dimensions of the world. This enables that practice of *upāya kaushalya* or 'skillful means' encountered in chapter 5, which permits the morally expert Buddhist to adapt to the uncommon moral reality of a given situation in ways that may require the modulation or even suspension of the moral conventions of *śīla*. We saw this flexibility in Aristotle as well, where practical wisdom enables prudential judgments when standing moral conventions are inadequate. Likewise, Confucians eschew rigidity and praise flexibility in adapting to uncommon dimensions of moral life—as in Mengzi's case of the man who must not let ritual propriety stop him from laying hands on his brother's drowning wife.

Classical virtue traditions value flexibility as a way to honor moral *truth*, that which is based on reality rather than delusion. It grants no license to yield to falsehoods or morally alien norms. Here we confront the boundary between classical and liberal notions of tolerance, and the 'hard problem' for cultivating a flexible public technomoral character. As we have noted, emerging technologies from global transportation networks to ICTs increase our exposure to a plurality of conceptions of the good life, many of which are incommensurable with one another in part or in whole. Mere exposure to such conflicts is nothing new. The existence of incompatible systems of moral life was a well-known and widely discussed classical phenomenon; Confucian literature reflects on the *junzi's* potential engagement with 'barbarian' tribes, and Aristotle reflects on the starkly different value-conceptions expressed in the political constitutions of Athens and Sparta—not to mention the vast moral distance he perceived between Greek and non-Greek ways of life. The truly novel challenge of our contemporary technosocial environment is the extent to which 21st century technology networks—the speed, range, and power of which have grown by many orders of magnitude in mere decades—ensure the radical and virtually *irreversible interdependence* of these morally incommensurable cultures and communities.

The result is a *qualitative* shift in the problem of moral and cultural diversity. Although our species has been dangerously slow to realize it, the stakes are no longer limited to our ability to form temporary alliances against hostile political actors, nor to our short-term economic productivity and access to capital in a global financial system. The extent of our new interdependence even goes beyond the increasingly global character of scientific, literary, and artistic endeavors, or our access to worldwide information networks and an encompassing 'Internet of things.' Today our access to even the most basic necessities of biological existence—breathable air, clean water, minerals, protein-rich foods, dry land, and a livable climate—depends inextricably upon the technosocial practices, beliefs, and attitudes of global others.

Given this shift, the disposition to be flexible only with those with whom we share a distinctive conception of the good life is no longer viable. Not even the liberal mantra 'live and let live' captures the newly flexible disposition we need to cultivate if we are to continue to flourish as a species; for that outdated mantra assumes that we still have the practical luxury of leaving others to their own devices, and just staying out of their way. Instead, we face a changed world in which our individual and local fates are largely conditioned by what humans *collectively* and *globally* practice. The good life cannot be reliably secured for any of us without the global coordination of practically wise courses of technosocial action that unfold in a stable and intelligently guided manner over the comparatively long-term, in many cases on timescales significantly longer than the average human lifespan. Yet we continue to live out our individual lives in local communities grounded in differently sedimented histories of moral meaning and value. What role can a new, *technomoral* virtue of flexibility play in all of this?

At the start of this section, we defined technomoral flexibility as a *reliable and skillful disposition to modulate action, belief, and feeling as called for by novel, unpredictable, frustrating, or unstable technosocial conditions.* We now add that today, this entails something not seen as virtuous by classical traditions: an intelligent and habituated disposition to forbear those cultural value-systems and beliefs that do not significantly impair the ability of their possessors to cultivate and exercise practical wisdom in matters of *global* technosocial concern. In chapter 6 we have been building up a constellation of virtues that can together constitute *global public character*, and technomoral flexibility is a critical part of it.

Thus the *capacity for global technomoral agency* is one plausible criterion for deciding which differences in cultural norms warrant mutual forbearance in the interests of global civility and flourishing, and which norms cannot safely be tolerated because they are objectively inimical to such agency and thus to the flourishing of our species and planet. For example, any local conception of 'feminine virtue' that is incompatible with the education or civic visibility of women is globally intolerable on this criterion. Women shape the technosocial practices and habits of every culture—thus their systematic exclusion from deliberative processes that aim to prudently guide such practices is *globally* self-defeating. Likewise, a cultural norm that enables only a small wealthy or technocratic elite to acquire and exercise the virtues of public character is globally intolerable; for again, it is entire peoples, not simply the powerful few, who carry out in daily life the technosocial practices and habits that must now be intelligently steered. Global technosocial policy created without robust stakeholder input and deliberative participation will be grossly ineffective at best and harmful at worst, not to mention profoundly unjust. We can find evidence of this on a smaller scale in the litany of ineffective or counterproductive laws regulating the Internet that have been passed by legislative powers lacking either the technical expertise or the expansive user knowledge base needed to craft a prudent law governing the Internet.

There are of course many other, arguably stronger, moral reasons not to tolerate the systematic disenfranchisement of women or the poor. My point is simply that the emerging need for the cultivation of *global* technomoral agency and wisdom gives us all a new incentive to push back on those who would inhibit that cultivation among their people. In contrast, the parties to a cultural disagreement about whether individual data privacy is an inviolable human right (a view widely held in the European Union) or a selfish luxury (a common view in some East Asian countries) can afford to acknowledge and at least temporarily forbear these philosophical differences in the interests of good faith negotiations of international agreements and global technology standards that necessarily affect the personal privacy of their peoples. Here, flexibility makes room for diverse cultural interests and values to have a voice in the debate, without the implication that any compromise ultimately reached decides the underlying value question or declares invalid the sincerely held value commitments of any particular group.

We must therefore aim to cultivate *technomoral flexibility*, a child of the liberal virtue of tolerance that aims to enable the co-flourishing of diverse human societies. Yet unlike a version of liberal tolerance that merely counsels us to 'live and let live,' technomoral flexibility requires that we cultivate a global capacity to actively deliberate together and agree upon prudent courses of technosocial action, even with people whose conceptions of the good life vary substantially from our own. This will produce inevitable conflict, insofar as we may disagree on where the human ship should be heading, so to speak. The defining feature of our present technosocial reality, however, and the one that distinguishes it from the liberal dilemma, is that no one's community can choose to 'get off the ship.' Our immediate fates are now inextricably interlinked with those of the other passengers. In each of the remaining chapters we will have an opportunity to reflect on the challenges of technomoral flexibility in an age of new media, surveillance, robotics, and human enhancement technology.

6.11 Perspective: Holding on to the Moral Whole

[Related Virtues: Discernment, Attention, Understanding]

The notion of moral perspective as a virtue has no precise classical analogue. Proper moral perspective is implied by the more comprehensive virtue of practical wisdom (*phronēsis*) in Aristotle, as well as by the virtue of wisdom or 'intelligent awareness' (*zhi*) in Confucian ethics, and the concept of wisdom (*prajñā*) in Buddhism. Yet it also has associations with the perceptual faculties of moral attention we discussed in chapter 4. Moral perspective can be roughly defined as a reliable disposition to *attend to, discern, and understand moral phenomena as meaningful parts of a moral whole.* The technosocial applications of moral perspective will hold our interest here, but because moral perspective is holistic by definition, its exercise cannot be limited to narrowly technosocial phenomena.

Comparative Buddhist scholar Nicolas Bommarito has developed a helpful account of moral perspective as an outlook that grounds familiar virtues such as patience and modesty. In his discussion of the perceptual dimension of Buddhist patience (*ksānti*), Bommarito describes moral perspective as entailing not a personalized outlook on the world and its value (in the way an English speaker might say "well, that's my perspective,") but rather a morally proper "sense of scale."[73] For example, this scale is engaged when I react patiently to the delayed boarding of my flight because reading on my laptop about the one that just crashed into the ocean has "given me some perspective." Moral perspective helps us to estimate the relative importance of competing values and desires and offers us "a sense of our place in a larger context," as when a soldier might say, "When I see that photo of my unit, it gives me some perspective and I feel I'm part of something more important than myself."[74]

Moral perspective allows me to see how my own desire at a given moment (say, my desire to board 'my' flight sooner) is more properly scaled within a holistic picture of desires and values that I hold in view. It is this 'holding in view' that is the exercise of moral perspective. In our example, a virtuous person would be able to hold in view her own standing desire not to be a selfish, impatient, or rude person, as well as the desires and values of the equally frustrated gate agent and fellow passengers, and the mechanics and crew who rightly value safety over on-time departures. Moral perspective is an essential disposition of a virtuous person; I cannot be a patient, honest, compassionate, or just person unless I act in these ways fairly reliably, and it does not seem that I will be able to do so unless I can reliably maintain moral perspective when I think and act. As Bommarito emphasizes, keeping moral perspective is a lot harder than retaining moral *knowledge*. I can intellectually *know* and *believe* that my needs and desires are not all that special or important in the 'big picture,' but that doesn't mean that I will actually be able to see the world in that light when my desires are activated.[75]

Yet my perspective *does* need to be cognitively successful—that is, responsible to the *reality* of the moral whole it keeps in mind. For example, Bommarito notes that a person who is characteristically meek and never puts her needs above those of others is not virtuous, because she fails to keep in proper perspective those needs and desires of hers which *really are* important in her moral environment.[76] Such a person might risk her own physical safety just to avoid inconveniencing or irritating a close relative, and this would be the vice of servility, not virtue.

At the end of chapter 5 we discussed those Buddhist exercises in imaginative 'perspective-taking' that foster the moral practice of appropriately extending empathic concern. We also referred to other, secular techniques of narrative and artistic expression as ways of stretching our sense of the moral whole; of seeing and 'holding in view' more of the important relations, obligations, and interdependencies between ourselves and others; of conceiving and feeling the moral meaning of others' needs, desires, and interests alongside our own. Thus a reliable disposition to hold an appropriate moral perspective is, in addition to enabling other virtues, a virtue that can be deliberately cultivated in its own right.

Still, what is a *proper* moral perspective? Doesn't this presuppose a single correct view of moral reality, one true scale by which the importance of ours and others' desires and interests should be judged? Isn't this obviously incompatible with the global ethical pluralism we have endorsed? Yes, but really no. Yes, because there is far more widespread agreement about moral scales than we acknowledge. The clinical sociopath's moral scale will be agreed to be wildly off by the vast majority of informed judges. So will the pathological narcissist's. So will the person who throws violent tantrums at gate agents whenever 'her' flight gets delayed, who quietly and knowingly dumps gallons of toxic sludge into a local stream rather than taking it for safe disposal, or who knowingly builds a grade school with substandard concrete in order to skim a few extra bucks. While the consensus will never be perfect, even locally, there is a well-established global sense of proper moral perspective and a large number of human behaviors and judgments widely regarded as contrary to it.

And yet, despite all this agreement, there are individual, local, and cultural differences in moral perspective that cannot be elided, and that we arguably should not merely tolerate, but welcome. At the finest levels of discrimination there is no single true scale of desires, values, and interests, nor need there be in order for us to have widely shared, even globally resonant conceptions of virtue. All we need for the latter is an understanding of what habits are generally conducive to human self-cultivation, which material and technosocial conditions are most essential to our flourishing together as a global and interdependent human family, and which among our various visions of the good life are consistent with that end. What, then, is the distinctly *technomoral* significance of the virtue of moral perspective? Our growing interdependence in a networked world entails that we will confront more and more choices the implications of which are global in scope. The need for proper moral perspective becomes ever more apparent as our actions and their consequences magnify in scale; as the potential impact of our technosocial choices widens, the desires, values, and interests that come into view become more numerous and more challenging for us to rank. Such considerations are of far greater moral significance than ever before, up to and including the immediate and long-term survival and vitality of ours and other species—even the ability of the planet to support life at all. In the 21st century, the job of keeping proper moral perspective is both more difficult, *and more important*, than it has ever been.

Moreover, emerging technologies offer ever-new ways to both undermine and enrich our moral perspective. From the rise of the ubiquitous 'selfie' enabling grinning self-celebrations at funerals and, notoriously, even concentration camps, to self-tracking apps that allow you to obsess over everything from the number of steps you take in a day to the number of people who 'like' your cat photos, technosocial life can make our perspective seem suffocatingly small. Yet it can also be used as a tool to enlarge our moral perspective with humor—as with the widely popular Twitter hashtag lampooning #firstworldproblems—or with intimacy, as with the many uses of social media and even digital surveillance to make the experiences, desires, and interests of marginalized peoples and classes globally visible and more immediately visceral. Our aim in Part III will be to take a closer look at how greater moral perspective might improve our use of these and other technologies, and in turn how the ethical use of emerging technologies might aid us in our efforts to enlarge and enrich our moral perspective.

6.12 Magnanimity: Moral Leadership and Nobility of Spirit

[Related Virtues: Equanimity, Courage, Ambition]

The Aristotelian virtue of magnanimity, or *megalopsuchia* (greatness of soul or spirit), has been the object of much modern critical scorn. It is often associated with an entitled sense of political and personal superiority, and the idea that a virtuous person ought to believe and feel himself 'above' the troubles and concerns of others. Understood in this sense, we might see magnanimity as inimical to the technomoral virtues of justice, empathy, care, civility, and perspective we highlighted above, and thus a better candidate for a technomoral *vice*. So what should the reader make of its presence here?

First, we should note that while the characterization given above might well have described some members of the Athenian *polis* that Aristotle greatly admired, this would be the result of a failure by Aristotle to correctly apply his own concept. First, his concept of magnanimity or 'greatness' is an explicitly *moral* one, thus excluding political and economic status as conditions of its satisfaction, except insofar as in a particular society these happen to be contingent prerequisites for one's moral development (as Aristotle thought some degree of material and political fortune were). The point here is that an inflated sense of superiority, or one that is based on one's political or economic status rather than one's character, is by definition contrary to the virtue of magnanimity. Only a *morally excellent* person who has cultivated virtue to an exceptional degree (is unusually just, honest, wise, etc.) can be a *megalopsuchos* ('great-souled man').

Furthermore, the sense in which the 'great-souled' or magnanimous person is 'above' the common person is chiefly concerned with their lack of *pettiness*—their unwillingness to defile their virtue by scrabbling in the dirt over trivial advantages, honors, titles, prizes, or other ego-boosting trifles. The great-souled person does not ignore these things because he wishes to be above others, rather *he is above others just because he tends to ignore these things.* The things the great-souled person values are more valuable. The magnanimous person is the one who has a sense of nobility and self-worth founded in a lifetime of moral and social efforts rather than relatively meaningless zero-sum contests of ego. The magnanimous person can afford to be generous in spirit where others are not. He can absorb a petty insult without having to repay it. He can warmly greet the person who has pretended not to notice his arrival. He can let the other car swoop into 'his' parking space at the mall without responding like a rabid dog.

Of course there is no doubt that Aristotle's conception of magnanimity is riddled with indefensible gender, ethnic, and class bias. Property-owning Greek men could be great-souled. Women, Persians, and slaves could not. It is also true that ancient Greek culture would have fostered specific acts of magnanimity that many contemporary citizens of more egalitarian societies would rightly find obnoxious or obscene. But does that make a *contemporary* re-conception of the 'great-souled' or magnanimous person incoherent? I hope my readers will admit that, no matter how unenlightened Greek thinkers may have been on the subject of who can or cannot qualify as magnanimous, they were right in this respect: there is much to admire in a person whose ego is neither a giant sucking black hole of need, nor packed to the brim with a bunch of morally vacuous honors and 'successes.'

A similar sense of moral nobility is embodied in the Confucian ideal of the *junzi*, who is repeatedly described as above pettiness and who neither overestimates nor undervalues his moral worth. In Buddhism the notion is more problematic, as the very concept of 'self-worth' or being 'great in soul' is metaphysically misleading and hence morally destructive. Yet there is a *behavioral* equivalent of magnanimity in the Buddhist image of the exemplary person, who is certainly 'above' petty concerns and ego contests, and who has cultivated the moral resources to absorb the pettiness of others without being ruffled or moved from calm equanimity (*upekṣā*).

What, then, is the relevance of this classical ideal for 21st century life? Who talks or thinks about being 'noble' any more? This is precisely my point. Magnanimity enables and encourages moral ambition and moral leadership, two things sorely lacking in our contemporary technosocial milieu. Moral ambition can be described as the ability to 'think big' in one's moral aims. The magnanimous, those with *justified* moral ambition, are able to go beyond what most of us can afford in the moral realm (often little more than 'I'm going to try to be slightly less of a selfish jerk today.') The magnanimous can pursue and lead others in moral projects that require enduring courage, deep wisdom, expansive empathy, extraordinary care, and tolerance for great frustration and conflict—because they have successfully cultivated these virtues as resources for such projects.

Of course most who view themselves as great moral leaders more likely suffer from a self-aggrandizing messiah complex, one reason that many of us regard anyone who identifies himself as a moral beacon with understandable suspicion. Nor is anyone born or destined to be great-souled, although some people's native psychology or life circumstances might offer more favorable conditions for extraordinary moral development. Still, those who have managed to cultivate themselves well can take on, in concert with others, projects that would be impossibly daunting for the morally average person to lead: fostering peace and goodwill between warring groups with a lengthy record of mutual atrocities; transforming a whole culture's prejudices about a despised ethnic group, women, or LGBT persons; ending practices of slavery or the poaching of wild animals. Persons of ordinary moral character can and should contribute much to these aims, and the vast majority of us involved in such efforts *are* just morally ordinary people doing our best, and getting out when the frustrations and disappointments become too much for us to bear. Yet it is hard to see how such projects can be reliably sustained and coordinated over the long-term without the moral energies of the truly magnanimous.

The magnanimous are those who have rightly earned the *moral trust* of others, who can inspire, guide, mentor, and lead the rest of us at least somewhere in the vicinity of the good. It is not an unrealistic ideal. There have been many such people in history, some famous, most unfairly forgotten, people who were still imperfect but far better than most, and who helped others to be better and *do* better. Yet does our contemporary technosocial world foster and support the ambitions of genuine moral leaders? Can we name specific living individuals who we trust to lead us *wisely* and *morally* through the thicket of global technosocial dilemmas facing humanity? Who can reliably guide us in making prudent choices about the expansion of global digital surveillance tools, armed robots, brain-computer interfaces, or the replacement of millions of human laborers with automated software systems? I will let that question hang in the air, rather than attempt to answer it. If the reader sees clear evidence that global technomoral leadership is in great or even adequate supply, then I am happy to be wrong. But if I am not wrong, then it is time to ask how human flourishing in the 21st century can benefit from a revival of magnanimity as a technomoral virtue, and how we might be able to achieve it.

6.13 Technomoral Wisdom: Unifying the Technomoral Virtues

The greatest of the virtues is the one we will say the least about, for it encompasses the totality of what has already been said. Whole books have been written on Aristotle's concept of practical wisdom or *phronēsis*, but as with Kongzi's notion of *ren* in its broader use, moral wisdom is best understood as a term for 'complete virtue,' representing the successful integration of a person's moral habits, knowledge, and virtues in an intelligent, authentic, and expert manner. Thus a person of moral or practical wisdom *just is* a person who reliably puts into practice the seven moral habits articulated in Part II, and who has used those habits to cultivate and integrate the virtues essential to flourishing in their given moral world. Such a person is not perfect but nevertheless exemplary—flawed but far better than most, finding more pleasure and joy in living well and rightly than most of us get from living badly or thoughtlessly, and exhibiting flashes of authentic moral beauty in action that give the rest of us pause.

Technomoral wisdom is therefore a virtue in a different sense from the others. It is not a specific excellence or disposition, but a *general condition* of well-cultivated and integrated moral expertise that expresses successfully—and in an intelligent, informed, and authentic way—each of the other virtues of character that we, individually and collectively, need in order to live well with emerging technologies. Each of the other eleven technomoral virtues find their highest expression when integrated in the actions of a person with technomoral wisdom.[77]

Having already laid down the justification, conceptual framework, and content for an account of the technomoral virtues, our concern in Part III will be narrowly practical and applied: identifying how the technomoral virtues individually, and technomoral wisdom in general, can aid us in coping with the rapid emergence of new technological powers from an increasingly opaque and unpredictable technosocial horizon.

PART III

Meeting the Future with Technomoral Wisdom, Or How To Live Well with Emerging Technologies

7

New Social Media and the Technomoral Virtues

IN PART III we investigate how various forms of emerging technology, depending upon how we choose to develop and engage them, may enable or frustrate our efforts to individually and collectively become virtuous: to make ourselves into the sorts of human beings able to live truly *good* lives. Here our framework of technomoral habits and virtues is put to work.

Since humans and technologies evolve in tandem, we will be asking not only how these emerging technologies shape our moral character for better and for worse, but also how cultivating technomoral virtue can help us to develop better technologies and wiser technosocial practices. Our aim is constructive: we will not hesitate to criticize suboptimal human-technology relations and effects, but the goal is to think about how those relations can be changed for the better, and how our technologies, communities, and *selves* can be made better in the process. For the ultimate engineering task is the fragile, endless, and sublime human project of using the culture we produce to make ourselves into the beings we wish to become. The main topic of this and subsequent chapters is whether and how emerging digital, robotic, artificially intelligent, or biomedical technologies can assist us in this task.

7.1 New Social Media and the Good Life

The scope of this chapter is limited in its precision by the constantly shifting forms and increasingly permeable boundaries of the digital media technologies that are its subject matter. Our focus, however, will be the innovations most responsible for the rise of 21st century digital culture: mobile media devices and software applications, online social networking platforms, and the encompassing architecture of Web 2.0.[1] In the early 2000's Web 2.0 standards transformed the Internet from a large collection of relatively static personal, institutional, and commercial webpages to a highly interactive and dynamic multimedia environment. Social networking and media sharing sites such as Facebook, Twitter, YouTube, and Instagram dominate this new landscape, along with Google, Apple, Amazon, and other corporate titans. But the distinctive feature of Web 2.0 is that it is *users* who create and share the vast majority of its content: the videos, music, photos, stories, mashups, satires, reviews, comments, complaints, opinions, hashtags, jokes, and memes that make up our new digital media culture. Just as essential to Web 2.0 as digital content is the distinctly *social* character of new media consumption. It is the constant stream of new connections and exchanges between billions of individual content creators and sharers that gives the Internet its present character as a *global* social network.

The unrestrained, even irrational early hype trumpeting the Internet's power to cure all the world's ills (*or*, in the eyes of its most careless critics, to destroy everything true, beautiful, and good) has thankfully given way to more realistic and nuanced assessments of its many benefits, dangers, and trade-offs. Most researchers today prefer a critical posture reflected succinctly in the title of leading Internet researcher danah boyd's book on teens and new social media: *It's Complicated*.[2] Yet there is still a strong conviction among many that the social Internet is a powerful force for human well-being, a silicon-paved path to the good life. Among those championing this view most loudly, we should not be surprised to find those technology leaders and innovators whose pockets are lined by public faith in the Internet.[3] But resolutely positive assessments of digital culture's impact on human well-being have also been voiced by many scholars presumably less invested in the verdict.

In their 2012 book *Networked: The New Social Operating System*, sociologist Barry Wellman and technology researcher Lee Rainie championed the new "networked individualism" as a force destined to engender a more educated, creative, productive, and happier society. Their core research metric is *social capital*—a sociological construct used to quantify the amount of social support from which an individual can draw in order to solve problems and meet her needs. Empirical research suggests strong links between social capital and psychological well-being or 'life satisfaction.' Yet the empirical measures of well-being privileged by sociologists and psychologists are simply *not* measurements of the same type of well-being with which virtue ethics is concerned, the type of human flourishing that Aristotle called *eudaimonia* and that is the motivating aim of ethical life.

Empirical studies of well-being often employ instruments such as the Diener Satisfaction With Life Scale (SWLS) or the Diener Flourishing Scale (FS). These are simple 5-to-8-item questionnaires that take no more than one or two minutes for a typical subject to complete. On the Flourishing Scale, for example, subjects rank items such as "I lead a purposeful and meaningful life" on a scale of 1 to 7, from "strongly disagree" to "strongly agree."[4] As useful as these instruments may be for many purposes, they do not tell us whether these subjects are *actually* living well or flourishing—at best they tell us whether the subjects *believe* themselves to be living well, based on a cursory review of a small number of criteria. That one is living well in the ethical sense is a fact that one can be mistaken about. A successful sociopath, a violent and unrepentant serial abuser, or an oblivious narcissist may readily score their own lives highly on such scales without actually meeting any of the requirements of a good life. If this isn't obvious, consider the fact that most parents' greatest desire is that their children flourish and live well, yet most would be horrified at the thought of having raised a self-satisfied and unrepentant sociopath.

Living well in the ethical sense, then, is an external and holistic condition that is not entailed by subjective reports or feelings of personal happiness and life satisfaction. Although most virtue ethicists regard living well in the ethical sense to strongly *support* such positive feelings, even to be the most reliable means of attaining them, it is psychologically possible for a person to be satisfied with a terrible, ignoble life. A person can also be *dis*satisfied, at least temporarily, with a life that is objectively noble and choiceworthy.[5] This means that the sorts of empirical data often used as evidence of a strong correlation between new social media and the good life are instructive, but not adequate for our inquiry. We need to look more deeply at the relationships between particular new media practices and the moral habits and virtues that enable and even *constitute* living well in the fullest possible sense.

7.1.1 New Media Habits and Rituals: The Virtues of Communicative Friction

My career as an emerging technology ethicist started with new social media for a simple reason: No technology that I had ever encountered had a comparable power to so rapidly and profoundly transform people's communicative habits, and given the communicative structure of moral life, it seemed obvious that these shifts had clear implications for human character development, and especially for the cultivation of our moral capacities. I began to write about how new social media habits might impact the development of what I called 'communicative virtues' such as honesty, patience, and empathy.[6]

Compared with traditional media, new social media foster communicative acts that are more rapid, flexible, concise, varied in style and content, and fluid in tone, pace, scope, and structure. Yet they also promote a dubious ideal: *frictionless* interactions that deftly evade the boredom, awkwardness, conflict, fear, misunderstanding, exasperation, and uncomfortable intimacies that often arise from traditional communications, especially face-to-face encounters in physical space.

Of course, such frictions still *do* arise in digital media connections. Yet when they do, the emerging norms of new media increasingly afford me or my interlocutor a quick and easy escape. We text 'gotta go,' log off Facebook, or close an app, tab, or window and quickly open another, safer, more attractive, and less demanding connection. And who can resist availing themselves of such escape routes, given that doing so is socially accepted, even increasingly expected?

We shall see later in this chapter the extent to which the new digital norm of frictionless sharing can imperil civic life. But intimate contexts of family and friendship are also impacted, in ways that may profoundly challenge character development. This risk came into sharp focus during the controversial 2013 Facebook Home ad campaign, which featured persons using the app to evade undesirable interactions at home, at work, and on an airplane. Technology ethicist Evan Selinger criticized the ads as "social engineering spectacles" that implicitly advocate "transforming vice into virtue."[7] Is this fair? Consider the 'Dinner' ad targeted by Selinger. It approvingly depicts a savvy teen's use of her phone's Facebook Home app to tune out a tiresome relative's story at a large family dinner. The ad asks us not only to sympathize with the teen's plight (and who doesn't?), but also to admire her bold solution. While her tablemates suffer and squirm, praying for deliverance, she smiles serenely into the glowing screen in her lap.

Yet what might she have done if using Facebook at a family dinner was simply beyond the social pale? She might have mustered the courage to interrupt and change the subject. She might have engaged in sympathetic eye-rolling with a fellow family hostage across the table. She might have started a quiet side conversation. She might have just sucked it up and patiently waited it out. What is important to recognize is that selecting well from these options calls for the skillful deployment of a range of moral habits, including attention (What is the offender droning on about? How engaged are others at the table?); relational understanding (Who is the offender? What is her position or status in the family? Is this a typical family dynamic?); reflective self-examination (Am I being a jerk here?); discerning and prudent judgment (How much deference does the offender deserve? When is the moment to act? What's the most defensible strategy? What technique will cause the least harm or insult?); and moral extension (Who should my solution benefit besides myself?). The person who has cultivated these habits and learned to respond reliably *well* in such situations possesses admirable virtues: self-control, social courage, empathy, care, flexibility, moral perspective—perhaps even practical wisdom. Where does our young Facebook Home user find *her* chance to practice these moral habits and cultivate these virtues?

Of course it would be silly to blame new media for the timeless ills of teenage boorishness, social haplessness, or self-absorption. Yet it is anything but silly to ask which new media habits will help teens *grow out* of that condition successfully, and which of today's new media habits tend to delay or impede that maturation. Empirical studies of the effects of new social media on youth lend increasing support to what is called the 'rich get richer' thesis: young people who are already well on their way to mature social competence and who have acquired above-average skills of social discernment, empathy, care, self-control, social courage, and confidence tend to *thrive* and gain important benefits from new social media practices.[8] On the other hand, young people who lack social competence, or who report high levels of social anxiety or loneliness, not only have less positive experiences online but in many cases seem to suffer increased anxiety and loneliness as a result of their new media habits. This 'rich get richer' hypothesis is entirely consistent with our account of moral self-cultivation thus far. It is not technologies *themselves* that determine whether or not we flourish socially, but rather the habits, skills, and virtues we have cultivated, with or without their help. Once we acquire those habits, skills, and virtues, we are not only better able to avoid harm in new social media environments, we are able to use new media in ways that further enrich our well-being.

Yet how did the already socially 'rich' *get* that way? Where did they acquire the social skills and warranted confidence that allowed them to flourish in the same new media environments that further impoverished their socially 'poorer' peers? While this topic has yet to be adequately researched, I venture the hypothesis that youth who flourish online acquired the lion's share of their social competences in environments that 1) did not permit or reward quick and easy escapes from social friction, challenge, or discomfort; 2) allowed considerable freedom to experiment with different communicative strategies for managing and coping with communicative friction; and 3) tempered that freedom with virtuous constraints—norms or rituals of communication designed to habituate them to social respect, competence, care, and wisdom.

Returning to the Facebook Home case, the girl in the ad is not forced to try out *any* of the techniques that we considered for negotiating the boring, alienating experience of the family dinner—she can simply nullify the experience by means of her social media escape pod. The problem is not the technology alone, but the technology *plus* the lack of an appropriate virtue-inducing constraint—such as a rule or strong social expectation that *one does not use Facebook at family dinners*. It is illuminating to consider that many high-tech leaders impose a surprising level of restraint upon *their* childrens' technology and new media usage at home, banning electronic screens from the family table, child bedrooms, and/or on school nights.[9] Such constraints are not plainly *not* the refuge of the technophobic; if anything, *they are the rituals of the technologically wise*.

As Selinger points out in his scathing commentary on the Facebook Home campaign, communicative norms that foster vice don't only impoverish the young and immature:

> Ignored Aunt will soon question why she's bothering to put in effort with her distant younger niece. Eventually, she'll adapt to the Facebook Home-idealized situation and stop caring. . . . Selfishness is contagious, after all. Once it spreads to a future scene where everyone behaves like Selfish Girl, with their eyes glued to their own Home screens, the Facebook ads portend the death of family gatherings. More specifically, they depict the end of connecting through effort.[10]

Selinger notes that the ad campaign idealizes a world where we tune out anyone whom we did not ourselves *select* as a social connection on the basis of shared interests and their personal entertainment value. Aside from the deep disrespect for family bonds that this evokes, it is a profoundly *immature* moral viewpoint. Indeed it reflects a value-system suited to Facebook's own origins: the narrow social aims of college undergrads away from home and, for the first time, free to assemble their own social network from a pool of strangers on the *sole* basis of their own personal desires and tastes. This is a liberating experience and an important stage in a young person's development. But its generalization as a universal norm of social life would be devastating. A frictionless world where every social bond and duty was conditional upon the ongoing ability of others to keep us stimulated and pleased would diminish us all.

In particular, it would render meaningless technomoral virtues such as *care*—the skillful, attentive, and emotionally responsive disposition to labor to meet the needs of other members of our technosocial world. The moral world of adulthood, *in any culture*, is one in which relational bonds of family, friendship, work, and civic life require us to care for, understand, and engage deeply with people who we at least initially find to be boring, weird, irritating, alienating, or exasperating. Some of those people we will learn to enjoy with ease. Others will always require more effort to be around, but we will come to respect, admire, and even love many of them anyway. A world where this kind of care and flexibility is not expected or even desired is one that very few of us would want to live in, and that none of us *should* want. It is a world without true friendship or love.[11] As Selinger points out, the Facebook Home campaign shields us from this horror by ensuring that in each ad, there is only one lucky soul using the app; everyone else in the room is still doing the heavy lifting of keeping the social fabric whole. But the campaign doesn't target just *one* family member, or one airplane passenger, or one person in the boring business meeting. The ads target us *all*—and therein lies their fundamental dishonesty.

This is why it is essential that we pay closer attention to the ethical impact of new social media on human character in different contexts, avoiding mindless celebration *or* demonization of these technologies. One important context involves the individual. As we have seen, socially well-habituated individuals are more likely to use new media to enrich, rather than to diminish, their flourishing. Another kind of context is the concrete situation. If our Facebook Home girl had a close friend undergoing life-saving surgery, we would not question her need to periodically check for updates on her phone, assuming she had explained to her family the legitimate reason for her distraction. A third kind of context is cultural—for the specific moral norms, habits, and rituals that structure one's environment help to determine the moral meaning of one's actions in that space. For example, philosopher of technology Pak-Hang Wong argues that today's new media architectures flatten relational structures in ways that are inconsistent with Confucian values centered on family roles and hierarchies. He claims that the personal freedom to manage one's own social media presence, without constant supervision and approval by family elders, might be genuinely unethical in a Confucian home—even though most Western virtue ethicists would find such oversight and control unduly intrusive.[12]

Context also drives ethical issues with new media that are increasingly hard to regulate, such as privacy, copyright, and cyberbullying. As Helen Nissenbaum notes, respecting privacy online requires acknowledging the *contextual integrity* of situations in which the disclosure of personal data is a factor. What users reasonably expect to happen with their data in one context may be entirely different in another, even where the covering laws are the same. Exclusive control over the use and reproduction of creative digital property is in some cultural contexts an inalienable right of the author, and in others a selfish inclination to hoard the fruits of one's creative efforts.[13] Speech that is hateful cyberbullying in one online context may be edgy humor in another. These contextual challenges cannot simply be resolved by imposing fixed rules or principles of a universal order. They require effective moral communication, sincere cooperation, and prudent, flexible negotiation, often across cultural lines. It is time for a closer look at the technomoral virtues that can best facilitate such exchanges in personal, local, and global contexts.

7.2 New Social Media and the Virtue of Self-Control

In chapter 6, we saw that the technomoral virtue of self-control is *an exemplary ability in technosocial contexts to choose, and ideally to desire for their own sakes, those goods and experiences that most contribute to our and others' flourishing.* It is a reliable disposition to discern and eventually to author *right* desires, those compatible with the good life. And as we saw in chapter 3, self-control begins in childhood with socially imposed habituation to right action. Ideally, this eventually gives us the moral continence to act only on desires that we correctly discern to be good, and to restrain ourselves from acting on vicious desires that often arise spontaneously. As moral understanding, experience, and motivation grow, restrained continence can mature into virtuous temperance, a state in which we become increasingly free to act on our desires, because our dominant desires have themselves become good.

Perhaps the greatest 21st century challenge to self-control is a culture in which consumption itself is the most valued activity of citizens. But new media deepen this challenge by means of the vastly expanded range of goods now available for our consumption through digital channels, and the associated fragmentation of social consensus about which goods and experiences are worthy of our consumption. These developments have a clear benefit: instead of cultures curated by a handful of state-owned or corporate gatekeepers, we have a newly enriched and constantly changing global tapestry of offerings at our fingertips. No longer can monolithic powers so easily restrict what citizens see, hear, discuss, or buy, or dictate which goods, topics, events, or creative acts are worthy of their attention. This is a benefit we should not wish to surrender. The trade-offs, however, have been considerable. Moreover, our digital liberation from cultural hegemony is itself endangered by multinational media consolidation and the deliberate design of 'sticky' digital media delivery systems that exploit neurological and psychological mechanisms to undermine our self-control.

Concerns about new media and self-control are often expressed by psychologists in terms of Internet 'addiction' or compulsion. While empirical research is ramping up, reliable studies of the true scope, severity, and causes of online media 'addiction' remain in short supply.[14] Early indications of a problem are not so hard to come by. A quick Google or Amazon search reveals a rapidly expanding cottage industry of books, blog posts, editorials, and software tools devoted to helping consumers diagnose, treat, or mitigate the harms produced by their new media addictions. Apps such as Freedom, Anti-Social, and SelfControl allow you to lock yourself out of part or the whole of the Internet for a fixed period of time, serving as modern-day Faraday cages for new media addicts like myself who are self-aware enough to know that they have a problem. These are less radical versions of the solution Nicholas Carr was compelled to adopt to finish his 2010 book on the Internet's impact on our brains, *The Shallows*: move to an isolated mountain town with no cell service or broadband connection.

What is interesting is that neither myself nor Carr, nor the many other scholars and creative professionals who seek such measures, can easily restrain *ourselves* from excessive and unproductive new media consumption. This is remarkable in a population of people who earlier in life proved themselves disciplined enough to write lengthy technical dissertations or other creative works requiring intensive concentration and self-discipline. Even Carr, whose book details the ways that new media change our brains in ways we do not choose, admitted that he slipped back into his new media habits once his book was done; voluntarily (or perhaps not?) surrendering the mental clarity, focus, and power he says he regained during his digital media sabbatical.[15] Perhaps Carr and I are in fact not self-disciplined at all; perhaps we are viciously incontinent or self-indulgent. But I venture that many readers know from personal experience that some new media habits pose an unusually strong challenge to our faculties of self-control.

Neuroscientific explanations of Internet addiction tell a story about operant conditioning, reward schedules, and dopamine receptors.[16] But the underlying social mechanism is rarely discussed: the techniques of those who develop, market, and measure the success of new media technologies. From the popular software industry site *Re/Code*, here is a remarkably frank admission of the deliberately addictive design of new media, in a post from Suhail Doshi, the CEO of Mixpanel—an analytics firm that sells developers tools for measuring how users interact with their software. The title of the post is "Mixpanel: How Addictive is Your App?":

> Whether you're building a game, a social network or a CRM tool, your ideal customers are the people who engage with your product at least once every day — *better still if they're using it constantly*. We've taken this a step further by introducing a new analytics report — *Addiction* — which will be available to users soon, and which tells companies how frequently people use their apps throughout the day. For example, if you've built a game, and a customer plays once in the morning, once at lunch, and once on the commute home, they've engaged with your product during three separate hours: As such, they'll fall into the three-hour bucket. (For the record, about 22 percent of actual game players fall into the three-hour addiction bucket, which makes sense — only a very lucky few can spend the entire day gaming.) . . . *Social apps have a stable, consistent and thoroughly addicted user base*, with 50 percent of people engaging with social networks for *more than five hours a day*, and even a small percentage logging time during *every waking hour*. (emphasis added)[17]

If it's not already obvious that an addicted user base is the explicit design goal for many app developers, Doshi reminds his audience that if users aren't logging on multiple times a day, they are doing something profoundly wrong, but if "most users are lighting up your Addiction report by using your app for 10 hours every day, you're doing something very, very right."[18]

If self-control is as essential to human well-being today as it was in Aristotle's, Kongzi's, or the Buddha's time, then we have a serious ethical problem on our hands. On the one hand, the pleasures of new media culture are vast and unprecedented; not many that have tasted them, even among the highly educated and culturally literate, would wish to surrender them. Thus new media seem to survive John Stuart Mill's test for discerning true goods—after all, he tells us, "What is there to decide whether a particular pleasure is worth purchasing at the cost of a particular pain, except the feelings and judgment of the experienced?"[19] On the other hand, many new media pleasures are consciously designed to be delivered to us in ways that undermine our cognitive autonomy and moral agency. They make it harder, not easier, for us to choose well. Even a utilitarian such as Mill understood the role of character, especially the virtue of self-control, in sustaining our ability to not merely know, but actually *choose* the good life:

> Capacity for the nobler feelings is in most natures a very tender plant, easily killed, not only by hostile influences, but by mere want of sustenance; and in the majority of young persons it speedily dies away if the occupations to which their position in life has devoted them, and the society into which it has thrown them, are not favourable to keeping that higher capacity in exercise. Men lose their high aspirations as they lose their intellectual tastes, because they have not time or opportunity for indulging them; and they addict themselves to inferior pleasures, not because they deliberately prefer them, but because they are either the only ones to which they have access, or the only ones which they are any longer capable of enjoying.[20]

If I were a more temperate person, I would *still* choose to use new media technologies—but I would do so consciously and selectively, not at times when it disrupts family life, not at times when I need to focus for an extended period of time on a single creative or intellectual effort, and not at times when I need to attend closely to my physical and moral environment. The question is: How do I and others become temperate or even adequately *continent* in our use of new social media? There is no simple answer, but an adequate solution must go beyond the habits of the individual consumer. I, like many others, can already manage to grit my teeth and activate my Freedom app when I am at home writing alone, and perhaps over time such individual practices can make us all more continent in this regard. Perhaps someday I will be able to turn on Freedom without gritting my teeth. Perhaps one day I will stop trying to open my browser every 15 minutes *even though I know that I am still locked out*. It helps that I have a small but supportive community of fellow writers with the same struggles, who directed me to the app in the first place. Our struggle, however, is greatly intensified in social environments where there is no such support: such as conferences where not only do fellow scholars openly use social media while presenters are speaking, but where we are asked by conference organizers, or even speakers themselves, to tweet the event in real time!

Likewise, setting a no-phones rule at your own table is easy enough, but what does one do at a dinner out with friends or colleagues that is already lit up by six different glowing, beeping screens? At that point imposing your own self-control seems pointless, since there is no one present enough to receive your undivided attention, and even self-defeating, as it leaves one stewing in a morally unhelpful mix of resentment, boredom, and holier-than-thou superiority. Of course, one can just seek new dinner companions, but most of us can't avail ourselves of this nuclear option easily or often. Instead, satisfactory solutions to digital media 'stickiness' will have to involve *collective* cultural agreements to seek healthier digital norms or social rituals, which will be specific to particular contexts but globally concerned with promoting the cultivation of technomoral self-control. An early example of this was the spread of the restaurant 'phone stack'; but such remedies remain thin and inconsistently practiced.

Another part of the solution will require engaging more assertively with software industry professionals on the social and moral impact of their designs. As Evgeny Morozov notes in his recent book *To Save Everything, Click Here*, this is a challenging hurdle to clear, given the false but widespread and resilient belief that "The Internet" in its current form (and by extension, the structures of the diverse technologies of which it is really composed) is a fixed given—an Archimedean point with which it is either futile or disastrous to tinker. He asks us why is it that when we find ourselves harmed by the digital media environment, the only responses we can think of are to block our access to our own digital tools, or flee their reach?[21] Or as I have asked elsewhere: Why must we choose between debilitating tools, and digital lockboxes to keep us away from them? Why not demand *useful tools that do not debilitate us*?[22]

While ethics in the software industry is a hot-button issue, too many new social media investors, developers, and engineers still believe that ethics is not their business—or even that it is inimical to innovation and productivity. In response to a media question about whether his company had ever thought of hiring an ethicist to consult on their decision to conduct secret experiments on their users, Christian Rudder, co-founder of the online dating site OK Cupid, joked "To wring his hands all day for $100,000 a year? . . . No, we have not thought of that."[23] Such attitudes in the software industry need to change, and soon—not just for users' sake, but for the sake of the health and growth of digital media culture itself. Too many inside the Silicon Valley 'thought bubble' are oblivious to the fact that ethics matters, not because academics or media critics say so, but because people will *always* want good lives, and will eventually turn on any technology or industry that is widely perceived as indifferent or destructive to that end. It only took a few decades for cigarette companies to go from corporate models of consumer loyalty and affection to being seen as merchants of addiction, sickness, and death, whose products are increasingly unwelcome in public or private spaces. Only extravagant hubris or magical thinking could make software industry leaders think they are shielded from a similar reversal of fortune.

The cultivation of self-control requires more support from industry and social norms, but it also remains within the reach of our own moral practices, especially the habit of *moral attention*. Indeed, *executive* attention, the ability to notice what in our physical or mental environment we are thinking about and consciously modify or redirect that thinking, is an essential part of self-control, involving the ability to delay gratification, moderate our emotional impulses, and restrain reflexive, unthinking actions. If new social media habits challenge our self-control, it likely has much to do with our attentional capacities in technosocial environments. How do new social media shape our habits of paying moral attention, and the virtues such habits foster?

7.3 Media Multitasking, Moral Attention, and the Cultivation of Empathy

Much has been written on the topic of new media and attention; unfortunately the news is not very encouraging. A growing body of empirical evidence suggests that new media practices, especially the ubiquitous habit of *media multitasking*, have profoundly negative effects on attention, along with other cognitive faculties including problem-solving, task switching, knowledge transfer, working memory, and memory consolidation. Many of these effects appear to endure for some time after multitasking exposure has ceased.[24] This has obvious implications for learning and productivity, but less often noted are its *ethical* implications.

Attention subserves our moral capacities no less than our narrowly cognitive ones. Reviewing the research on attention and empathy, psychologist and science writer Daniel Goleman notes that while sociopaths can often reason quite competently about human emotions, they lack the affective sensitivity to others' emotions that stimulates emotional empathy from the bottom up. Goleman describes affective empathy as a neurological process that begins with a form of attention, in which the anterior cingulate 'tunes in' to another's emotional state and activates the anterior insula and amygdala to produce genuine affective pairing or co-feeling.[25] This brain activity makes possible the feeling of 'being moved' which, as we saw in chapter 6, is essential to the virtue of *empathic concern*. Unfortunately for media multitaskers, their resulting distractibility may be a significant obstacle to this experience, for there is evidence that the neural mechanisms of attention that produce empathic moral concern require considerably more time to activate than others.[26]

Moreover, empathic concern relies heavily on the ability to read emotional cues from the faces and bodies of others, since it is this embodied perception that activates the bottom-up circuitry of affective empathy.[27] Simply being told of a person's emotional state does not produce the same affective response in us as does directly *perceiving* that state. But how can I discern the emotional cues on my friend's face if, as he describes his weekend for me over coffee, I am simultaneously glancing at the texts, Twitter, and email notifications popping up on my phone? How can I read the emotional state of my patient if I spend the majority of the appointment attending to my handy medical records app, designed like most apps to maximize attention capture and user engagement? Looking up every few minutes to make a second or two of eye contact is unlikely to be sufficient either to accurately read another's emotional state or to activate the higher-order moral circuitry of empathic concern. Moreover, there is evidence that in addition to empathy, bottom-up processes of moral attention facilitate other important virtues, such as flexibility, care, and perspective.[28]

One way of responding to these worries returns us to the virtue of self-control. If I can muster the discipline to put my phone or laptop away in the relevant circumstances, I should find it easier to be more attentive and more appropriately empathic. But we said above that executive attention is a *prerequisite* for self-control; so how do we first break out of the vicious circle of distraction? Fortunately, the executive ability to notice and manage our own mental states can be cultivated from a young age, by techniques that encourage children to envision and anticipate longer-term outcomes, show them how to direct attention away from immediate attractors, and teach them to experiment with modifying their own perspectives and emotions.[29] These are in fact not unlike the mindfulness trainings central to Buddhist practice. Thus one constructive response to a new media age is to devote considerably more educational and parenting resources than we do now to the active cultivation in children of techniques of executive attention, and in particular, to reward children for practicing those techniques *in new media environments*. It goes without saying that media multitasking adults can likely benefit from these same techniques, but the rewards are likely to be greater the earlier these habits are formed. Indeed, such attentional habits in childhood have been shown to be as strongly correlated with adult well-being as IQ, family wealth, or social class, and in some cases more so.[30]

Yet another technique shown to restore depleted attentional capacity is contact with nature; this is known as the "attention restoration thesis."[31] When compared with work spaces, coffee shops, or even a 'relaxing' walk down an urban street, spending time around trees, flowers, animals, and water seems to have far more restorative effects on our cognitive powers. Social media addicts can likely benefit greatly from habits of nature-walking and park-sitting to replenish our bottom-up attentional capacities. Finally, as noted earlier, cultural norms can be powerful constraints. What if it were socially *anathema* not to look others in the eye when we speak with them, or to make them compete for our attention with five other digital channels? The most extreme media habits are antisocial enough to make one a deserving subject of kind but firm interventions. A person who shows *no* interest in reforming their habits sufficiently to make them morally responsive to others should arguably be considered undateable, unhireable, even unfriendable. This is a harsh remedy, however. Rather than have to punish those whose moral capacities are already damaged by their new media habits, would it not be better to ask how new media practices themselves can be made more compatible with human flourishing?

For example, what about creative moral uses of mixed media technologies that include video and narrative? Such techniques are heavily relied upon by nonprofit organizations and activists to focus attention on, and stimulate empathic concern for, human and animal suffering that would otherwise go unrecognized—soldiers maimed by war, children in famine-ridden countries, women in violently patriarchal societies, disabled persons denied social accommodations, captive whales and elephants, abused dogs and cats in shelters. It is too easy to forget that new media afford communications much richer than 140-character tweets. Still, the sheer number and range of these calls for compassion on our various media feeds can produce an empathic overload that leads to withdrawal and apathy. Alternatively, such appeals risk being reduced to viral social memes with a shelf life far shorter than the moral need—remember KONY 2012? Some critics of new media 'slacktivism' likewise challenged the moral depth and discernment of the viral (and arguably water-wasting) 'ice bucket challenge' for ALS research in 2014—a short-lived meme that nevertheless raised over $115 million.[32] Empathy, like any virtue, must be carefully modulated and intelligently calibrated according to the facts of the situation. We must be empathic at the right time, in the right way, to the right degree, and toward the right beings. Too much *or* too little empathic concern will leave us emotionally and morally crippled, and concern expressed foolishly will miscarry.

Perhaps our social media channels could be modulated to better *focus* our attention on empathy-demanding information, such that as we actively take up a particular moral concern, the delivery of competing solicitations is slowed for a period of time. Or perhaps new media sites and apps could actively reward consumption of *extended* moral narratives, incentivizing habits of reading that counteract those fostered by a steady media diet of listicles and slideshows. Indeed, our current distractibility (well captured by the popular acronym 'TL; DR'—'too long, didn't read') can be tempered by some of the same technologies that feed it today. For example, a site might permit comments only from those whom, the software judges, took the time to thoughtfully read a longform piece (using a digital measure known as 'dwell time'). Or a site might offer discounted subscriptions or exclusive content for readers with an average dwell time above a certain threshold. Rebuilding these attentional habits is key to enabling empathic concern, since empathy and moral perspective-taking are strongly facilitated by longform narratives, especially fiction.[33] In these formats, solicitations of empathy are embedded in a broader human context in which even profound suffering or joy is tempered by humor, wisdom, beauty, and the mundane. In addition to being less emotionally overwhelming, these enrich our moral perspective on the causes and lived meaning of suffering and joy as elements of a larger moral whole. The fact that current new media practices disfavor attention and empathy-sustaining formats is much lamented by media critics, and rightly so—but this is not a law of nature. It can be changed, especially if our flourishing *depends* upon it changing.

We must also realize that there is no 'collective moral intelligence' living in the cloud that can substitute for genuinely attentive and empathic human perceptions. Contrary to the vision of Web 2.0 enthusiast Clay Shirky and others, billions of half-read webpages and hastily skimmed e-books do not magically add up to a wise or even informed body.[34] So whatever we do, let us not concede our most basic moral capacities to the contingencies of the market or the illusion of a hive mind. As noted by the authors of *The Online Manifesto,* the EU-funded project on digital society, our attentional capabilities are in fact as essential to our personhood as our bodies:

> To the same extent that organs should not be exchanged on the market place, our attentional capabilities deserve protective treatment. Respect for attention should be linked to fundamental rights such as privacy and bodily integrity . . . the default settings and other designed aspects of our technologies should respect and protect attentional capabilities.[35]

Markets, as human institutions, are legitimized *only* by their promotion of human flourishing. To sacrifice the latter to the former is neither pragmatism nor realism—it is insanity.

7.4 New Social Media and Virtuous Self-Regard: Humility, Honesty, and Perspective

Attentional capabilities enable empathy with others, but also awareness of *ourselves* in the world. We saw in chapter 4 that moral self-cultivation presupposes effective habits of self-appraisal, yet these are easily distorted by new media practices. On the one hand, our friends' carefully self-edited streams of personal and career triumphs on Facebook can make our own real, but inconsistent records of achievement seem paltry. Such comparisons have been shown to be a significant cause of emotional distress in social media users, especially those who already suffer from low self-esteem.[36] Yet social media tools and practices can also artificially *inflate* our sense of personal worth and importance. People predisposed to narcissism are among those most inclined to maladaptive uses of social media, but few of us are wholly immune to social media's enabling of self-preoccupation.[37] It has become routine to ridicule the 'here is a photo of my breakfast' crowd, and new words such as 'humblebrag' are coined for social media practices seen as transparently ego-driven. Yet many users remain compelled to post photos of their breakfast and to humblebrag, so much so that few of us still bother to be all that bothered by it. Narcissism in a new media age is a moving target.

New social media seem to act on our self-regard as a hall of funhouse mirrors—all manner of distortions are possible, but the common thread is that these media in their *current* forms fail to deliver reliable feedback for self-appraisal. That said, psychologists have long known that humans are generally *terrible* at self-assessment—with or without new social media. Some self-distortions can be psychologically protective, but it is not easy to find the mean between the positive and negative poles of self-delusion. An honest respect for the truth about one's own character, and an accurate perspective on the value of one's life within a larger moral whole, are rare conditions of virtue. Yet *precisely because* honesty, humility, and moral perspective with regard to one's own life are such challenging virtues to cultivate, we need to resist media habits that make these valuable qualities any harder to cultivate than they already are.

Technologies are not stone tablets delivered from on high. They are malleable human creations that can be reshaped in the service of living well if our collective will demands it. So what shall we demand? As a starting place, we can look to those methods traditionally used to foster virtuous self-appraisal in other domains of our lives. In most societies, people expect and thrive upon honest feedback from loving family members and teachers. Trusted friends, lovers, and colleagues can offer the same benefit, though not always with the same authority. Exposing ourselves to moral narrative can help; comparing ourselves with fictional or historical others may put our own virtues and vices in proper perspective, as can reading the philosophical meditations of those working through their own projects of moral self-cultivation and reflection, as one finds in the writings of Marcus Aurelius, Boethius, and many others. How can new media be adapted to similar purposes?

We have already discussed the need for new media to be more conducive to extended moral narratives; but consider that many existing social media *already* function as meditations on the challenges and rewards of self-cultivation. Self-reflective blogs on the experience of being a morally imperfect parent are legion, and inspire many readers to have the courage to continue to try every day to get the job of parenting right, despite the certainty that we will continue to do very many things wrong. Such blogs also do something traditional literary meditations do not—namely, allow the authors to give and receive real-time encouragement and enjoy ongoing, honest feedback from their readership. Although comment spaces are almost always sprinkled with banality and random abuse, many bloggers develop strong mutual relationships of moral support with their dedicated readers. Moreover, blogs such as Allie Brosh's beloved *Hyperbole and a Half* can inspire millions of people to persevere under psychological hardships not articulated by the likes of Marcus Aurelius, conditions in which just getting dressed and cleaning some things, or returning a halfway convincing smile, are hard-won moral victories. What is often lumped in with 'new media narcissism'—a blogging culture in which people of no great fame or achievement share their mundane struggles and victories—also contains many more types and exemplars of moral cultivation than readers of past ages could ever have hoped to access.

Yet social media are also rife with apps designed to *disengage* one's online activity from the creation of an enduring and stable moral narrative. Apps such as Secret and YikYak use anonymity to enable sociality without relationship accountability. Snapchat and Sobrr allow users to send photos, videos, or messages to friends or nearby strangers that 'self-destruct' after a certain period of time; these apps use the power of ephemera to guarantee users that the history of their interactions through the app will not be preserved. The co-founder of Sobrr, Bruce Yang, proclaims that the app "promotes sharing only the most refreshing contents and leading a casual relationship with no strings attached."[38] There are many contexts in which it is perfectly appropriate, even beneficial, to enjoy superficiality, unaccountability, and ephemerality in social behavior. Yet as we can infer from our Facebook Home example, and Secret and YikYak's reputations for fostering hateful abuse and cruelty, new media platforms too often obscure the *contextual limits* of that appropriateness in a good life.

While seeking new media tools more conducive to moral self-cultivation, we must not forget that our virtues can also be cultivated independently, then leveraged as cultural pressures for more morally supportive technologies. As Mengzi says, "When the Way prevails everywhere, use it to pursue your personal cultivation. When the Way does not prevail, use your personal cultivation to pursue the Way."[39] The seven habits of moral self-cultivation outlined in Part II are still with us, and nothing in the world of new social media stops us from reclaiming their importance for our time, as Aristotle, Kongzi, and the Buddha did in theirs. In the next section we shall see why this is essential not only for *personal* flourishing, but also for preserving and enlarging the *civic* potential of new social media.

7.5 New Social Media, Civic Virtue, and the Spiral of Silence

The civic potential of new social media has been championed since the dawn of Web 2.0, and while reality has since dampened the most naïve predictions of a global Internet utopia, many still embrace these technologies as humanity's best hope for expanded global democracy, freedom, enlightenment, and community. Among them are Google leaders Eric Schmidt and Jared Cohen, who breathlessly assure us that: "With so many people connected in so many places, the future will contain the most active, outspoken, and globalized civil society the world has ever known."[40] Not only will the new global citizen be active and outspoken, she will be informed, enlightened, and constructively engaged in building better civic institutions:

> The data revolution will bring untold benefits to the citizens of the future. They will have unprecedented insight into how other people think, behave and adhere to norms or deviate from them, both at home and in every society in the world. The newfound ability to obtain accurate and verified information online, easily, in native languages and in endless quantity, will usher in an era of critical thinking in societies around the world that before had been culturally isolated. In societies where the physical infrastructure is weak, connectivity will enable people to build businesses, engage in online commerce and interact with their government at an entirely new level. . . . Citizen participation will reach an all-time high as anyone with a mobile handset and access to the Internet will be able to play a part in promoting accountability and transparency.[41]

Perhaps the best-known prophet of a newly empowered and enlightened global media commons is writer and social media theorist Clay Shirky, who tells us that thanks to new social media, "we are living in the middle of a remarkable increase in our ability to share, to cooperate with one another, and to take collective action, all outside the framework of traditional institutions and organizations."[42]

Yet in his book's epilogue, Shirky confronts the reality that online collective action in the civic domain has fallen far short of fulfilling this promise. He admits that beyond the organization of coordinated protests, relatively few sustained civic projects of significant scale have resulted from the use of these tools, which seemed in the body of his book to be invested with nearly magical powers to enable spontaneous social change for the better. In the epilogue, he concludes that the solution is to supply citizens with yet *another* kind of tool, namely, new laws for incorporating online collectives so that they may benefit from formal civic standing.

This reflects a fundamental error in the thinking of Shirky and other heralds of a digital civic renaissance such as Kevin Kelly, Jeff Jarvis, Eric Schmidt, and Nicholas Negroponte: the assumption that you will get a thriving civil society *simply by supplying citizens with the proper tools*. 'If cooperative and fruitful civic action online is not forthcoming,' so the thinking goes, 'then we obviously have not yet given our would-be citizens of the global digital *polis* the right instruments with which to construct it.' The mantra seems to be: *if we equip them, they will build it*. What is missing is any attention to the particular civic characters, motivations, and capacities of the 'them' in question. The collective desire, skill, and virtue necessary to cooperatively wield new media technologies for civic aims are all (wrongly) assumed by this model.

Evgeny Morozov, who is among the harshest critics of such models, has argued that the ideology of technological solutionism reflected in these visions will fail for two reasons. First, it treats technologies as neutral tools for producing things and solving problems, rather than as extensions of the human value contexts in which they operate. Giving people a new media platform cannot 'solve' the 'problem' of civic apathy if the platform's affordances and values are shaped by the same political conditions as the problem. The ideology's other failure is its treatment of people as essentially identical and interchangeable nodes in a network of rational agents, rather than as the complex, diverse, and constantly developing creatures that we are.[43] The ideology seems to be that if technocrats just wire up all the human nodes in the right way and enable information flows between them in the proper directions and at the necessary speeds, the 'system' will magically self-optimize for the relevant values (in the case of Western technocrats things like efficiency, freedom, transparency, democracy, etc.).[44] But we are not identical nodes in a network. We don't all respond to or exchange information in the same fashion or with the same abilities, and most importantly, we aren't all trying to optimize the same set of values, or balance values in the same way.

Nor will the obstacles to a flourishing global commons be removed simply by raising greater awareness of them. Internet scholar Maria Bakardjieva calls for expanding critical research to illuminate the details of new media's sometimes beneficial and sometimes "impoverishing effects on the lives of users," enabling users to "navigate the Internet both tactically and strategically in ways that defy oppression and advance emancipation."[45] Bakardjieva is right to call for a more granular approach to new media ethics, one that eschews global judgments about the effects of new media in favor of a more empirical and particularist approach. For example, we might study some of the more 'tactical and strategic' uses of cellphone video amplified by social media to expose police brutality and racial injustice in the United States. During the protests that followed the 2014 police shooting of Michael Brown—an unarmed black teen in Ferguson, Missouri—such techniques were effective in reinvigorating, at least temporarily, a long-neglected discussion of these simmering civic issues in traditional media outlets, schools, and homes across the country.[46] Yet sustaining civic awareness through new media in a manner that produces substantial and lasting social change will require more than a heightened awareness of how the Internet can be used for civic purposes, and more than improved new media tools; it will require *deliberately cultivated skills and virtues*.

If people are going to use social media for cooperative civic purposes to reliably and effectively promote human flourishing, then these will need to be not just *any* kind of people, but people with strong *civic virtues* of the technomoral variant. Contrast the most effective civic uses of new media in the aftermath of Ferguson, those that exposed the deeper structural patterns of racial and economic injustice in the county of which Ferguson is a part, with the actions of the Anonymous hacker(s) who tweeted the name of an officer they had recklessly misidentified as responsible for the Ferguson shooting. The hacker responsible for the misidentification was abetted by other 'doxxers'—people who collect personal information on individuals for online release—exposing an innocent man to grave risk of vigilante violence.[47] Such actions not only show a morally blameworthy lack of judgment, they also weaken public faith in the reliability and moral legitimacy of social media activism, undermining their own professed cause. Civic practices of using new media to promote justice and accountability cannot stabilize in a moral vacuum; they require intelligent and skillful guidance by people of technomoral virtue.

We need the technomoral virtues in order to ensure that civic cooperation and flourishing are mutual and sincere aims of those involved, and that those who cooperate for this purpose have the habits and skills to enable their joint success. How might these virtues also aid us in realizing the untapped civic potential of new media technologies? Consider the technomoral virtue of honesty, which we defined as a *respect for truth in technosocial contexts.* How is it related to the civic use of new media? On the one hand, such media seem to be custom-made to expand truth seeking and sharing. Web 2.0 technologies enable citizen journalism, foster vast knowledge-building projects such as Wikipedia, and allow anyone to generate and share data on an unprecedented scale. New social media have fostered new norms of institutional transparency and 'open government.' Truth can also be served by allowing a wider range of perspectives to have a public voice, and Web 2.0 technologies promote this goal. But of course, this is all too simple. Honesty as a virtue is a *discerning* respect for truth, not a mindless fetish for transparency or context-free data dumps.

Moreover, the relevant facts about many topics (quantum physics, global climate, evolution, high-frequency trading) are not equally accessible from all perspectives, so if adding more perspectives in these cases merely dilutes the views of those few who have reliable knowledge, the result will not be truth but confusion. Consider the global spread via social media of false and exceedingly dangerous beliefs about a causal link or correlation between autism and vaccines, which persists despite an overwhelming global scientific consensus that the totality of years of research shows no such link. Not only do these false beliefs persist, but the millions of parents who have acted on the social media advice of entertainment figures and conspiracy theorists have put others at grave risk by not vaccinating their children against fatal or disabling infectious pathogens. Is this the fate of truth-seeking 2.0? Or can we admit that cultivating respect for truth in the 21st century will require more than just handing everybody a louder microphone?

With respect to the volumes of 'big data' being mined from the new media 'social graph' and our associated online activities, it is far from obvious that more data always yields more knowledge, or more reliable access to truth. For one thing, data's relationship to knowledge and truth is still ruled by the enduring principle of 'garbage in, garbage out.' How much can a social scientist, politician, or public health researcher reasonably infer about human values, beliefs, needs, and preferences from inputs to a database that merely reflect superficial social media behaviors such as 'likes,' 'clicks,' and 'follows'? Furthermore, large datasets can and usually must be manipulated to yield any useful analysis, and these manipulations can mislead and obfuscate as much as they reveal. Every assumption, every highlighting or discounting of a variable is a human value judgment that does not come to the surface when the result is presented as the mechanical output of an unbiased, infallible algorithm. Finally, in our obsession with mining truth from the mountains of data embedded in the social graph, we risk a disastrous reduction of the meaning of truth *to* data. We forget that truth always has a context. We cannot respect truth by stripping it of all reference to the concrete worldly situations that *make* it true.

A further challenge to truth in social media environments is the cluster of concerns related to 'filter bubbles,' 'cyberbalkanization,' and 'echo chambers.'[48] Filter bubbles are the predicted effect of personalized browser search algorithms, such that, based on your online behavioral history, when you search for any given term—for example, 'vaccines,' 'climate,' 'Democrats,' 'guns,' or 'drones,'—you will be shown only those results that the algorithm predicts you *want* to see. This sounds like a good thing until we consider that respect for truth among an informed citizenry would likely be served better by inviting people to visit sites that they do not *already* want or expect to see. The worry is that personalized search algorithms magnify our existing cognitive biases and blind spots, and by ensuring that we encounter more and more perspectives that echo our own, reinforce inflexible belief patterns inconsistent with a civic environment that fosters respect for truth. This effect has not yet been empirically demonstrated to follow from personalized search alone, but in combination with other types of online personalization, the worry remains significant.

This is because social media platforms such as Facebook and Twitter also use sophisticated algorithms to determine what shows up on users' feeds; contrary to many users' belief that they are seeing an uncurated, real-time stream of all their friends' posts and shares, they are actually seeing a highly selective collage deliberately arranged *by* the platform to maximize their engagement *with* the platform. The problem is, what will maximize your engagement is likely to involve activity that closely aligns with your own existing values, beliefs, and affiliations. There is no 'dislike' button on Facebook, and no 'Unfavorite' on Twitter, so the permitted forms of engagement ensure that most users see limited content that disturbs, confuses, or disgusts them.

For example, the social media images and posts you see regarding conflict in the Middle East largely depend on your existing political and religious views and alliances. If you are a strong supporter of Israel, your feeds are curated in a way that deemphasizes pro-Palestinian posts and images, and vice versa.[49] Whether you see any content on this topic *at all* depends on your demonstrated interest in current events, religion, or politics. Nor are Facebook and Twitter alone in this practice; virtually all new media platforms, including the online presences of major news outlets such as the *Wall Street Journal* and the *Washington Post*, have developed such customization tools. These personalization algorithms and their effects are not disclosed to the public. To find out just how powerful they are, *Wired* contributor Mat Honan experimented with 'liking' every single item on Facebook for two days; he was stunned to find his feed overtaken by increasingly polarized and uninformed content:

> This is a problem much bigger than Facebook. It reminded me of what can go wrong in society, and why we now often talk at each other instead of to each other. . . . We go down rabbit holes of special interests until we're lost in the queen's garden, cursing everyone above ground. . . . But

maybe worse than the fractious political tones my feed took on was how deeply stupid it became.[50]

It seems, then, that the new Eden of social media has yet to fulfill its promise. Or perhaps it has; after all, wasn't Eden its own filter bubble of blissful ignorance? Does this mean that social media technologies are by nature the enemy of truth and civic wisdom? Hardly. For one thing, not all social media platforms are alike; their algorithms, users, and design interfaces vary, and the environments for truth vary with them. Facebook is not Twitter, Twitter is not Instagram, and Instagram is not Vine. A platform such as Reddit is hardly a harmonious Eden of bland 'likes'—it can be a chaotic and morally bankrupt place, but it can also foster intellectual controversy and challenge. Of course technomoral honesty, like all virtues, is a mean between extremes. Overexcited Redditors can be more irresponsible in their relationship to truth than the most incurious and complacent Facebook users, as when a 'crowdsourced' Reddit search for the Boston Marathon bombers went horribly awry, targeting innocent citizens as collateral damage.[51]

Yet Reddit's management was quick to issue a thoughtful and self-searching public apology. It is noteworthy that unlike Facebook, Reddit has an explicit ethical code ['Reddiquette'] that asks users to, among other things, "Remember the human," "Moderate based on quality, not opinion," "Look for the original source of content," and "Actually read an article before you vote on it."[52] Such distinctions matter, whether or not they are enforceable or have the immediate results we would like. Reddit, Facebook, and Twitter are not fixed givens but fluid human creations whose norms, structures, and affordances are constantly being tinkered with. The question is, *what moral ends will inform that tinkering?* The point here is that new media do not offer a single path to the truth *or* a uniform threat to it; they are varied and *ongoing* human experiments with it and other values.

Once again, social media technologies are not neutral tools for seeking the good life, such that the only ethical question is whether we have got the right tool in our hands. Just as important as the tool is the question of *whose hands* the tool is presently in, and what the present character of that person allows them to build with it. What are the virtues of the people who most closely approach in practice the ideal Redditor? What sort of online platform could those persons create with an even *better* technomoral character? What are the vices of those who embody the ugliest side of that community? How do they inhibit the flourishing of the site and its best possibilities? Remember, the virtues and vices of humans wielding technologies shape how the technologies themselves evolve; thus let us ask not what social media technologies are doing *to* us and our world, but what our technologies are doing *with* us, and what we ought to do *with* them.

Among the early hopes for new social media was that they would enrich the quality of scientific, cultural, and political discourse, and foster a newly global, open, vibrant, and engaged public sphere. There were of course early skeptics; philosopher Albert Borgmann was among the first to claim that no true civic life can flourish in the social shallows of the Internet, where persons are necessarily drawn to commodify themselves as "glamorous and attractive personae" rather than citizens committed to working out together our "reasons for living."[53] As the dust has settled, however, most scholars have begun to acknowledge that the reality of civic life online is far more complicated than either the techno-optimists or techno-pessimists predicted.

On the one hand, online community is everywhere, in a dizzying and constantly evolving array of forms. Yet open, civil, and *reasonable* online discourse is a rarity. In the United States and United Kingdom, large majorities of users decline new media's political and civic opportunities.[54] One factor is the scourge of anonymous trolls dedicated to disrupting or preventing reasoned online discourse by means of deliberate provocation, misinformation, and abuse. Despite site hosts' efforts to restrain them, trolls have been successful in persuading millions of sincere participants to flee online comment boards, and many hosted discussion communities have had to be shut down. The cure would initially seem to be a technological one: simply restrict or eliminate anonymity in online civic fora, and hold people's local identities personally accountable to their fellow citizens for what they say. Certainly this might be part of the solution; our technologies have specific affordances that condition our behavior, and in many contexts, anonymity in online communities with civic aims is an 'unaffordable affordance' to trolling behavior.

However, things are not so simple. Paradoxically, making people socially accountable for what they say online can have an effect *contrary* to the aims of robust civic discourse. The Pew Research Center, which has historically conducted research that dispels negative prejudices about social media and the Internet, has released a worrisome study suggesting that social media platforms not only fail to reliably *foster* open civic discourse, they often *inhibit* it.[55] The study found that social media users are generally much *less* likely than non-users to be willing to discuss a controversial political issue with others, online *or* offline. Among the specific findings: "86% of Americans were willing to have an in-person conversation [about the issue]", but just 42% of Facebook and Twitter users were willing to post about it on those platforms."[56]

Reinforcing earlier worries about echo chambers, filter bubbles, and cyberbalkanization, the study found that, compared with non-users, social media users in the United States were more likely to believe that their friends, family, and acquaintances shared their own opinions, and more reluctant than non-users to discuss a political or civic issue if they felt their audience might not hold the same view. With remarkable consistency across the study metrics, the researchers found that social media users seem to be disproportionately inclined in civic life to fall victim to the phenomenon known as the 'spiral of silence.' This is the tendency of holders of a minority viewpoint to become increasingly inclined to self-censor in order to escape the social penalties for disagreement, which in turn artificially boosts the perceived strength of the majority opinion across the community, further quieting dissent. The study authors hypothesize that in online contexts,

> this heightened self-censorship might be tied to social media users' greater awareness of the opinions of others in their network (on this and other topics). Thus, they could be more aware of views that oppose their own. If their use of social media gives them broader exposure to the views of friends, family, and workmates, this might increase the likelihood that people will choose to withhold their opinion because they know more about the people who will object to it.[57]

One of the most disturbing features of the study is that this effect appears to spill over into offline civic contexts such as the family dinner table, the workplace, or a night out with friends. Thus it appears that social media practices may have a 'chilling effect' on civic discourse more broadly. Overall social media platforms were not seen as more welcoming or safer spaces for civic discourse: of 14% of responders who would not discuss a political issue in person, only 0.3% of them were willing to post about it online. Social media platforms may therefore distort our perceptions of others' values and interests, making political and civic issues, especially those that draw controversy, appear to us to be far *less* important to our fellows than they actually are.

This is a potentially serious obstacle to the cultivation and maintenance of technomoral virtues central to civic flourishing, including *justice, courage, civility, flexibility,* and *moral perspective*. My ability to reliably respond to injustices–especially those involving controversial issues such as race, gender, class, and religion–will be hampered if those platforms through which I learn about my world cause me to fear and avoid public expressions of concern about these very injustices. How can I cultivate and express technomoral *civility*, the disposition to cooperate successfully with others to pursue civic goods in technosocial contexts, if those same contexts obscure how many of my fellows share my interest in civic issues? How can I cultivate the tendency to make appropriately *flexible* civic judgments if I am shielded on social media from the true diversity of beliefs, values, and aims of my fellow citizens and stakeholders? How can I possibly cultivate an accurate *perspective* on the world as a moral whole if those social media platforms that promise me a wider view of that world actually deliver a view that is politically and morally denatured in the ways we have described?

Here we see the importance of designing technologies that afford habits between vicious technosocial extremes. For example, a civil person will resist aggressive and unyielding impositions of her political worldview, but *also* resist the habit of polite docility that encourages silence on issues of critical public concern. Furthermore, civility is expressed differently at different times and places, in a situationally intelligent manner. What is virtuous civility at a wedding, where the mildly ignorant prejudices of one's table companions are suffered in favor of social harmony (unless, perhaps, they are directed at other guests), may well be vicious if one tends to remain silent when those same common but harmful prejudices are voiced at town meetings or in classroom discussions. For millennia humans have negotiated nuanced and flexible practices of civility in offline spaces. The same situational intelligence must now be embodied in *technomoral* civility online. Yet we have a lot of learning to do in this new arena of civic virtue.

For example, when in 2014 the Chancellor of the University of Illinois, under pressure from donors and trustees, rescinded an offer of tenure to scholar Edward Salaita on the basis of his 'uncivil' use of Twitter to criticize Israel and its policies, she ignited a firestorm of debate about the potential conflict of a 'civility litmus test' for scholars with the norms of academic freedom and intellectual honesty.[58] Neither Twitter nor the classroom are appropriate places for the kind of bland civility one might maintain at a wedding, and the academic community was vocal and well-justified in its objections to what appeared to be the Chancellor's rhetorical abuse of the concept of civility for political ends. Yet this does not mean that Twitter or academia should be 'civility-free' zones. Just as we would be right to deem a scholar civilly unfit for the academic profession if he or she were in the habit of using their fists at conferences, or shouting down students with profanities and threats, we must recognize that on Twitter one can behave in a sufficiently uncivil manner as to warrant having one's tweets deleted or one's account blocked. Indeed, Twitter itself has adapted, if slowly, to evolving social expectations that users will be protected from threatening abuse and cyberstalking.[59]

What are we to make of all this? More importantly, what should we *do* about it? We must resist the easy temptation to lay the blame for civic docility *or* incivility at the feet of new social media platforms, as if their impact is independent of our moral agency. Yet it would be equally misguided to let them off the hook as neutral tools with no impact upon that agency. Technologies neither actively determine nor passively reveal our moral character, they *mediate* it by conditioning, *and* being conditioned by, our moral habits and practices.[60] New social media can shape our civic habits for better or for worse. At present, they too often shape it for the worse, by promoting both vicious extremes of abusive incivility and docile civic quietude. But let us resist the natural tendency to focus upon those for whom social media networks appear to be a barrier to civic flourishing; let us focus instead on the virtues of the few who seem to have *overcome* this barrier.

Let us return to the Pew study discussed above. Among those 42% of users who were willing to post online about a political issue, we can assume a large number did so only because they anticipated robust agreement from a like-minded audience. Let us set them aside. Who are the minority who ventured their opinions in the online public sphere even when they knew they would be challenged, and what is their civic character? Assume that some were abusive trolls—insincere or uncivil enemies of cooperative civic engagement. Set them aside too. Certainly some nonnegligible subset of our original group remains. How did *they* come by the distinctive habits and virtues that allow them to use physical spaces *and* social media platforms effectively for civic purposes? What habits and practices, on *and* offline, cultivated in them the technomoral excellence to courageously resist the 'spiral of silence' and speak up on important and controversial matters of global justice, security, and community? Such people play a significant role in launching those waves of political and moral consciousness that periodically sweep through social media and mobilize civic action. We can benefit from closer study of the character of such individuals, and, more importantly, how, where and from whom they began to acquire the technomoral virtues of *honesty, courage, civility,* and *perspective* they seem to exercise more readily than most. We might also learn how *they* would design or modify social media platforms to be more conducive to civic flourishing than they are at present.

7.6 Technomoral Wisdom and Leadership in a New Media Age: Looking Forward

Among those who recognize the intersection between new media practices and virtue is media scholar Nick Couldry, who has repeatedly called for a new emphasis on the role of character in global media ethics.[61] Arguing that accuracy, sincerity, and care are not just ethical norms of media practice but *virtues* to be cultivated in them, he writes:

> If we agree that media . . . are integral to the life conditions that humans now encounter, that is, lifeworlds of complex interconnection across large scales, then media are plausibly part of the practices that contribute to human excellence. Conducting the practice of media well—in accordance with its distinctive aims, and so that, overall, we can live well with and through media—is itself part of human excellence.[62]

Couldry's core *media virtues* of accuracy, sincerity, and care are consistent with a broader framework of technomoral virtue such as the one suggested in this book. Moreover, he reinforces the distinctively *global* character of emerging technomoral problems. As he notes, global interconnectivity forces the entire human family to pose to one another the following question: "How can we live *sustainably* with each other through media, even though media unavoidably expose us to our moral differences?"[63]

To answer Couldry's important question, living sustainably and well with emerging media technologies, especially social media, will require more than just finding the right media tools to put into human hands. We must also pay attention to the technomoral *character* of the diverse stakeholders wielding those tools. We must look at the different ways in which new media practices condition the habits through which technomoral character develops, such as moral attention, relational understanding, reflective self-examination, and prudential judgment. At the next level of analysis, specific technomoral virtues (e.g., self-control, honesty, empathy, care, justice, civility, flexibility, and perspective) need to be identified and empirically studied in action across a range of local and global media contexts and cultures.

Of course, studies of new media practices will uncover patterns of vice more often than virtue; virtue is always rarer, and takes more time to emerge and stabilize in social practices. Yet when confronted with the problem of fostering virtue in morally confused or apathetic times, neither Aristotle, Kongzi, nor the Buddha saw the practical obstacles as insuperable. Each offers us a compelling lesson in moral hope. None allowed the abstractions of moral theory to cloud the need for action, or the perfect to become the enemy of the good. Instead, each turned to the concrete question of how moral education and habituation could be strengthened in their respective worlds, in the contexts of family, political, religious or monastic life. They and their followers then moved to put these views into *practice*, with powerful and lasting effects, both on local moral cultures and the many other global cultures with which each eventually interacted. To begin to foster global *technomoral* virtue, we must follow their lead.

First and foremost, we must create more and better practical spaces for *technomoral education*, places where people may apply the habits of moral self-cultivation to the contemporary challenge of living well with emerging technologies, and the inevitable surprises awaiting us in our technosocial future. Technomoral education must eschew passive learning of fixed rules and the associated 'compliance mindset' in favor of active habituation and practice across a variety of technosocial contexts, fostering habits of technomoral reflection, the study of technomoral exemplars, and the skills of moral discernment and judgment needed to adapt and flourish in new and evolving technosocial circumstances.

We also need to become more adept at recognizing worthy targets of moral admiration and emulation in media practice—individuals and groups who already promote human flourishing, locally *and* globally, through their use of these technologies. Such models not only inspire and guide others toward living well with existing new media, but help us to develop *better* media technologies and practices. Here we need the rarest of technomoral virtues: *magnanimity* and *wisdom*. Among the most commonly bemoaned effects of the new media age is a loss of mature moral expertise and leadership. While men like Walter Cronkite and Edward R. Murrow were humanly imperfect media leaders, they embodied for many the media vocation's nobler ideals, and inspired many other minds to follow. Today, radical changes in the economic model of the industry have led to widespread collapse of public trust in the media, and in our age of increasing information-dependence, it is difficult to overstate the global social price of this collapse.

New social media and the broader Web have yet to fill the moral vacuum left by the traditional venues whose departure they hastened. Hans Jonas lamented in 1979 that "we need wisdom the most when we believe in it least."[64] Indeed, many today find more reliable moral leadership in media satirists than in those who seriously profess the vocation. Unless this changes soon, our hopes of meeting the challenge of living well together in a globally networked media society will be greatly dimmed. We must ask ourselves how we as new media consumers and producers can start to deliberately foster, identify, and encourage ambitious moral leaders with the courage, nobility, and wisdom to use and shape these technologies for our common good.

8

Surveillance and the Examined Life

CULTIVATING THE TECHNOMORAL SELF IN A PANOPTIC WORLD

SURVEILLANCE TECHNOLOGIES ARE nothing new. For as long as there have been creative minds, people have used them to find ingenious ways of watching, listening, and tracking one another, and their motives have been as various as the means. The implications of the topic for questions of ethics, justice, and human nature have also magnetized the philosophical imagination. Among the most famous thought experiments concerning surveillance are Plato's Ring of Gyges, which imagines how human character might be corrupted by perfect immunity from surveillance. At the other extreme is Jeremy Bentham's notorious design for the Panopticon, a prison designed to allow constant surveillance of its inmates in order to pacify them with the ever-present *possibility* of being watched. In the 20th century, Michel Foucault famously reflected in *Discipline and Punish* upon Bentham's Panopticon as an illustration of a broader philosophy of social control: *panopticism*.[1] Panopticism employs a range of political, material, psychological, and economic techniques to establish maximally effective and far-ranging forms of social discipline with minimal obtrusiveness, expenditure, force, and risk.

8.1 Virtue in the Panopticon: Challenging the New Cult of Transparency

Many decades into the digital computing revolution, networked devices for pervasive and unobtrusive surveillance are now well established in countless millions of homes, schools, cars, workplaces, parks, playgrounds, government buildings, shops, restaurants, airports, and sporting arenas. They are in phones we carry in our pockets, in medical devices implanted in our bodies, embedded via RFID chips in our credit cards and clothing tags, in the invisible geofences around our children's schools, in the high-powered satellites that orbit us, and even in camera traps and webcams installed in the most remote swaths of terrestrial wilderness, allowing monkeys and rare mountain pallas cats to unwittingly record 'selfies' for our awe and amusement.

Today's means, motives, forms, and uses of surveillance are, as they have always been, as varied as the human imagination. What is genuinely new is the way in which massively networked data storage banks and powerful algorithms now allow us to integrate, aggregate, compare, and extrapolate from the output of these diverse and globally distributed surveillance tools, producing an unprecedented and seemingly unfathomable ocean of discretely retrievable data about people, places, things, and events. What is more, most of this data is not originally created by surveillance mechanisms as such. Most of it becomes material *for* surveillance simply by means of the potential commercial and security value of every piece of digital information your life generates. Since, relative to its potential value, data is cheap to collect, transmit, and store, your credit card purchases, turnstile exits, emails, web searches, phone contacts, social media 'likes,' personal photos, family health risks, prescription history, favorite vacation spots, college essays, movie rentals, music listening history, driving habits, and dental hygiene are most likely all stored *somewhere*, and potentially linkable to any other piece of information about you and others you know. This is the phenomenon known as *dataveillance*.

Dataveillance complicates Foucault's account of a panoptic society in several ways. First, much of the data we surrender appears to be freely given. Even if we discount data unknowingly surrendered under opaque digital 'terms of service' or end-user license agreements (EULAs), many of us routinely post our physical locations, consumption habits, political and religious affiliations, voting intentions, health conditions, and romantic histories with the conscious belief that the overall benefits of sharing these data outweigh the privacy risks. We often fail to see, or care, that data which individually seems trivial and harmless can, when aggregated by powerful algorithms, be profoundly revealing of our selves. Is this the ultimate triumph of the panopticon's invisible social discipline? Or an indicator that dataveillance is driven by spontaneous, bottom-up forces rather than political control? Additionally, in a world where our exposures to surveillance are so varied and many, it is not clear to what extent surveillance still functions as an effective means of control; perhaps we will become increasingly numbed to surveillance's inhibitory effects.

Moreover, the lines between watchers and watched have never been so fluid. Fears of pervasive government surveillance, aka 'Big Brother,' endure; indeed, in the wake of Wikileaks and Edward Snowden's revelations about the NSA and GCHQ, they are more acute than ever. Yet global networks of 'hacktivists' and citizen watchdogs are using their own digital tools to watch the watchers from below. Also growing are practices of *lateral surveillance*: ordinary citizens watching and recording one another's activities for any number of purposes, from the relatively benign to the criminal. When we add to these the growing movement of *personal or self-surveillance* discussed in the next section, it becomes clear that the lines of power, discipline, and control that Foucault tried to articulate in his account of a panoptic society have become increasingly tangled and diffuse.

Some employ the term *sousveillance* to encompass this contemporary culture of expanding, reflexive, and manifold forms of watching and being watched.[2] The newest hardware innovations associated with the emerging sousveillance culture are wearable personal computing and recording devices. Examples include fitness and health trackers such as the Fitbit and Jawbone, the Apple watch, and Google's controversial Project Glass. Driving this culture are powerful advances in algorithms for geolocational tracking, facial and speech recognition, and aggregating and processing large datasets. A sousveillance society's predominant cultural value is *transparency*. One of the earliest forecasters and most avid defenders of 'the new transparency' is futurist David Brin, whose 1998 book *The Transparent Society* called for a radical extension of liberal enlightenment philosophies of social and political openness. Arguing that the interests of justice and equality can only be served by a culture in which their violations can be openly observed by anyone, Brin admiringly cites J. Robert Oppenheimer's remark that "We do not believe any group of men adequate enough or wise enough to operate without scrutiny or without criticism. We know that the only way to avoid error is to detect it, that the only way to detect it is to be free to inquire. We know that in secrecy error undetected will flourish and subvert."[3]

Oppenheimer's view echoes the oft-quoted sentiment of U.S. Supreme Court Justice Louis Brandeis that "Sunlight is said to be the best of disinfectants," but all too often it is forgotten that the author of this quote was the Court's most effective advocate of individual privacy, arguing in 1928 in the dissent to *Olmstead v. United States* that privacy, or "the right to be let alone," is "the most comprehensive of rights and the right most valued by civilized men."[4] Brandeis's remarks do not contradict one another, but acknowledge important asymmetries in power between individual citizens who require privacy as a means of preserving what liberty they have, and those political institutions and groups which require regular 'disinfection' by public inquiry. A similar point is made by Brin critic Bruce Schneier: "Forced openness in government reduces the relative power differential between the two, and is generally good. Forced openness in laypeople increases the relative power, and is generally bad."[5] Brin responds, however, that this falsely presumes a society in which there is reliable information about who has the power that warrants watching, and who does not.[6] In an increasingly oligarchic society where a small number of private citizens amass vastly disproportionate shares of wealth and quietly convert it to unprecedented social and political influence on commercial and government institutions, Brin may have a point.

Yet is the ethical question about surveillance *only* about power and public corruption? Or does the growing 'cult of transparency' that drives the vision of many futurist and tech leaders challenge the good life in other, more personal ways? When Brandeis remarked that privacy was the right "most valued by civilized men," did he think this was simply because civilized men happen to be fondest of their political liberties? Perhaps becoming or remaining 'civilized' *requires* that I enjoy some creative license to experiment with thought and action in arenas that are not immediately subject to judgment by others, especially when those 'others' represent the cultural status quo. In her 2012 book *Configuring the Networked Self*, privacy law scholar Julie Cohen writes extensively on privacy's role in preserving spaces for free moral and cultural *play*, a critical element in the development and maturation of selves and communities. Likewise, and in a manner that intersects with Jaron Lanier's concerns about technological 'lock-in,' Evgeny Morozov warns of the profound risk of choosing technologies which, rather than encouraging deliberation and experimentation with new moral and political patterns of response, reify standing norms by enforcing and rewarding rote moral and political habits.[7] Surveillance technologies that work too well in making us act 'rightly' in the short term may shortchange our moral and cultural growth in the long term.

If this seems to overstate the risks of a sousveillance culture, consider that many defenders of a transparent society anticipate and explicitly welcome its salutary ethical impact on private, not merely political conduct. Brin's favorite example of the power of transparency is the way in which the visibility of other diners in a restaurant helps to ensure that they refrain from reading our lips, or leaning in to overhear our conversations. The moral ideal is that of having 'nothing to hide,' and thus nothing to lose from a transparent society. As Google CEO and Chairman Eric Schmidt famously said in a candid news interview, "If you have something that you don't want anyone to know, maybe you shouldn't be doing it in the first place."[8] Aside from its outrageous conflation of discretion and vice, this philosophy overlooks the fact that information about me is *also* usually information about the others with whom I share my life, and thus to focus only on the question of whether *I* have something to hide is a profoundly solipsistic attitude to privacy concerns, one incompatible with the virtue of moral perspective.[9] Still, there are many who would gladly impose this cost on others in order to promote a supposedly greater end: a perfectly good society, or as close as technology will let us get. David Friedberg, CEO of a big data analytics firm owned by Monsanto, recently summed up the emerging mantra of a morality of transparency: "What happens when every secret [is exposed], from who really did the work in the office, to sex, to who said what, is that we get a more truthful society. . . . Technology is the empowerment of more truth, and fewer things taken on faith."[10]

Implicit in this statement is the unquestioned privilege of truth over other moral values, including trust, respect, compassion, humility, and flexibility. Better to always have the 'truth' about who on the team "really did the work" than to permit the guy who was up all night soothing a colicky child to quietly slack off at work one afternoon without risk of exposure. Better to have the 'truth' about the name your husband called you in the heat of anger than to allow him the protective shield of a friend's discretion, or the liberty to substitute a gentler word in confessing the tirade to you later. Here what Aristotle, Kongzi, and the Buddha celebrated as practical wisdom, the ability of the virtuous person to perceive the moral world in richer, subtler tones than others, is turned on its head: the technocrats would have the moral world rendered in the stark black and white of binary thought—truth or falsehood, faith or fact, good or bad.

Thus even the moral value of truth is distorted here by a profoundly antihumanistic form of what Morozov calls *information reductionism*, rightly scorned by Nietzsche as the delusion that there is a "'world of truth' that can be mastered completely and forever with the aid of our square little reason."[11] One can grant the existence of objective truths while recognizing that reality, and *especially* moral reality, will always overflow any concrete human representation of it. The moral truth about a situation is always richer and more complex than our first glimpse reveals, and the ethical response to that limitation is not simply to gather more 'bits' of morally salient information but also to cultivate better ways of seeing, questioning, thinking about, and listening to it.

These are exactly the talents missing in the morally rigid and passively conformist 'village honest man' who Kongzi held to be the 'enemy of virtue.' It goes without saying that the village honest man cannot remediate his deficient moral vision simply by donning a pair of Google glasses. Moreover, his (or her) rigid, binary moralism often ends in tragedy. Consider the indiscriminate data dumps by the hackers of the Ashley Madison website in 2015, in which the exposure of an unethical business model and legions of would-be marital cheaters was carried out with breathtaking hubris and recklessness—endangering the lives of closeted gay and lesbian citizens living under oppressive regimes, handing foreign intelligence agencies mountains of leverage for blackmail of military and government workers, and leaving a trail of unspeakable devastation in the collateral damage of suicides and family dissolutions. *This* is what a transparent society looks like when untempered by humility, justice, care, empathy, moral perspective, or wisdom.

This is why the moral and political challenges facing humanity cannot be met simply by equipping societies with more sophisticated tools or more advanced STEM skills, independently of the virtues of the people who will wield them. Just as you do not make me a good surgeon by handing me a fancy surgical robot, or a good soldier by sitting me at the controls of a remote UAV, you do not produce more honest or discerning human beings simply by supplying them with more powerful technologies for surveillance. Technology ethicists Evan Selinger and Woodrow Hartzog further illuminate this flaw in the logic of sousveillance culture by pointing to the many empirical studies showing that people do not, in fact, reliably respond to the supply of more information by making more ethical or rational judgments. Just as often, in ways that are highly context-dependent, they respond to increased information flows by becoming more passive, less responsible, overwhelmed, and more likely to fall back on dangerous cognitive biases and shortcuts in ethical judgment.[12] For example, the notorious 'backfire effect,' which psychologists have demonstrated in media, political, and healthcare contexts, causes people with entrenched but false beliefs to respond to 'corrective information'—that is, reliable facts that undermine that belief—in an entirely counterintuitive and nonrational way. Such persons tend to respond to corrective information flows not by abandoning or revising their false belief, but by becoming even *more* confident in its truth![13]

Newly data-driven powers of surveillance can also project dangerous illusions of technological neutrality. Consider the increasing reliance of law enforcement and other authorities on facial recognition software and biometric surveillance in public spaces, in combination with predictive algorithms for individual behavior and risk profiles. 'Big Data'-driven 'predictive policing' is often cited as the future of law enforcement and anti-terrorism efforts, and a more objective, less biased methodology for crime prevention.[14] Yet the datasets upon which predictive algorithms are trained reflect many of the same blindspots, flawed assumptions and biases of the humans that generated and labeled that data—biases all too easily disguised and reinforced by the power of data anonymization and aggregation.

Many defenders of the cult of digital transparency fall victim to a distinctive cognitive bias of their own, namely *techno-fatalism*, which preempts rational consideration of otherwise compelling social, political, or ethical critiques of emerging technological development. Brin expresses this bias repeatedly in *The Transparent Society,* telling readers to surrender the naïve fantasy of regulatory or cultural constraints on emerging surveillance technologies, or any effective steering of their course. This is a bizarre stance for a champion of Enlightenment liberalism, given that tradition's hostility to unchecked powers. But Brin seems to have decided that 21st century technology, although *massively dependent* on political subsidy *and* regulatory and consumer support, has somehow acquired a quasi-magical power to defy the public will:

> Those cameras on every street corner are coming, as surely as the new millennium. Nothing will stop them. Oh, we can try. We might agitate, demonstrate, legislate. But in rushing to pass so-called privacy laws, we will not succeed in preventing hidden eyes from peering into our lives. . . . As this was true in the past, so it will be a thousandfold in the information age to come, when cameras and databases will sprout like crocuses—or weeds—whether we like it or not.[15]

Such fatalism in emerging technology discourse is patently false as a claim of logical necessity, and it fails to be compelling even on historical grounds; after all, humans have managed through intensely cooperative efforts to keep germline engineering as well as nuclear, chemical, and biological weapons from proliferating like unregulated 'weeds.' Yet if enough people are convinced by such rhetoric to surrender any hope of intelligently steering the course of emerging technologies, it becomes truth.[16] Moreover, without the virtues of technomoral character outlined in chapter 6, our attempts to steer the sousveillance culture and other emerging technosocial developments in ways compatible with our long-term flourishing *will* most likely fail—not out of fatal necessity, but out of careless indifference to our ongoing development as moral agents.

But this is not a counsel of despair. For two-plus millennia the human family has, in diverse cultural traditions spanning the globe, developed sound practices of moral self-cultivation. When properly nourished by individuals and institutions, and applied in the relevant domains, they yield remarkable, if humanly imperfect and inconsistent, results. No one can reasonably doubt that Confucian moral education—when habitually, sincerely and thoughtfully enacted—has been successful in cultivating *stronger* moral dispositions toward filial care and loyalty, even if these practices remain subject to extramoral influences *and* error. Only a wholly uncharitable mind could doubt that millions of Buddhists have, over the centuries, managed by means of their practices to expand their capacities for flexible tolerance, moral perspective, and compassion, even if these capacities remain limited and imperfect. And few would doubt that Aristotle's philosophy, in those communities where its influence on moral education endured, shaped the virtues of civic affiliation and political friendship so cherished, if imperfectly realized, by Greco-Roman and later European cultures. If devoted, sincere Confucians have cultivated stronger filial respect; if devoted, sincere Buddhists have cultivated enlarged compassion and generosity; and if devoted, sincere Aristotelians have cultivated more active civic friendship, then why should we now doubt our own potential to cultivate greater *technomoral* virtues in ourselves?

The emergence of a digital sousveillance culture thus invites two questions. First, how might this culture influence, for better or for worse, the cultivation of technomoral virtues such as honesty, empathy, justice, and self-control? Second, and flipping the first in a way that should by now be familiar, how can cultivating the technomoral virtues outlined in chapter 6 help us to improve our chances of flourishing together in an increasingly panoptic technosocial environment? Let us begin by delving into an aspect of sousveillance culture that poses an especially strong challenge to traditional practices of moral self-cultivation: namely, the quest for the *quantified self*.

8.2 Technologies of the Examined Life

In the 4th century BCE, the Greek philosopher Socrates asserted that "the unexamined life is not worth living for a human being."[17] In that same century, the Chinese philosopher Mengzi quoted Kongzi (Confucius) as saying, "If I examine myself and am not upright, although I am opposed by a common fellow coarsely clad, would I not be in fear? If I examine myself and am upright, although I am opposed by thousands and tens of thousands, I shall go forward."[18] How might a 21st century sage pursue this age-old ideal of the 'examined life'? Will she employ the very same methods of conscious reflection, meditation, and critical questioning used since classical antiquity? Or will she lean on new technologies to better understand who she is, and how well she is living? Consider this alternative vision of the examined life, proposed by contemporary futurist Kevin Kelly:

> Through technology we are engineering our lives and bodies to be more quantifiable. We are embedding sensors in our bodies and in our environment in order to be able to quantify all kinds of functions. Just as science has been a matter of quantification—something doesn't count unless we can measure it—now our personal lives are becoming a matter of quantification. So the next century will be one ongoing march toward making nearly every aspect of our personal lives—from exterior to interior—more quantifiable. There is both a whole new industry in that shift as well as a whole new science, as well as a whole new lifestyle. There will be new money, new tools, and new philosophy stemming from measuring your whole life. 'Lifelogging' some call it. It will be the new normal.[19]

In chapter 4 we saw that a habit of reflective self-examination is an essential part of the practice of moral self-cultivation, a practice that cuts across cultural and historical boundaries.[20] In addition to enabling us to periodically take stock of our character and assess the moral trajectory of our lives thus far, habits of self-examination can help us to respond appropriately in novel and unfamiliar circumstances, which increasingly define 21st century life. While a person of mature character moving through familiar moral territory can summon her virtues automatically and effortlessly, a practically wise person can also work her way through *unfamiliar* moral territory, in part by calling upon her explicit knowledge of her own moral competencies and limitations as they relate to the practical challenge confronting her.

For classical philosophers, the imperative of self-examination answered a practical rather than a theoretical question—namely, *how can we best secure the good life for ourselves and others?* Today a growing number of people embrace new technology-driven habits of self-examination in the hope of attaining the good life by means of the 'quantified self.' The quantified self is a picture of one's own existence produced by perpetual surveillance via biometric and mobile sensors, which yield unprecedented volumes of streaming data about one's psychological and physical states and activities. The movement's motto? "Self-knowledge through numbers." Is this simply the next phase of humanity's historical quest for the examined life? Is it an entirely *new* vision of the good life and the path leading to it? Or does it embody a profound misconception of the human self and its potential to flourish?

The Examined Life and the Ethical Ideal

As we have seen, a person leading an examined life historically sought reflective self-knowledge not as an end in itself, but as a means to something else: the *cultivated* self. A cultivated self is one that has been improved by conscious, lifelong efforts to bring one's examined thoughts, feelings, and actions nearer to some normative ideal. The examined life was thus part of a broader practice of self-care, what Plato called the 'care of the soul.' In classical thinking, this care involves philosophical habits of self-awareness that enable a gradual realignment of one's actions, values, emotions, and beliefs with the Good. We explored three cultural variants of this element of self-care in chapter 4.

Yet leading an examined life is a struggle for most. We are frequently blind to our own failings, unable to distinguish between true and false or higher and lower goods, or distracted from virtuous aims by nonmoral needs and desires. As a result, says the Confucian philosopher Mengzi, "the multitude can be said never to understand what they practice, to notice what they repeatedly do, or to be aware of the path they follow all their lives."[21] If Mengzi is correct, then few of us can hope to live well without some kind of philosophical intervention. If our path to a good life is blocked by these obstacles to moral self-cultivation, then we require one or more practical remedies that can help us move over, around, or past them. Reflective self-examination is one such remedy.

Of course, the habit of self-examination must itself be carried out wisely, with appropriate flexibility and moderation. We should not indulge the kind of obsessive self-examination that produces paralyzing anxiety and endless self-recrimination, and/or a narcissistic overestimation of the worldly significance of one's own moral being. That said, developing this habit in a measured way is an essential part of preserving individual and collective human moral character in new technosocial domains, and will help us to identify and cultivate the technomoral skills and virtues we need in order to flourish in increasingly unstable and opaque practical environments. Unfortunately, making this virtue a lasting habit requires more than simply recognizing its importance. As with all moral practices, it is vulnerable to disruption, neglect, corruption, and replacement by counterfeit. To see how this might be a particular danger in today's technosocial environments, let us recall what is most essential to self-examination as a *moral* practice.

8.2.1 Technologies of the (Cultivated) Self

In his 1982 lectures of the same name, Foucault called such practices "technologies of the self" (*techniques de soi*): methods by which members of a given culture "effect by their own means or with the help of others a certain number of operations on their own bodies and souls, thoughts, conduct, and way of being, so as to transform themselves in order to attain a certain state of happiness, purity, wisdom, perfection, or immortality."[22] 'Technologies of the self' commonly associated with philosophical practice include the Socratic habit of dialectical questioning; the careful study and emulation of exemplary persons (e.g., Aristotle's *phronimoi* or the Confucian *junzi*); immersion in philosophical and moral education; the cultivation of narrative habits of confession, letter writing, and meditation; reflective examinations of conscience and memory; and the testing of one's decisions according to general moral principles.

In talking about these philosophical practices as "technologies of the self," we hearken back to the ancient Greek concept of *techné* as a craft by which a product (here, the cultivated ethical self) is gradually constructed and shaped. Yet to be authentic, techniques of self-care must be consciously and reflectively embraced by the moral agent being cultivated. While others may help, the agent must actively examine and evaluate her own conduct. Even Foucault, who emphasizes the construction of the self by impersonal cultural, historical, and political forces, holds that the very thing such forces ultimately produce is a self-conscious agent who assumes moral responsibility for herself.

In *Discipline and Punish*, Foucault uses Bentham's design for the Panopticon prison tower to show how social, economic, and political structures can effectively internalize the normalizing power of surveillance, such that the subject of power willingly creates *herself* within the confines of her culture's dominant ideals. Foucault's account reflects a postmodern suspicion of such processes, which bring individual thought and behavior into alignment with cultural or political norms. Such suspicion is hardly groundless; we need not rehearse the many ways in which normalizing powers can inflict not only bodily but psychological violence upon human persons.

But it would be rash to assume that technologies of the self are *inherently* oppressive. To do so I would have to deny that actively striving to configure myself for a flourishing life could be an authentic choice, and this seems to unduly demean human agency. Furthermore, technologies of the self do not function as pure constraints on my development—they also reveal richer possibilities for living that may otherwise go unseen and unrealized. For example, by exposing myself to the moral questioning of others, or by comparing my actions with those of someone whose character I deeply admire, I might discover that I am far less forgiving of others than I imagined myself to be, and that I hold onto petty resentments far longer than is reasonable. In discovering this, I reopen the possibility of becoming a person less consumed by anger and judgment, freer to love, and to bond more deeply with other imperfect souls. My life is not *restricted* by this result of self-care—it is greatly enhanced and enlarged.

Thus while the accepted aims and means of self-cultivation are shaped by powerful cultural forces, these need not impose any single universal image of the 'cultivated self,' or any one best technique for pursuing it. To justify the use of philosophical technologies for an examined life, we need only assume that for any individual, the decision to pursue a developmental path by such means *can* be authentic and justifiable. Yet not every technology of the self is philosophical. The Quantified Self movement, embodied in Kevin Kelly's assertion that life "doesn't count unless we can measure it," aims to use emerging technologies to transcend the limitations of philosophical and other qualitative methods of self-examination.

8.3 'Smart' Surveillance and the Quantified Self

Adherents of the Quantified Self movement, as well as an increasing number of average consumers, employ mobile, wearable, and/or biometric sensors such as the FitBit and Jawbone devices, smartphone apps such as Moves and Chronos, video cameras, and a range of other devices to measure, track, analyze, and store volumes of recorded data concerning an ever-expanding list of personal variables. Weight, blood pressure, muscle mass, sleep patterns, mood, physical activity, energy levels, creativity, and social and cognitive performance are just a few variables of common interest. Among the more active and vocal adherents of the movement is statistician Konstantin Augemberg, author of a blog called *Measured Me* (www.measuredme.com) where he posts the results of his efforts to quantify and track everything from alertness and sleep efficiency to happiness and life satisfaction. Among his most ambitious projects is a quest to quantify 'living well,' under which he subsumes measures of physical activity expended, healthy eating, diversity and engaged 'flow' of his daily experiences, and time spent 'recharging' with sleep and meditation.

The Quantified Self movement claims thousands of followers across the world, with hundreds of local meetup groups and an annual global conference at which enthusiasts share information on the latest gadgets for self-surveillance and tracking. Yet these numbers are dwarfed by the millions of casual consumers who compose the market for affordable and user-friendly self-tracking devices. Is this just another phase in the evolution of technologies of the self, in Foucault's sense? Or does the move from philosophical to digital methods of self-examination represent a more profound moral shift?

A satisfactory answer would require deeper inquiry, but let us weigh some preliminary considerations. As Foucault explains at length in his 1982 lectures, the concept of the examined life is historically fluid. Classical, medieval, and modern visions circulated between the ideal's first incarnation as a form of self-care that maintains the overall excellence of one's life *activity*, and an alternative conception of self-care as a way to preserve the integrity of the *soul*. Foucault observes that by the 20th century, the classical vision of self-care as a tool for setting youth on the right path to political life had evolved into a medical-psychological model, in which the task of self-care is seen as a lifelong duty to take continual stock of one's *moral health*.

Nevertheless, certain core notions endure across these historical changes. One common thread is the idea that a good human life presupposes my ability and intentional choice to habitually *reflect* upon and *attend* to my own moral development or trajectory. A second commonality, implicit in the first, is the requirement that I take steps to actively cultivate or steer that trajectory in the desired direction. This is so even if I reject the idea of a fixed end or destination toward which my life must aim. For even if I may freely invent the form of life to which I aspire, that choice implies a commitment to actively shape my own person into the sort of being that can achieve such a life. Lastly, the idea of self-care has historically included a duty to be concerned with those aspects of my life and activity that are central to my moral character (i.e., my virtues and vices). All of these common habits of moral self-care appeared in our examinations of the practice of moral self-cultivation in chapters 3, 4, and 5. While Platonists, Confucians, Buddhists, Aristotelians, Stoics, medieval Christians, and moderns all had very different ideas of what one's moral constitution ought to be like, none would have been satisfied with habits of self-surveillance that left the moral virtues largely unexamined.

While one might simply reject the notion that a good life is an examined one, let us assume for the sake of argument that it is and see what follows from it in relation to the Quantified Self movement. Let's start with the requirement to pay conscious attention to one's developmental trajectory. It may seem that the movement not only satisfies but exceeds this requirement by encouraging a kind of hyperattention to, even obsession with, that trajectory. A quick perusal of blog posts, websites, and essays produced by Quantified Self enthusiasts conveys an atmosphere of fevered excitement about finding ways to track ever more personal variables, with ever greater degrees of mathematical precision and reliability. Yet as the number of tracked variables expands, we must ask whether the act of 'attending' to oneself becomes more difficult and how many of the factors being tracked are actually meaningful. As any philosopher of perception or cognitive scientist knows, attention is as much about the ability to *screen out* information as it is about taking it in; in fact, the former capacity enables the latter.

Consider the *Measured Me* project mentioned earlier, which posted weekly 'lifestream' updates for years before taking a hiatus in Fall 2014. Its author's published personal dataset for October 2012 included sixty-six daily datapoints, taken for twenty-one personal variables ranging from 'mental energy' to 'charisma' and 'self-esteem' (along with six variables for the weather). The previous month's dataset included still other variables, such as calories consumed/expended and personal 'entropy,' a measure of how "chaotic" a particular morning, afternoon, or evening was. Now to be fair, these are early experiments in self-surveillance and quantification, rather than a honed practice of self-cultivation. But it is worth asking whether the latter can be served by habits that aim primarily at expanding the *range* of self-surveillance rather than selecting for and attending to the most salient features. Indeed, the obsessive quality of many Quantified Self habits evokes the philosopher Mengzi's warning against hyperactive, overly self-conscious efforts at self-improvement, which he compared to the self-defeating habit of the farmer who tried to help his plants grow faster by pulling at their sprouts.[23]

This brings us to the second core element of an examined life: the decision to use the results of self-examination to cultivate or steer one's own personal development in a desired direction. This is a form of philosophical 'perfectionism,' what Stanley Cavell has described (taking Emerson's phrase) as a fundamentally human quest for my "further, next, unattained but attainable self."[24] There is little evidence that this is a central aim of devotees of the Quantified Self. It is true that they seek correlations between tracked variables precisely because these could aid prediction and control of one's personal states. For example, if I discover that my happiness or resilience tends to decrease sharply as local temperature or humidity rises, I might think to move to a different climate and see if my well-being improves.

Yet few Quantified Self enthusiasts express a clear sense of what *kind* of self they wish to cultivate overall; nor is the 'cultivated self' an explicit and recognized goal of the movement. Indeed the perceived value of the practice appears more epistemic than developmental, as indicated by their motto, "Self-knowledge through numbers." The Quantified Self website claims that the movement's aim is "to help people get meaning out of their personal data." But this is profoundly vague; what *sort* of meaning? To what end, if any? While Quantified Self enthusiasts are developing techniques that could perhaps someday be placed in the service of a philosophical practice of self-cultivation, this is not presently their goal. Since it isn't aimed in that direction, the movement in its current form does not offer nor even promise a new road to the cultivated self.

Does a quantified life display the third core feature of an examined life? Can it attend to the specifically *moral* features of one's character and activity? On this score, we find the greatest tension between a quantified life and an examined one. The moral dimensions of the self are among the most difficult to translate into numbers. In contrast with variables such as calories, muscle mass, or sleep hours, the moral features of my person seem to be difficult to formalize into a tidy list of variables, much less variables that can be assigned discrete values. Yet as Kevin Kelly's quote makes clear, the Quantified Self movement is committed to a form of self-surveillance that is both encompassing *and* scientific, where 'scientific' entails 'quantifiable.' Its enthusiasts readily assign numbers to philosophically rich concepts associated with the good life, such as 'happiness' and 'living well.' Yet even these are not interpreted in moral terms; rather, Quantified Self enthusiasts use them to describe a subjective sense of well-being and engaged 'flow' that is quite compatible with morally weak character or practice.

These preliminary analyses suggest that Quantified Self practices do not currently constitute technologies of the self in Foucault's sense, nor do they fulfill the aims of an examined life needed to cultivate virtue and promote sustained human flourishing. It remains to be seen how the movement will evolve. It may prove to be a technological fad with transient appeal. Or it may grow and endure as a distinct cultural practice of self-surveillance unrelated to the aims of an examined life. In that case, however, it would likely be viewed as an *alternative* to moral technologies of the self, as it is difficult to see how the habits of self-tracking promoted by the Quantified Self movement can coexist happily with the philosophical and spiritual habits of an examined life. After all, each practice demands a considerable investment of time and mental energy; can one imagine a Quantified Self devotee faithfully cataloguing and analyzing dozens of daily datapoints on their behavior and mental states, while also exercising the daily habits of narrative self-examination practiced by the likes of Marcus Aurelius or Emerson?

Still, why not just conclude that philosophical technologies of the self are hopelessly outdated, destined to be replaced with more 'objective,' 'scientific,' and user-friendly tools for living the examined life? Indeed, a central aim of the movement is to rely more and more on technological devices to record, store, and analyze these datapoints for us. Its advocates expect that within a few short decades, nearly everyone will use such devices to collect unprecedented amounts of data about ourselves in real-time, including data that are inaccessible to conscious reflection. Furthermore, such data will be far more precise and less vulnerable to subjective distortion than the contents of reflective self-examination. Feedback from artificially intelligent life coaches will be able to tell us what adjustments we need to make to our behavior or thinking. Technological practices of self-quantification may thus appear to offer a *superior* replacement for moral technologies of the self. So why reject this thinking?

The answer is straightforward enough: this thinking commits a profound category mistake. The most accurate and comprehensive recording of your past and present states would not constitute an examined life, because *a dataset is not a life* at all. As Aristotle reminds us, my life includes my future, and thus the examined life is always a project, never an achievement. It is the future toward which we live, and for the sake of which we examine and cultivate ourselves. We prize the examined life not for the 'data' that reflective philosophical practice yields, but for the transformative nature of the practice itself and the dignity it confers upon those who take it up. The examined life is worth living because it embodies those chosen habits of mind and conscience that constitute a person who takes responsibility for her own being, and in particular her *moral being*. This capacity may or may not be uniquely human (can a dolphin reflect upon what kind of dolphin it is, and how far it is from the dolphin it wants to be?); but it is the capacity that most uniquely defines a moral agent, and a person in the fullest sense.

8.4 Surveillance and Moral 'Nudges'

Converging with sousveillance culture is another emerging technosocial alternative to the conscious practice of moral self-cultivation. This is the phenomenon known as *nudging*. The idea behind 'nudges' is to use technology to foster prosocial behavior in ways that require little or no conscious effort from users. Some nudges are built into physical environments: just as speedbumps promote safe driving, one might design a building's stairwell to be more accessible and attractive than the elevator as a means to promote residents' exercise.[25] Today, digital surveillance and self-tracking technologies afford far more personalized and responsive forms of nudging. Apps on your phone can now remind you to text your girlfriend, call your mother, and eat more fiber. Your 'smart' pill bottle can ask you to take your diabetes medicine. Your ToneCheck software can tell you to write a nicer email to your boss, and your Apple Watch can tell you to calm the hell down during a family argument. Ten or twenty years from now, it is hard to imagine a domain of practice in which you won't be able to have an artificially intelligent behavioral monitor if you want one. In many places, such as the workplace, it is unlikely you will be offered the choice.

These are astonishing possibilities, and used selectively and in moderation, they could contribute a great deal to human flourishing. Unfortunately, we are not today cultivating ourselves, or educating our children, to attain the technomoral virtues needed to steer such practices wisely and intentionally. It is essential for human agency that our moral practice, whether supported by philosophical or digital technologies of the self, remain our *own conscious activity and achievement* rather than passive, unthinking submission. Imagine if Aristotle's fellow Athenians had been ruled by such perfect laws that they almost never erred in moral life, either in individual or collective action. Imagine also that Athens' laws operated in such an unobtrusive and frictionless manner that the citizens largely remained unaware of their content, their aims, or even their specific behavioral effects. Our fictional Athenians are reliably prosocial, but they cannot begin to explain *why* they act in good ways, why the ways they act *are* good, or *what* the good life for a human being or community might be. For they have never given a moment's thought to any of these things, yet they go on every day behaving reliably justly to one another. Would we want to say these hypothetical Athenians are moral beings? *Do* they live well? Do they flourish in the way that we would most want our own children and friends to flourish?

How do our hypothetical Athenians differ from the 'pod people' or 'Stepford Wives' described in all manner of modern science fiction dystopias, where moral conformity is achieved through a process of social transmission that operates independently of deliberation and choice? In such fictions, the pod people often retain intellect or consciousness—they converse with one another about everything from sports to business to war to childrearing. It is what we *cannot* imagine them talking about that makes them dystopian horrors. They cannot talk about how they themselves might become better people, what obstacles, trade-offs, and choices they face in doing so, or what that project of moral self-cultivation would mean to them or the world.

The real Athenians were not pod people, and fortunately neither are we. Nor will apps, wearable computing, or 'smart' environments make us into mindless moral zombies. Yet this does not make them harmless. From the risks of 'helicopter parenting,' in which the agency of children and young adults is tightly constrained to prevent them from making mistakes or missing opportunities, to the seductive pull of political and religious groupthink, humans *already* struggle to habitually exercise their own moral agency, a struggle that is millennia old. If surveillance and nudging technologies are marketed and embraced as yet one more social license to relinquish this struggle, so that our moral lives may be quietly and seamlessly molded into the shapes programmed by Silicon Valley software engineers and technocrats, we may not become pod people—but will we become *more* or *less* like the human beings we wish to be?

8.5 Flourishing with Emerging Surveillance Technologies

The practices of moral self-cultivation we articulated in Part II—and human flourishing itself—face serious challenges from the rapid emergence of a global sousveillance culture. If one were a techno-pessimist, one might conclude that the technologies driving this phenomenon promise only to magnify asymmetries of political and economic power; to diminish the space for moral play and authentic development; to render trust in human relations superfluous; to reduce embodied moral truth to decontextualized information; and to replace examined lives with datasets. On top of this, we hear from futurists and technocrats that the course of our sousveillance society is already out of our hands, and that our only hope is to adapt to whatever shapes new surveillance technologies impose upon on our lives.

Fortunately, the latter claim is entirely false, and not one of the pessimistic scenarios we have mentioned is fated. Nor is there any reason to think that the only alternative to those scenarios is to simply reject all new surveillance and 'smart' technologies, or to resist every one of the cultural transformations they engender. Of course, effective resistance very probably *would* be impossible in the absence of more widespread global cultivation of the technomoral virtues that enable prudent civic deliberation and collective action in contemporary life. But let us imagine that through improved technomoral education and practice such virtues *were* to be more widely cultivated. In that case, their use in prudent technosocial action would be far more discerning, subtle, and flexible than any indiscriminate refusal of new technologies. Possessors of technomoral virtues would recognize, for example, the ways in which emerging surveillance technologies can be redesigned, experimented with, applied differently, restricted to appropriate contexts, made accountable and responsive to social and political critique, and used as *extensions and reflections* of our moral virtues rather than substitutes for them.

Let us consider an example. Even in a society where technomoral education for civic virtue is reasonably effective, political corruption and abuses of power will exist. The need for some degree of transparency in government and other centers of power is real, and new technologies of sousveillance can be helpful means of holding those powers accountable—*if and only if* they are used wisely. As Morozov notes, it is naive to think that government and other large civic institutions would be improved by recording every detail of their operations and making them available to all on demand. Not only would the glut of data produce more confusion among citizens than clarity, but it would likely change institutional behaviors for the worse. If representatives and leaders knew that every word and act would be made available for public inspection and criticism, they would be discouraged from reaching difficult but necessary compromises, voicing uncertainty, taking important risks, and making the painful trade-offs that prudent governance requires.[26] Just as privacy secures an important space of free play and experimentation for individuals, it also secures this for institutions.

Thus human flourishing requires cultivating *technomoral honesty*, a respect for truth in its diverse appearances across the range of technosocial contexts, rather than the flattened, grossly reductive account of truth offered by the defenders of a 'transparent society.' An institution, politician, or citizen who records *or* shares information without discernment or contextual sensitivity is *not* virtuous but vicious. Imagine a politician who records sensitive details of diplomatic negotiations and then disseminates the unedited footage on social media, with disastrous results for global civic flourishing. This is *not* an honest politician, because she is not exemplary in her treatment of political truth. Compare this contextual role with that of an academic, where in general the norms of exemplary conduct will be more conducive to plain truth-telling. Yet even here, people must be discerning in the treatment of information; an honest scientist must know her audiences and venues, and there are times when excessive precision or transparency will obscure the truth rather than reveal it. Should we imagine that it would be exemplary or truth-serving for a scientist to embed in her articles a computer-generated transcript of every mundane minute of laboratory activity leading up to publication, including that morning when she, due to an error in preparing some slides for dissection, spent hours analyzing and remarking upon a completely contaminated sample?

Still, there are circumstances in the life of government, academia, business, and other institutions—even the private lives of citizens—that justify opening the books for inspection, so to speak. We need the technomoral wisdom to make more intelligent and *practically discerning* use of the new technologies that can make opening the books in such circumstances easier. For example, we might insist upon the maintenance of encrypted electronic recordings of certain legislative sessions or high-level government meetings, but require these records to remain encrypted unless the courts declare a compelling public interest in the information. We might expand the rights of private citizens to safely record their interactions with law enforcement, but look toward designing technologies for doing so that are less intrusive than having multiple cell phone cameras held up in the face of every officer trying to do her job. We might require that wearable recording devices audibly announce when a photo or video is being captured. We might discourage or ban their use in public recreation or wilderness areas.

To balance the interests of child safety with the developmental importance of parent-child trust, those who wish to use RFID tags or other geo-enabled devices to track the locations of their children could be encouraged to have the data stored in third-party systems that are inconvenient to access, lowering the temptation to monitor children in non-emergency circumstances. Children in state custody or new foster homes could be supplied by courts with digital means of documenting and reporting abuse or neglect. To tackle the cognitive biases and blindspots that commonly impede accurate moral self-assessment, future self-surveillance technologies could be designed to provide new forms of *moral biofeedback*, allowing us to better discern unflattering, destructive, or antisocial habits and traits that evade traditional introspective methods of self-examination. We might also look to design the outputs of self-surveillance technologies in ways that resist reductive quantification; instead of yielding sets of raw, decontextualized numbers or graphs, we might design such technologies to display their results in a more descriptive or holistic form that requires integration within a moral narrative of the good life.

Of course, such proposals are conditioned upon cultural and technical feasibility, and some may just be bad ideas; but the point is that *without* technomoral virtues, we will lack the ability to determine individually *or* collectively which goals and means of surveillance are wise and worth pursuing. In particular, the *global* goods of human flourishing described in chapter 2 will be virtually impossible to secure in these virtues' absence. We have already spoken of technomoral honesty, which is of clear relevance to the prudent guidance of surveillance technologies, but let us consider a few others:

Living well in the 21st century will require enhancing our capacities for deliberative *self-control* in technosocial contexts. Which specific surveillance and nudging practices will aid us in this effort? Which are more likely to induce greater moral passivity in the manner foretold by Orwell's *1984* or Huxley's *Brave New World*, weakening our ability to deliberate upon and mold our own desires? Which modes of education and cultural expression can strengthen our self-control in technosocial contexts, allowing us to intelligently steer rather than passively adapt to the emerging sousveillance phenomenon?

Additionally, could cultivating *technomoral humility* help us to acquire more modest and realistic expectations of how rationally and effectively humans respond to surveillance practices? How might surveillance technologies be redesigned with this in mind, and used to help us to grasp the limits of, and form more reasoned hopes for, our informational capacities? Could such practices help us to more honestly confront, accept, and adapt to a technosocial future that will continue to outstrip human mastery?

With respect to the virtue of *technomoral justice*, how can we use surveillance technologies to enhance our awareness of unjust asymmetries in the global benefits and risks of technosocial developments, and the impositions of technosocial power on the basic rights, dignity, or welfare of others? Could we then put this new awareness to work in educational and cultural practices that foster more just uses of surveillance technology? Can more discerning applications of surveillance technologies foster greater *technomoral empathy* with the suffering and joys of others outside our immediate view, and better moral *care* and loving service of their needs? Can discerning surveillance practices foster more *flexible* responses to local and global technosocial change and opacity? Or more holistic and less fragmented *moral perspectives*?

Of course, considerable technomoral virtue is *already* required in order to frame and execute ethically constructive applications of surveillance technologies. We must ask ourselves where such exemplary virtues can be found today. If they are widely lacking in the relevant institutions and cultures, then the most pressing question for us is this: How can we begin to design and implement *now* educational and cultural projects to enhance the cultivation of technomoral virtue? For this is the only way to ensure that new surveillance technologies are wisely deployed to promote human flourishing in an increasingly opaque future—as opposed to the indiscriminate, extreme, and reductive uses of these technologies all too often favored by defenders of a new, radically transparent surveillance society.

9

Robots at War and at Home

PRESERVING THE TECHNOMORAL VIRTUES OF CARE AND COURAGE

TODAY WE STAND on the threshold of welcoming robots into domains of activity that will expand their presence in our lives dramatically, extending well beyond the primarily industrial contexts in which robots have for decades been working alongside or in place of us. The integration of machine robotics with biomechanical engineering and artificial intelligence (AI) software has vastly expanded the range of human functions that robots will be able to perform in the coming decades. Many predict a 21st century labor incursion of robotic and AI systems that compete with or even outperform human workers in jobs that even a decade ago seemed unthinkable targets for automation.[1] If these predictions are right, then we may soon be served (and many of us disemployed) by a wave of artificially intelligent legal assistants, accountants, pharmacists, drivers, lab technicians—perhaps even teachers, artists, and composers.

Investment in these technologies is driven by a host of factors, not the least of which is that robots and software 'employees' don't need vacations, health care, or workers' rights. Nor will they sexually harass each other, yell at clients, or question their managers. Their sheer computational power, speed, and accuracy in many tasks will also be increasingly impossible for humans to match. Robotic "surgeons" already conduct or assist in medical procedures with a precision well beyond the capabilities of human hands, while IBM's Watson computer is hyped as soon to be the "best diagnostician in the world."[2] Sophisticated robotic drones already assist or take the place of human soldiers in critical and dangerous military operations. Millions of consumers have embraced the commercial rise of personal and social robots that promise to do more than clean our floors and mow our lawns—today they promise to entertain, educate, bathe, feed, play, and chat with us, and do the same for those we love.[3]

These developments naturally give rise to a broad range of philosophical questions. Some are metaphysical (Could artificially intelligent robots become conscious or self-aware? Could they have emotions or moral agency?); others are economic (What do robots mean for the future of human labor and capital?) or political (Can/should robots have rights?); some are existential (What purpose will human lives have in a hypothetical future when our intelligence has been superseded by machine intelligence?); but arguably, most are ethical questions (Is it ethical to take a robot lover? To develop autonomous 'killer' robots? To hire a robot nanny?).

9.1 Robot Ethics as Virtue Ethics

Fortunately, the study of robot ethics (sometimes abbreviated as *roboethics*) is already well underway.[4] Although robotics and artificial intelligence are distinct technologies, their convergence in today's marketplace means that roboethics is an increasingly eclectic field, addressing ethical questions that arise from AI research and development as well as the study of human-robot interaction (HRI). Our study here is guided, as the reader will by now expect, by concerns about robots and the cultivation of *virtue*. This is a good thing, because much work in roboethics operates within deontological (rule-based) or utilitarian ethical frameworks. As we noted in Part I, such frameworks can struggle to accommodate the constant flux, contextual variety, and increasingly opaque horizon of emerging technologies and their applications. This is not to say that such approaches are never useful in robot ethics; it is only to say that they remain incomplete.

Allow me to illustrate the point with utilitarian ethics. Utilitarians recognize just one fixed and universal moral principle: the duty to promote the greatest overall happiness or net pleasure. So, when judging the ethics of robot childcare, a utilitarian must determine whether robot nannies will produce the greatest overall happiness, over the long term, for parents, children, and other affected stakeholders. Whether it is ethical to build or buy a robot nanny is supposed to be answered by this calculus. Let's simplify the problem. Imagine that the utilitarian only has to judge the ethics of using a robot nanny *herself*, not robot childcare as a broad social practice. However, as Sherry Turkle and other researchers have discovered, human-robot interaction demonstrably *changes* how humans feel about robots in ways that are not easily predicted, and that evolve in complex ways over time.[5] Thus a person who finds the idea of using a robot nanny deeply repugnant or alienating cannot know whether she will still have those unpleasant feelings after a month, or a year, of using one. The sum of happiness she and her children will get from a robot nanny will be affected by the robot's presence in ways she cannot readily predict. The point is that whatever utilitarian calculation one makes at a given point in time is radically destabilized by the power of social robots to alter our likes and dislikes in ways that continually surprise us. And how do we judge which of our sentiments we should trust? Those we have when a social robot first enters our lives? Or the very different feelings we may come to have later?

Moreover, the capacities and behaviors of robots *themselves* are continually evolving; how much happiness I and my child get from Nannybot Version 1.0 tells me very little about how much happiness the robot might bring us by the time it has auto-upgraded to Nannybot Version 6.0. Of course, this might just mean that I constantly have to revise my calculations of the robot's expected utility, but this will not be a simple task, since the robot's powers will change *qualitatively* with each substantial upgrade, and again, I cannot predict how future interactions with the robot will alter my, or my children's, likes and dislikes. How then can I ever reliably assess the robot's expected utility to my family? In this fluid moral landscape, what we need is an ethics that relies on the presence of flexible, discerning, and practically wise human agents, who actively cultivate the very traits needed to judge wisely and well in such unstable conditions. This, of course, is where virtue ethics begins. Whereas utilitarians and Kantians offer us a fixed formula or recipe to use in every moral judgment, virtue ethicists would rather we learn *how to become good judges*.

Moreover, strictly deontological and utilitarian approaches to roboethics tend to shortchange or even bypass the question of what sorts of goods, and what activities and contexts of receiving them, are *in fact* good for humans to enjoy. This can be starkly illustrated by the case of robot abuse or torture, one that is not merely speculative, as we already have some striking real-world examples.[6] Consider a future subculture of humans with a habit of torturing social robots—they get immense pleasure from the robots' screams, writhing, and desperate pleas for mercy. We can assume that these robots are not sentient—they only mimic pain and don't 'truly' suffer. However, since humans who interact extensively with social robots commonly project robust mental states onto them, assume that the torturers vividly perceive that their robots *do* suffer greatly.[7] Let us also imagine, simply for the sake of the argument, that this sadistic habit will not spill over into the robot owners' treatment of their human fellows—a consequence that would otherwise make it easy for almost *any* moral theory to label this habit as unethical.[8]

If we come to it from a Kantian perspective, the scenario described above does not appear to violate the categorical imperative. That moral rule requires only that a practice be universalizable, and that it not treat a rational agent as mere means to an end. A universal practice of robot torture is not impossible to conceive, nor are these particular robots genuinely rational, autonomous beings with dignity, deserving of being treated as 'ends in themselves.' From a utilitarian standpoint, torturing these robots seems not only permissible, but *ethically positive activity* for the sadists, for it produces pleasure without causing anyone pain. These conclusions reveal why such ethical theories are incomplete. Neither framework has the resources to ask what human practices or traits of character are noble, admirable, and worthy of our aspirations. From the perspective of virtue ethics, people who spend most of their free time getting immense pleasure from torturing robots, even if they increase the net pleasure in the world by their own harmless enjoyment, are not living well or flourishing, because they are not by this activity cultivating any of the character traits, skills, or motivations that constitute *human* excellence and flourishing. Nothing about the torturers' favored activity makes them *better* human beings, and because a good human being embodies character traits and patterns of activity directly contrary to sadism, our robot abusers are less noble for their sadistic habits even if no living persons are injured by those habits.

This is not to deny utilitarian and deontological standpoints a place in roboethics. We must continue to ask how 'killer robots' can ever be consistent with the principles of just war theory and international humanitarian law; whether drone warfare is increasing or decreasing civilian suffering; whether the number of lives potentially saved by autonomous cars means that we must surrender the pleasures of driving; and whether robot caretakers violate the privacy rights or dignity of the elderly. But there are other important ethical questions about robots that only virtue ethics readily allows us to pose: How are advances in robotics shaping human habits, skills, and traits of character for the better, or for worse? How can the odds of our flourishing with robotic technologies be increased by collectively investing in the global cultivation of the technomoral virtues? How can these virtues help us improve upon the robotic designs and practices already emerging? Among the many types of robots and human-robot practices that demand our ethical attention, in this chapter we shall focus on just two, *military robots* and *carebots*, along with two excellences of character of special relevance to these: the technomoral virtues of *courage* and *care*.

9.2 Autonomous Military Robots: Courage and Hope for the Human Future

The rapid global development and deployment of advanced military robotics, especially autonomous and semiautonomous robotic systems, has greatly outpaced attempts by public policy experts, international human rights advocates, and ethicists to bring moral and legal clarity to the phenomenon. The technological capacity arrived before anyone could predict and assess its implications; one would be hard pressed to find a better illustration of our condition of acute technosocial opacity. Fortunately, vigorous scholarly and civic responses to the phenomenon are catching up, ranging from mass media and political critique; to books, journals, and conferences aimed at both popular and scholarly audiences; to organized global coalitions such as the Campaign to Stop Killer Robots.

While there is considerable fluidity to the concept of a 'robot,' and even more fluidity to the concept of an 'autonomous robot,' we can define the class of semi- or fully autonomous military robots *broadly* as:

1. mechanical systems designed and/or used to support and/or carry out tasks associated with, or constitutive of, military operations,
2. with the power to sense, process, and act upon information in the physical environment in ways that alter said environment, and
3. that have the capacity to calculate, select, and initiate particular courses of action independently of human judgment or oversight.

Such robots differ from familiar *non-autonomous* military robots such as PackBots and Talons, commonly used to destroy improvised explosive devices (IEDs). Human soldiers manipulate these with joysticks or other remote controls, although attempts to make autonomous versions of such robots are underway.[9] The best-known *semi-autonomous* military robots are unmanned aerial vehicles (UAVs or 'drones'), which may be armed or unarmed. These can typically fly, navigate, and/or track targets with some degree of independence, but require human oversight and do not target or fire autonomously. Unmanned ground and underwater vehicles have similar semi-autonomous capacities. Examples of robustly autonomous military robots already in use include automated sentry robots such as South Korea's Samsung Techwin SGR-1, which can select and shoot at targets without human oversight (though these do not currently operate in autonomous mode); 'fire and forget' systems such as Israel's Harop/Harpy 2 anti-radar UAV; and antiaircraft/antimissile systems such as the U.S. Phalanx and Israel's Iron Dome, which can in principle identify, track, and fire upon targets autonomously. Countless other variants of autonomous military robot are being designed, developed, tested, and deployed by militaries around the globe. Many also have civilian applications, such as the delivery of medicine, food, or other emergency supplies to otherwise inaccessible locations.

Advances in conventional robot mechanics are accelerating thanks to rapid convergence with advanced artificial intelligence research, as well as biomechanical research involving the integration of robotic and human or other animal systems. As a result, military ethicists are starting to ask not just how robotic systems can be responsibly *used by* humans as tools of war, but also how robots themselves will alter, cooperate, or compete with the agency of human soldiers. Military-funded researchers are working hard to develop robot soldiers with artificial *moral* intelligence, robots more restrained and trustworthy in battle than human soldiers will ever be.[10] Modern militaries already rely upon increasingly powerful software algorithms for ranking the value of potential human and property targets, assessing the collateral damage of strikes, and evaluating strategic options—all formerly the exclusive province of human judgment by officers presumed to possess the experience, perspective, and practical wisdom needed to make such high-stakes ethical decisions.[11] Due to the rapidly increasing speed and precision of algorithmic 'intelligence,' we are told that the human soldier will soon be "the weakest link in the military arsenal."[12] Many believe that the demands of military efficiency, force protection, and mission success make the global development and use of fully autonomous weapons a foregone conclusion, in the absence of a firm and enforceable international commitment to renounce them.[13]

For fairly obvious reasons, then, the global debate about military robotics has focused almost exclusively on robots with the capacity for autonomous lethality. As important as this debate is, it has diverted public attention from the many nonlethal and potentially *morally beneficial* uses of military robotics. Consider Vecna Technologies' BEAR (Battlefield Extraction Assist Robot) prototype, which could evacuate wounded soldiers from the field without requiring additional human troops and medics to expose themselves to mortal risk. Here we have a military robot with clear ethical applications and few apparent ethical downsides. Yet the BEAR has been starved for funding, forcing its makers to turn their attention to other, less noble projects, as Vecna's Chief Technology Officer Daniel Theobald admits: "had we gotten a whole bunch of R&D dollars from [the military,] it probably would have kept our focus on that rather than big commercial products."[14]

We must recognize that as with all technologies, military robots are neither 'good' nor 'bad' in themselves; yet they *afford* or *invite* human choices and habits of varying virtue or vice. Moreover, *context* matters. Assigning an autonomous lethal sentry robot to help secure a nuclear waste disposal facility in a desolate, uninhabited region might under certain circumstances be a prudent option, but the temptation to install one at the periphery of an army base in an occupied territory with an already alienated and angry civilian populace is another matter. The point is that promoting indiscriminate global resistance to advanced robotics, even *military* robotics, would be morally blind and counterproductive. What we need instead is the technomoral wisdom to ensure that future robotic applications prioritize the *promotion* rather than the degradation of global human flourishing—including in military contexts.

While the ethical issues in this domain are thus not exhausted by concerns about 'killer robots,' the rise of autonomous robotic lethality is both imminent and morally grave, warranting our urgent attention. Let us see what our technomoral virtue ethic can add to the conversation. For now, let us set aside important deontological questions about whether lethal robots will ever be able to satisfy the principles of just war theory or the requirements of international humanitarian law.[15] We also set aside the utilitarian question, no less relevant, of whether autonomous lethal robots will ultimately prevent or cause more combatant and civilian suffering. Instead we will ask a different question: How might the development of autonomous lethal robots impact the ability of human soldiers and officers to live nobly, wisely, and *well*—to live lives that fulfill the aspirations to courage and selfless service that military personnel pledge?

We saw in chapter 6 that the virtue of courage takes multiple forms. People with physical courage have a context-sensitive disposition to appropriately modulate fear of physical injury or death, in a manner that reliably avoids the extremes of rashness and cowardice. Another type of courage is a context-sensitive disposition to appropriately fear *moral* injury or ruin. Both forms of courage fall under a more general description, still captured well in Aristotle's words: "[to fear] the right things and from the right motive, in the right way and at the right time,"[16] and both are often called for in military life.[17] Moral courage in particular we defined as a "reliable tendency to fear grave wrongdoing and a compromised character more than one fears the other dangers or injuries that one might invite by acting rightly," a virtue that requires the intelligent balancing of measured and justified fears with measured and justified confidence and *hope*. We said that courage presupposes that living well sometimes requires genuine sacrifice and acceptance of great loss. While virtues are typically conducive to subjective, psychological happiness, the specific virtue of courage entails one's willingness, when called upon, to give up things of real subjective worth in order to preserve one's happiness in the objective, *eudaimonistic* sense—that is, living well rather than badly.

Chapter 6 defined *technomoral* courage as "a reliable disposition toward intelligent fear and hope with respect to both the moral and material dangers, and opportunities, presented by emerging technologies." Consider the BEAR prototype again. A sensible way to explain its failure to attract sufficient investment, despite immense moral promise to save the lives and limbs of wounded human soldiers and medics, is in terms of *technomoral cowardice.* The U.S. Defense Department declined to gamble funds on an emerging technology that warranted great moral hope, *despite* their repeatedly demonstrated willingness to bet far greater sums on riskier and less morally promising technologies such as the RAH-66 Comanche helicopter, a massive boondoggle cancelled only after twenty-two years and seven billion wasted dollars.[18] Along with technomoral justice, civility, and magnanimity, modern moral leadership requires the private and institutional cultivation of *technomoral courage*: the disposition to wisely balance our technosocial fears and hopes, and risks and rewards, in the service of human flourishing. Educating for and rewarding technomoral courage in voters, lawmakers, contractors, and military leaders is therefore essential to realizing the legitimate promise of robotic and other emerging defense technologies. But how is this virtue *itself* being affected by the emergence of these technologies?

9.3 'Killer Robots' and Technomoral Courage: The 'Troubles We Do Not Avoid'

To begin—what does the global race to develop autonomous robot warriors say about our greatest fears, or our greatest hopes? The most obvious fear that it expresses is our profound and very natural fear of human death and mutilation, and, more specifically, our fear of exposing ourselves, our children, neighbors, and fellow citizens to such grievous harms. Better to have robots be shot out of the air or torn up by an IED than to see it happen to our own people. This is not our only legitimate fear, however. There is also our fear of *moral* injury, a concept all too familiar to veterans, their families and friends, and to the overwhelmed mental health providers who struggle to help veterans cope with the psychological aftermath of engaging in conflict that profoundly violates the norms of civilian life. Not only is this moral injury worth fearing, but virtue traditions tell us that great moral injury ought in general to be feared more than physical injury or death. If robots can do our killing for us, and if this could spare us and our fellows from the profound moral injury that often results from our participation in the horrors of war, wouldn't the virtue of technomoral courage *require* us to develop and deploy such technologies?

Of course, the moral reality is far more complex. First, robotic mediation does not always prevent moral or psychological injury. For example, the prediction that remote drone pilots would be spared psychological harm has been disproved. Drone pilots often track targets for days or weeks before witnessing their deaths essentially in real time, and they suffer stress disorders like PTSD to at least the same extent as conventional pilots.[19] Perhaps this might be prevented by taking them even *further* 'out of the loop.' If armed drones were *fully* autonomous, we would not need humans to trigger, approve, or possibly even witness the use of lethal military force, as drone pilots do today. Imagine mechanical creations efficiently and dispassionately visiting death around the globe, without having to trouble any of their creators in the least. If this spectacle gives you pause, then it is clear to you that perhaps the moral injury we should fear most from war is *not* the shock and trauma of witnessing war's horrors ourselves.

In reflecting on the importance of moral courage, Mengzi says, "That is why *there are troubles I do not avoid . . .*"[20] Is war one of the troubles that, *if* it cannot be prevented, a good human being will not avoid? One way of thinking about this is to consider the specter of profound injustice presented by a century in which the majority of the world's soldiers are among the youngest, poorest, and most disenfranchised members of their societies. Can a *good* human being avoid the 'trouble' of sending his Ivy League son or daughter to war to defend his own nation in a war he voted for, while parents on the other side of the tracks watch their sons and daughters go—and die—in droves? Another way to pose the problem brings it back to our concern in this chapter. Can a *good* human community avoid the trouble of paying its share of the global human price that war commands by sending armies of replaceable robots in their stead? Or does goodness, as I am clearly suggesting, require each of us to either work tirelessly to prevent and oppose war, *or* to courageously accept some share of the devastation and irrevocable loss that our failure to prevent war brings about?

Perhaps, you might think, the wars of the future will not involve moral horrors at all—perhaps they will be fought *between* robot armies, eliminating human suffering and evil from the picture altogether. Let us set aside the question of what the stakes could possibly be in such a scenario *unless* humans were threatened (i.e., how this would be *war* rather than an expensive and pointless international game of chess). We can set it aside because, in the concrete global reality in which we live, only a small portion of the planet's population has the wealth and resources to marshal robot armies, and thus the trend we see is increasingly *asymmetrical* warfare such as that seen in the global 'war against terror': drone strikes versus beheadings.[21]

In fact, asymmetrical warfare arguably *doubles* the types of moral horror for which the human family is accountable. On the one hand, we have a mechanical and impersonal technics of killing, one that fosters a military psychology in which the most grievous human suffering of innocent civilians can be reduced to the vocabulary and calculus of 'bugsplats.'[22] On the other side, we witness surges of increasingly gruesome and indiscriminate violence from those who pursue their low-tech methods of brutality without even the merest semblance of moral restraint. The oft-promised ability of drone warfare to minimize civilian casualties from airstrikes has yet to be empirically demonstrated by any neutral observer, but if the asymmetry that modern warfare fosters plays *any* role in feeding the warped psychology and recruitment successes of groups like ISIS and Al Qaeda, then the claim that robotic warfare will make innocent civilians *overall* safer from the horrors of war is plainly dubious. Thus the development of lethal military robots that promise to allow a minority of privileged human beings to detach even *further* from the physical, psychological, and moral horrors of war endured by the rest of humanity is deeply inconsistent with the technomoral virtue of courage, not to mention justice, empathy, and moral perspective.

It must be noted that moral concern for the cultivation and preservation of virtue operates *within* the military profession; it is not merely a critical pressure imposed from without. As I have argued elsewhere, the modern military profession is, like all genuine professions, defined by a special moral commitment explicitly *professed* to the public it serves.[23] The core profession of the modern military vocation is the ideal of "selfless service."[24] "Selfless service" expresses the moral virtues of *care* and *courage*, for military culture is rooted in our most basic moral understanding of our duty to care and make sacrifices for the security and well-being of those who depend on us most. A soldier is distinguished from a criminal, mercenary, or 'barbarian' only by virtue of this selfless service. In most contemporary contexts this entails fighting with courage, loyalty, honor, and other accepted martial virtues while accepting the norms of international humanitarian law.[25] That many soldiers fall short of this goal does not change its centrality to the modern military vocation. Relinquishing this commitment as incompatible with the technological expedience of autonomous lethal robots would destroy the integrity of the military *profession*, severing the always tenuous link between soldiers and the moral life of their communities, and the even more delicate link between warmakers and the human family writ large. The long history and moral significance of this link is reflected in centuries of recorded accounts of soldiers treated with mercy and even respect by foreign enemies, military and civilian, who recognized them as members of a shared and *noble* human vocation of selfless service to country.

Finally, we must address the role of *hope,* which attends all forms of courage, but including and *especially* military courage. Of course, military courage entails the hope of victory, or, if that cannot be had, the hope of honor in defeat. But there is a far deeper kind of hope that must be retained in the heart of military virtue. The highest ideals of the military vocation are haunted and challenged, as they must be, with the fundamental and undeniable *immorality* of warfare as a global human practice. What can make warfighting a morally necessary means to the immediate survival and security of a people is the very immorality of war as a human way of being in the world. Because we cannot always avoid those who would make war against us, our children, and civic fellows, we cannot forego all defensive wars without failing in our most basic moral duties of human care. But we do not by such means excuse ourselves from the stain of the worst kinds of human vice. Thus the military profession, in order to remain a moral vocation, must also have peace as its distal aim; it must retain *hope* for the end of war as a meaningful possibility.

What does our rush to invest massive resources in autonomous lethal robots say about our moral hope for peace? Consider the danger of technosocial 'lock-in' highlighted in Jaron Lanier's *You Are Not a Gadget*.[26] Lanier warns us to be alert to the gradual hardening of certain technical design choices that over time become increasingly difficult to undo, modify, or improve upon, as other systems, tools, and practices are built to work with that initial choice. The technosocial evolution of warfare has in fact been a gradual 'lock-in' of martial human culture. The more we invest our critical resources in manufacturing bullets, tanks, battleships, or, today, autonomous killing machines, the more we are invited to war by their standing presence. Expensive things beg to be used. So do things that reflect back to us our own growing technosocial cleverness. Only the unthinkable consequences of global nuclear destruction gave us the resolve to gradually decommission those great mirrors of human ingenuity, and even *this* task we have not allowed ourselves to finish. The mid-21st century may be humanity's last window for meaningful *hope* to relinquish a technics of perpetual war. Massive global investment in technologies of automated robotic lethality, already underway, may well ensure the closing of that final window of hope. Only an equally massive counterinvestment in technomoral practices that pursue the goods of global human community, justice, security, and wisdom will force it back open.

As we argued in chapter 2, early signs of an emerging, if comparatively weak, counterinvestment in these technomoral goods of *global* human flourishing are already visible. Thus despite the immense obstacles, this is not the time to surrender our moral hope. Rather it is time to cultivate renewed technomoral courage: the ability as a species to, at last, fear most that injury which is *truly* most fearful, and hold out hope for the greatest, perhaps now the *only* hope for the human family: an enduring and stable moral technics of peace.

Next we turn to a far quieter and more intimate setting of emerging robot presence in our lives, though one no less central to our moral identities and hopes for living well.

9.4 Carebots and the Ethical Self

'Carebots' are social robots designed for use in home, hospital, or other settings to assist in, support, or provide care for sick, disabled, young, elderly, or otherwise vulnerable people.[27] The kind of support they may provide varies widely, but most actual or potential functions of carebots fall into one of the following categories: *performing* or *directly assisting* in caregiving tasks (for example, bathing, dressing, turning, administering medication to, or transporting persons under care); *monitoring* the health or behavioral status of those receiving care, or the provision of care by human caregivers; and *providing companionship* to those under care.[28]

One primary motivation for the development of carebots is the emerging deficit of care providers in nations such as Japan, Germany, Israel, and the United States, where demographic, political, and cultural factors converge to strain these societies' capacities to adequately care for a rapidly expanding population of aged or otherwise vulnerable persons. Related to this are economic pressures upon individuals, private and public institutions to reduce the growing *costs* of care, social pressures to reduce growing institutional failures to provide *quality* human care, and recognition of the need to reduce the often-overwhelming physical and psychological *burdens* placed upon individual human caregivers.

Development and implementation of carebots is in a relatively early stage. Prototypes such as AIST's Paro Therapeutic Robot, a bionic fur-covered harp seal now classified in the United States as a Class 2 medical device to provide companionship in nursing home settings, have had notable success. Paro robots are in hundreds of care facilities and homes around the world, and their therapeutic effects on patients are the subject of ongoing empirical study. The Robear nurse robot developed by Japan's RIKEN-TRI Collaboration Center for Human-Interactive Robot Research (RTC) is among the more advanced carebot prototypes; it can lift a patient from a bed to a wheelchair, and vice versa. Since 2010 GeckoSystems International Corporation has been conducting limited in-home trials of its CareBot. Combining a mobile-service robot base with interlinked AI modules for navigation, tracking, scheduling, reminders (e.g., medications, food, etc.), and "chat," the CareBot is marketed as having commercial potential for both elder care and child care. In 2013, GeckoSystems told their shareholders that "[we] forecast, with a high level of confidence, that we can manufacture and market a $5000 personal robot, our CareBot, in the near term."[29] In addition to smaller firms such as GeckoSystems, robotic industry leaders such as Honda, iRobot, and Toyota are investing heavily in social robotics, including carebot development.

While it may be some time before practical, safe, and affordable carebots are available for widespread consumer or institutional use, their development is likely to continue to advance, given the grave deficiencies in care systems in many developed nations. Previous chapters have caused us to question the growing global appetite for 'techno-fixes' to what are, at bottom, social, political, and/or *moral* defects. Yet we have also warned against uncritical rejection of new technologies. The lock was, and remains, a 'techno-fix' for social and moral failures. Does it follow that the invention and use of locks was unethical? Naïve technophilia and reactionary technophobia are equally blind and unthinking responses to technosocial challenges. What we need is a measured, careful examination of the specific ethical risks and opportunities created by the imminent emergence of carebot technology.

One obvious concern is the *safety* of carebots should they be introduced. However, given the presence in most nations of social mechanisms for monitoring product safety (mechanisms which, one must grant, are far from perfect), let us assume for the sake of argument that safety concerns can be well-managed. Many issues remain. Many ethicists worry that carebots *objectify* the elderly and other patients as "problems" to be solved by technological means. Others highlight the effects carebots could have on the *capabilities, freedom, autonomy,* and/or *dignity* of those being cared for, effects that include possibilities both welcome and worrisome. Another question is whether carebots will enhance or reduce patients' *engagement* with their physical and social surroundings, and whether their meaningful contact with families and friends will be reduced. Particularly for carebots with tracking and monitoring functionality, carebots could also either preserve or intrude upon the *privacy* of patients. It goes without saying that we must also be concerned about the *quality* of physical and psychological care robots can realistically be expected to supply. Finally, carebot-human relations have the potential to be *deceptive* or *infantilizing*.[30]

Philosopher of technology Mark Coeckelbergh has helpfully separated questions of "roboethics" into three categories.[31] The first pertains to the minds and moral realities of robots themselves. Here we would place concerns about whether artificially intelligent robots could ever properly be considered moral agents, whether they might one day deserve "robot rights," and so on. The second pertains to the application of traditional ethical theories to human–robot interactions. In this category belong questions about robots' impact on the privacy, liberty, dignity, autonomy, health, and safety of humans. Coeckelbergh then suggests a new, third kind of roboethics—moving beyond what he calls "external" ethical criteria, and toward consideration of the "good internal to practice" that may be realized through robot–human interaction.[32] That is, he wishes to ask how such interactions change us, what they "do to us as social and emotional beings."[33]

As we saw in chapter 2, references to *goods internal to practices* belong to the language of virtue ethics. So do concerns about how robot-human interaction conditions the social and emotional *character* of human beings. Such concerns suggest that 'carebot ethics' is not exhausted by asking what carebots will mean for those cared *for*. The safety, happiness, rights, dignity, and capacities of the people cared for by robots must remain of central ethical concern—but not our only concern.[34] Nor is this subject exhausted by asking how best to "reduce the care burden" on caregivers and society in ways that liberate caregivers from rote, tiring, stressful, or dangerous tasks while decreasing the numbers of undercompensated, overworked, and ill-trained human caregivers.[35] Carebot ethics remains incomplete until we have also considered the possible impact of carebot practices on human habits, virtues, and vices, especially those of *caregivers*. That is, we must understand how carebots might affect our own abilities to flourish as persons capable of care, creatures whose moral status is deeply rooted in relations of caring *virtue*.

9.5 Technomoral Care in Relations of Human Dependence

Let us recall what it means to say that care is a virtue. 'Care' is an ambiguous word. Sometimes it refers to a mere attitude: ("My doctor really cares.") But 'care' can also be understood as an *activity* of personally meeting another's need, one that, if properly habituated and refined into a practice, can also become a manifestation of personal excellence. Such care meets all the requirements of a virtue. Caring for people *well* is not easy. We must learn how to care in the right ways, at the right times and places, and for the right people. Uncultivated caring activity often goes wrong despite our best intentions: we somehow make things worse, or we leap in too quickly and offend others' autonomy and dignity, or we wait far longer than we should to help. It takes a lot of effort and acquired moral intelligence to become even halfway good at caring, and even then we may screw it up royally. But people who care for others *well* are among those examples of human excellence that we recognize and respect most readily. In chapter 6 we defined the virtue of *technomoral care* as a "skillful, attentive, responsible and emotionally responsive disposition to personally meet the needs of others who share our technosocial environment." How might carebots degrade *or* enrich our ability to cultivate this virtue?

Anyone who has cared for an elderly parent, chronically ill spouse, or disabled child for any significant length of time knows that reducing the 'care burden' on caregivers is something often to be desired, even a social obligation, assuming it can be accomplished by ethical means. It will also be clear to most who have spent time in skilled nursing facilities or other underfunded and overburdened institutional settings that, as ethicists Jason Borenstein and Yvette Pearson state all too mildly, " . . . it is difficult to support the notion that current caregiving conditions ought to be preserved."[36] Yet the goal of 'reducing the care burden' is profoundly ambiguous. Do we mean that we wish to liberate caring practices from the sorts of physical, emotional, and financial burdens that can diminish the quality of our care of others, or that prevent us from continuing to care for them? Or do we mean that caring *itself* is a burden that we should wish to reduce for ourselves and other human beings when possible? The first goal seems an uncontroversial good. As we will see, the second goal is deeply problematic, making the ambiguity of talk about the 'burden' of care ethically perilous.

Given the widely acknowledged and grave deficiencies of the status quo in human caring practices and institutions, let us stipulate that carebots, properly designed and implemented, could improve the lives of both those being cared for and caregivers in ways that would be ethically desirable and, in the absence of better alternatives, ethically mandated. Yet before we pursue this possibility, we must reflect more deeply on the moral value of caring as a human practice, not only for recipients of care, but also for *caregivers*. We must resist well-motivated but one-sided portrayals of caregiving as a social requirement or burden that people (especially women) are unjustly expected to bear.[37] Even the provision of companionship and emotional support is all too easily characterized as a "task" in the same way as one describes the work of vacuuming floors.[38] Martha Nussbaum rightly notes that contemporary societies still tend to assume that women in particular will give care "for free, 'out of love,'" her scare quotes directing us to the injustice of caregiving as an unfairly distributed and uncompensated labor.[39] Yet if there are no further dimensions to caregiving practices that draw our attention, it would seem to follow that in the absence of injury or insult to those receiving care, we could have no reason to hesitate to surrender our caregiving practices to carebots, and to encourage others to do the same.

Intuitively, it seems this cannot be right, and few of the philosophers who have reflected on the ethics of robot care would endorse such a conclusion. Nevertheless, we must ask why the potential losses to caregivers who surrender such practices have not been more expressly considered. It is, of course, acknowledged by philosophers that caregiving may be freely chosen, indeed, this is presented as the ideal state of affairs. Borenstein and Pearson, following Nussbaum, assert that "those who provide care in non-emergency contexts should have the freedom to choose the extent, type, and manner of caregiving."[40] While putting it this way obscures the extent to which even nonemergency care is often a robust human obligation rather than simply one life choice among others, the reasons *why* one might freely choose to give care are not explored here, and these reasons matter a great deal to our inquiry.

Some such choices are motivated by ethical heroism, a decision to take on the burdens of care so that others do not have to. Others are motivated by a sense of personal and nontransferable duty, such that one would rather bear the burden of care than the guilt of shirking it. Still others are motivated by external rewards that outweigh care's burden: financial compensation, or expressions of gratitude and admiration from the recipient of care and other observers. But all of these reasons presuppose that the practice of caregiving *itself* is still only a burden, the free acceptance of which must be justified by something external to the practice. If this is right, then nearly all of the conditions that warrant freely giving care could be rendered superfluous by the introduction of carebots. We can transfer caring tasks to entities that will not experience them as a burden, and hence require no moral, social, or financial compensation. Yet there are moral goods *internal* to the practice of caregiving that we should not wish to surrender, or that it would be unwise to surrender even if we frequently find ourselves wishing to do so. For example, most ethicists would agree that "the elderly need contact with fellow human beings," and that if the use of carebots led us to deprive the elderly of this contact, this would be ethically problematic.[41] But rarely does one hear an ethicist ask whether caregivers need contact with the elderly! (Here we may substitute 'children,' 'the sick,' 'the injured,' the 'poor,' or any other likely subject of care).

We have falsely assumed that the provision of care is a one-way street. Let us recall from our earlier discussion of courage Mengzi's suggestion that a cultivated person has 'troubles [he or she] will not avoid.' Is *care* one of those troubles that a cultivated human does not avoid? Moreover, will a truly virtuous, exemplary person experience caregiving as *only* a trouble, one that entails heroically sacrificing one's own good, rather than receiving or sharing in a good? It seems this cannot be so, if there really *are* goods internal to the practice of caring. If there are, then I lose these goods if I give up the practice to a robot—even if that robot can do a better job of caring than I can. It is of course possible that caregivers will choose to *share* caring practices with robots rather than surrender them. However, given the physical, psychological, and financial costs of giving care, we cannot discount the fact that many humans will, once they trust the safety and skill of carebots, be tempted to *transfer* the lion's share of their caring duties to them.

What *are* the goods internal to the practice of caring? Let us consider what caring teaches us. First, it teaches us the moral meaning and importance of *reciprocity*. We learn in caregiving practices to see how reciprocity holds human relations together and allows other kinds of goods to flow across those connections. This may not be immediately evident if we focus on the asymmetry of need between the vulnerable person who receives care, and the one who gives it. Yet the elderly parents we hand-feed once fed *us*. The older sister whose radiation sickness we comfort perhaps nursed *us* through a childhood flu—and the moral meaning of that care endures even after the tables are turned. Even in institutional settings, nurses, doctors, and therapists are themselves, at times, the patients. We learn in a time of need that others are there for us *now*, and just as importantly, we learn *through being there for others* to trust that someday someone will be there for us once again. For once I perceive that I, who am *not* a moral saint but an often selfish and profoundly imperfect creature, can reliably give care to others, then I can more easily believe and trust that equally imperfect humans can and will care for me when the time comes. Of course, this trust in human care can be violated—indeed, the very fact of such violations drives carebot development. But those who lose all trust in the human bonds of care, or never acquire it, are profoundly *damaged* beings.

Caregiving practices also foster the internal good of *empathy*, the emotional connection that moves us to 'take trouble' for others, as Aristotle notes.[42] It is through giving care that I learn to recognize in concrete others those expressions of need and desire that can motivate empathic concern. Even as a child I come to this attitude through caring practices: looking after a younger sibling, feeding and watering a pet, fetching Kleenex for a flu-afflicted parent, or listening to the problems of a friend. It is through caring activity that I come to fully appreciate the *goodness* of the caring role and of caring *well*. As someone cared for, I can see the goodness of having my needs met—but I cannot fully grasp the goodness of *meeting* needs until I have taken on that role myself and experienced its challenges and rewards. Without exposure to caregiving practices, my grasp of the goodness of caring relations remains impoverished and one-sided. Moreover, it is only through repeated and intimate exposures to human need and care that I develop the abstractive capacity to "imagine the situation of another" outside my local circle of care, and to be moved to extend my concern to them. For those with whom I empathize in this way, I may indeed "take trouble."

Roboethicists Robert and Linda Sparrow have stressed the importance of empathy in caregiving as a means of showing the irreplaceability of human caregivers by carebots, claiming that "entities which do not understand the facts about human experience and mortality that make tears appropriate will be unable to fulfill this caring role."[43] Although I do not wish to challenge this claim, I would stress a different implication. It is primarily in fulfilling these caring roles that humans put their capacity for empathy *to work*, practicing and gradually cultivating empathic concern as a virtue of their character.[44] Without enough such opportunities, the development of this virtue may stall. The concern is that the availability of carebots as a substitute for human care may be all too appealing to those who have not yet become comfortable with the exposure to suffering that is involved in caregiving practices, leading them to abandon early on, or greatly limit, their exposure to caregiving. If carebots allow us to feel that our obligations to care are being met even when we have quietly removed ourselves from the suffering of our loved ones and fellows, is this only an injury to them? Or is it also an injury *to us*?

Like all moral habits and virtues, the capacity to care well is not something we are born with or that happens automatically, but something we have to cultivate and maintain in ourselves. Here we can draw upon the resources of care ethics discussed in chapter 6, as a way of more fully articulating the role of caring relations in moral excellence and human flourishing. As we noted there, care ethics shares with virtue ethics an orientation to the goods internal to practices, and the belief that morality arises from the enactment of such practices in concrete human relationships. While virtue ethicists and care ethicists differ on whether human flourishing or the caring relation is the ultimate good, they occupy considerable common ground when it comes to the question of how the ethical self is cultivated. Care ethicist Nel Noddings asserts that through our memories of being cared for, we acquire our first and ultimate understanding of goodness. The natural caring relation thus comes to function as regulative in human social behavior, motivating us to seek to restore it when it becomes deficient or disrupted. From this impulse we acquire our first ethical ideal, a notion of the "ethical self" as a personal commitment to meet others, especially those within our intimate circle, from within the caring attitude. Noddings notes that, "This caring for self, for the ethical self, can emerge only from a caring for others."[45] While the memory of being cared for spurs the development of the ethical self as a regulative ideal for my life, it is my committed practice of caring for others that sustains and enriches this ethical self.[46]

Noddings identifies two central criteria for a genuinely caring commitment: *engrossment* and *motivational displacement*.[47] Engrossment is an empathic orientation to the reality of the other, as opposed to my own. It allows the needs of the other to be foregrounded in my field of awareness, rather than as the background of my own needs and desires outside the caring attitude. Motivational displacement is the associated feeling that "I must do something," I must actively *address* the other's reality, "to reduce the pain, to fill the need, to actualize the dream."[48] These criteria, of course, create quite a high standard for genuine care. They exclude care offered by proxy, or though routine tasks that may "secure credit" for caring from others, but do not solicit or require my engrossment in the needs and sufferings of the other.[49]

To see why, it helps to reflect upon the moral psychology of the person who never becomes comfortable or skilled in caregiving. This person may refuse the caring role due to anxiety about being exposed to the pain, fear, or grief of the one needing care; or being made helpless in the face of the other's longing to be made strong or whole; or becoming intolerably angry, frustrated, depressed, humbled, or exhausted by the work that care entails. Most of us will find these fears recognizable—we have seen them expressed by others, and had them ourselves. When called, one takes on the caring role in spite of such fears, or one abdicates it—leaving it empty or to be filled by someone else. A person who consistently abdicates the caring role out of fear lives as a distant, "checked-out" parent, especially during a child's illness or disability. Such persons will find excuses not to answer phone calls from depressed or grieving friends. They will cite a hatred of hospitals as a reason not to visit the sick and dying. They may withdraw from a terminally ill spouse, or attempt to "care" for others in less intimate, risky ways—working overtime to help with medical costs, sending flowers and food baskets by courier, researching treatments online, planning a trip to Disneyland. These too are morally meaningful acts. But as long as one continues to avoid the practice of caregiving—drawing upon the full range of one's emotional, intellectual, and physical capacities to support someone who is suffering, dependent, or vulnerable—one's capacity to flourish as a moral being remains stunted and impoverished.

It is not fear that consigns such a person to this fate—we all have such fears. But the consistent abdication of caring practices allows those fears to be engraved into a person's emotional character, foreclosing possibilities for emotional and moral growth. Meanwhile, the person who takes up caring practices—even if initially out of guilt or social shame—is forced to meet these fears with courageous action. Caring requires courage because care will likely bring *precisely* those pains and losses the carer fears most—grief, longing, anger, exhaustion. But when these pains are incorporated into lives sustained by loving and reciprocal relations of selfless service and empathic concern, our character is open to being shaped not only by fear and anxiety, but also by gratitude, love, hope, trust, humor, compassion, and mercy. Caring practices also foster fuller and more honest moral perspectives on the meaning and value of life itself, perspectives that acknowledge the finitude and fragility of our existence rather than hide it.

Thus if I transfer my *full* obligation of care to a carebot, I have "done something" to care, but not in a manner that will sustain my self as a moral, caring being. Still, Noddings warns that in all caring situations, there is a risk that "the one-caring will be overwhelmed by the responsibilities and duties of the task and that, as a result of being burdened, he or she will cease to care for the other . . ."[50] Since ethical care is not a mere attitude but an activity, I can only care within the limits of the practically possible, and a host of environmental, social, or institutional conditions can constrain or impoverish my care. Engrossment in the need of another is impossible when one has had so little rest that it strains one's limits even to sleepwalk mindlessly through the motions of care—changing the bedpan, remembering to give medication, cleaning the wound. What more can one give?

The unmitigated stress, anxiety, and physical and mental exhaustion of unsupported caregiving often results in impaired judgment, depression, apathy, emotional withdrawal, or extreme emotional volatility—any or all of which can result in significant harm and fractured relations with others, including those for whom we struggle to care. As Mengzi and Aristotle understood very well, certain kinds of deprivation are deeply poisonous to moral practice. Environments that lack supportive resources for caregivers are likely to drain caregivers of emotional power, and starve empathic responses rather than cultivate them. Indeed, too many modern care facilities are 21st century equivalents of the ruined social environments Mengzi refers to in his famous parable of 'Ox Mountain': places so desolate and barren, so empty of human compassion, reciprocity, trust, empathy, and hope, that one might be forgiven for thinking that such virtues never existed.[51]

Rather than inviting us to be 'liberated' *from* care, if carebots can provide limited but critical forms of support that draw us *further* into caregiving practices, able to feel more and give more, freed from the fear that we will be crushed by unbearable burdens, then the moral effect of carebots on human character and flourishing could be immensely positive. Carebots in such contexts can *sustain* rather than *liberate* human caregivers, saving them from the degradation of their own ideal ethical self. Certainly caregivers need support to come in many more forms than smart and shiny robots—they need better support from extended family, friends, employers, lawmakers, healthcare providers, and insurance companies. They need better and more affordable care facilities. They need more financial security. They need access to effective caregiver education, training, and support networks. But in more than a few practices and contexts, caregivers might get significant 'moral support' from carebots as well.

For example, carebots could easily monitor vital signs, medication, and pain levels while caregivers sleep. They could warn of medication interactions or new symptoms that often escape human notice. They could answer caregivers' questions, link them to medical or legal resources and support groups, and keep track of financial and other documents. Carebots could detect signs of caregiver exhaustion, peak stress, and burnout, and recommend taking a break or rest. Caregivers might even take such advice more readily from a robot than from a human being who is capable of passing moral judgment on their 'weakness'—a fear that may inhibit caregivers from admitting to others when they have reached their emotional or physical limits, often with tragic results. Carebots could also help with tasks beyond the physical capacity of many human caregivers, such as moving heavy medical equipment, lifting and turning a disabled spouse in bed, helping parents in and out of wheelchairs, or supporting the weight of an unsteady friend as she walks to the bathroom.

Imagine how many physically petite, infirm, or elderly human caregivers might then be able to confidently and safely care for their loved ones in circumstances that were previously impossible for them to manage. Imagine how many people might be able to remain in their homes receiving care from loved ones, if the physical strains and dangers of care were mitigated by carebots. We have already said that living well, in ways that allow us to cultivate our ethical selves as caring beings, requires being able to successfully *take* care of those around us. It seems, then, that carebots could make the goods internal to caring practices accessible to many more human beings, and for far longer periods of time, than ever before.

As with all virtues, care is an appropriate and contextually sensitive disposition to the mean between extremes. As Aristotle said, of two extreme vices on either side of a virtue, there is usually one extreme that seems more ignoble to us. Relative to the virtue of care as loving service to others, the more obvious and ignoble vice is the carelessness of those who never learn to attend willingly or well to others' needs. But the reckless overextension of our abilities to care is *also* a vice. Whether from pride or cowardice, refusing to acknowledge that we need help when our caring skills, judgment, or strength are irretrievably degraded puts those we love most at grave risk of being harmed by us. An extreme, unmodulated disposition to care not only risks injury to others, but also unjustly perpetuates their dependence on us (hence the phenomenon of 'codependence' often seen in caretakers; we might also consider the harm to children done by extreme forms of 'helicopter parenting'). Moreover, unbalanced dispositions toward caring roles, especially in women, can serve as an excuse not to develop our *other* fundamental human capacities and virtues. Carebots can and ought to be developed and employed in ways that help us to avoid *either* extreme, supporting our self-cultivation as *successfully* caring beings, who provide care in the right ways, at the right times, toward the right persons, and to the right extent.

Yet is this the most likely outcome in our current technosocial climate? Earlier in this chapter we encountered the claim that humans are, or soon will be, the 'weakest link' in the technosocial network. If we are exhorted by carebot marketers or nursing home managers to 'let the robot do its job,' to hand over caring practices to a being which will never forget a pill, never lose its patience, and never let a loved one fall, we might be convinced that both we *and* our loved ones are best served by the full surrender of care. I may accept remote assurances of my loved one's well-being, or visit from another kind of distance, where I enter the physical presence of a loved one without ever allowing myself the experience of touching and being touched by her weakness, her sickness, her *need*. Yet Noddings is right, I think, that "I enslave myself to a particularly unhappy task when I make this choice. As I chop away at the chains that bind me to loved others, asserting my freedom, I move into a wilderness of strangers and loneliness, leaving behind all who cared for me and even, perhaps, my own self."[52]

If this is right, it has clear implications for how we assess the ethics of carebot uses. Many ethicists emphasize consulting the recipients of care about what levels of human/robot interaction *they* prefer.[53] Such preferences are highly salient, and any ethical assessment must take them into account. Yet imagine that a century from now, carebots are not only cheap, safe, and skilled, but good at keeping the elderly, children, and the convalescent happy—amused, comforted, and distracted from their needs or pains. Imagine that a great many people in this future *prefer* to be cared for by carebots rather than imperfect human caregivers, who still at times become impatient, dull, or angry in their company. How would honoring those preferences, and greatly restricting human involvement in caregiving practices, impact human moral development and flourishing in the long run? Alternatively, imagine a future in which most care recipients prefer, for aesthetic or tradition-bound reasons, *not* to have robots involved in their care, despite our having clear evidence that (1) limited uses of carebots for supportive tasks allow human caregivers to sustain and enrich their own caring practices to a significant degree; and (2) such uses have no negative impact on those being cared for *except* to offend their aesthetic and traditional preferences. Is it obvious that the sincere preferences of people on the receiving end of care would be morally decisive in either case? The virtue of technomoral care is a basic good in its own right, and not just for those on the receiving end. Any ethical assessment that fails to account for this good is profoundly inadequate.

What might technomorally *wise* proposals for carebot development look like? Aimee van Wynsberghe has developed a relational approach to carebot ethics, rooted in the ideas of care ethicist Joan Tronto, which offers specific design guidance to robot developers. Her approach emphasizes four norms of caring practice: *attentiveness, responsibility, competence,* and *reciprocity.* She asks roboticists to weigh how their designs impact these norms of *skillful* caring relations in a variety of contexts, with the goal of developing carebots that "safeguard the manifestation of care values," rather than displace them.[54] For example, van Wynsberghe emphasizes the importance of eye contact and touch in skillful caring relations. Describing a robotic patient lift, she notes that even if the robot is safe and efficient, a nurse 'lifting' a patient with its help is not engaged in skillful caring if the nurse's attention during the activity must be devoted to monitoring the machine controls instead of the patient. A superior design might be a mechanical exoskeleton that ensures a safe result while allowing the nurse to maintain eye contact with the patient, discuss his or her comfort level, and offer reassuring touch throughout the process.

Yet are we likely to see these nobler, wiser possibilities win out in the absence of a widespread and sustained cultural effort to preserve and cultivate technomoral virtue? Are today's roboticists taking pains to ensure that carebots foster the cultivation of excellence in human caring? Or will they overwhelmingly follow the path of least engineering and economic resistance, driven exclusively by efficiency, expedience, and the contingencies of the law and the market? If the latter is the most plausible answer, is a future of fully mechanized caregiving therefore *inevitable*?

It is *far* from inevitable—and not just because safe, cost-effective carebot development still has many technical hurdles to overcome. The global market will welcome this future only if we as a human family widely accept the dangerous idea that care—skillful and loving service to the needs of others—is no longer our ineradicable human responsibility. If this idea is accepted, however, there will no longer *be* such a thing as the human family, except in the thinnest genetic sense. Once the bonds of human care are not merely frayed but severed, once our ideal ethical selves as caring beings are forgotten, ethics *itself* dies at the root. We are not there yet, and there is still time—time to recover our hope for living *well* in the fully ethical sense; time to demand that our institutions reaffirm the practices of moral self-cultivation and the value of exemplary lives; time to demand, as other cultures have before us, the pursuit of human virtue.

10

Knowing What to Wish For

TECHNOMORAL WISDOM AND HUMAN ENHANCEMENT TECHNOLOGY

IN THIS AND coming centuries, nanoscale, biological, information, and cognitive (NBIC) sciences are expected to converge in ways that enable the development of powerful new technologies for human enhancement.[1] We can define 'human enhancement' as the technological improvement and expansion of various 'species-typical' qualities and capabilities of human beings in ways that differ importantly from existing *therapeutic* uses of medical technology. Enhancement can be sought by a variety of means, alone or in combination, including genetic, biomechanical, nanomedical, and/or pharmacological techniques. The precise scope and essential characteristics of enhancement, and even its coherence as a concept in contrast with medical therapy, are hotly contested.[2] What is not in question is the considerable market potential of these technologies, which may offer radical life extension and slowed aging; enhanced bodily strength, endurance, resilience, size, or appearance; enhanced memory, attention, sensory perception, judgment, mood, or wakefulness; and enhancement of our moral faculties or dispositions.

Some enhancements are already available, such as the controversial but common 'off-label' uses of pharmacological agents such as Adderall or Modafinil to enhance attention, wakefulness, memory, mood, or endurance in healthy people. Other enhancements are in development and likely to emerge in the near future, thanks to brain-computer interfaces, genome editing, and biomechanical implants to enhance both cognitive and physical capacities. Still other enhancements envisioned by futurists fall into the realm of wild speculation: for example, the digital 'uploading' of human consciousnesses into 'cyber-immortality,' or the creation of transgenic human-animal hybrids.

10.1 Competing Visions of Human (or Posthuman) Flourishing

The most enthusiastic promoters of these developments call themselves 'transhumanists.' Transhumanists expect the obstacles to human flourishing presented by age, death, disability, and disease, as well as 'normal' cognitive and physical limitations, to be partly or wholly transcended by means of emerging technologies. Many speak of a coming 'posthuman' era that will fully liberate us from the straitjacket of our biological heritage.[3] A diverse community of scholars, technologists, and futurists of various philosophical and political stripes, transhumanists are united by four widely shared convictions: 1) that enhancement technologies have the potential to greatly improve the quality of our existence; 2) that research and development of such technologies should generally be fostered rather than banned or discouraged; 3) that decisions about the wisdom of enhancing ourselves, or about what constitutes a genuine 'enhancement,' will often be best left to private individuals; and 4) that many or even most individuals will have good reasons to choose enhancement for themselves and/or their children.

Bioconservative ethicists, on the other hand, reject human enhancement on several grounds. Chief among them is the belief that the transhumanist vision poses a grave threat to human dignity, and depreciates the meaning and value of human nature.[4] Other concerns include worries about existing global socio-economic inequalities being magnified by a split between enhanced and unenhanced social classes, fear of the physical and psychological risks of enhancement, and theological concerns about 'playing God.' Like transhumanists, bioconservatives are motivated by diverse philosophical, political, and spiritual convictions. They differ, for example, in the degree and scope of their opposition to existing genetic and reproductive technologies such as in-vitro fertilization and pre-implantation genetic diagnosis. Yet they share the view that transhumanism, far from promising unprecedented levels of well-being, in fact threatens to undermine the most basic moral and material conditions of human flourishing embedded in our biological nature. Limitations of space preclude a full survey of the debate over the ethics of human enhancement.[5] There is, however, one fundamental question, largely ignored on both sides of this debate, which we need to ask: *what technomoral virtues would humans need to have in order to enhance themselves wisely and well?*

Of course, if human enhancement is always unethical *in principle*, regardless of the character and motivations of those who pursue it, then our question will be moot. However, as we will see, bioconservative arguments for an unconditional rejection of human enhancement technologies are deeply problematic, even philosophically incoherent. Too often, transhumanists treat the weakness of these arguments as a free pass to overlook thornier and more nuanced ethical worries about enhancement, including those motivating our question. To reach those worries, let us first clear the road ahead by exposing the problems with the extreme bioconservative position. Only then can we confront the more daunting ethical challenge for transhumanists, and assess the wisdom of pursuing human enhancement from within our present technomoral condition.

10.1.1 Bioconservatism, Human Dignity, and Virtue

Many arguments against human enhancement are of a socially and technically contingent, rather than principled, sort. For example, concerns about the biological or socioeconomic risks of human enhancement depend heavily upon the particular worldly conditions under which such technologies will be developed, tested, and disseminated. If those conditions can be suitably controlled, the risks of enhancement could perhaps be mitigated or greatly reduced. Yet most bioconservatives offer a single *a priori* principle as justification for a blanket rejection of any and all human enhancement proposals, regardless of the specifics of those proposals or their social implementation. Bioconservatives typically claim that any proposal for human enhancement involves a profound violation of the moral imperative to respect *human dignity*.

Yet it is far from clear what bioconservatives mean by 'human dignity.' Francis Fukuyama notoriously defines it as the possession of an ineffable "Factor X," an "essential human quality" defying description that lies behind "a person's contingent and accidental characteristics."[6] This is problematic for a number of reasons, not the least of which are its vagueness and reliance on a questionable and apparently *ad hoc* metaphysics. Fukuyama defends his view on the grounds that the cultural and political ideals of Western liberalism depend for their intellectual coherence upon the existence of this universal and essential human quality. One could easily dispute this assertion, but regardless, it is not clear how Fukuyama can defend a metaphysical claim *simply* by observing that it happens to sit better with his political and cultural ideals.

Leon Kass, in his essay 'Defending Human Dignity,' attempts to overcome the lack of specificity that he admits tends to plague bioconservative appeals to this concept.[7] He defines human dignity as fundamentally connected with human *aspiration*. Here dignity depends upon our species' conscious striving to realize our natural potentialities through excellent activity, in order to flourish in our given biological form. Thus he equates dignity with the very capacity for moral self-cultivation we explored in Part II. Yet Kass links this aspiration to fulfill our *natural* potential for excellence to another aim: namely, a "self-denying aspiration for something that transcends our own finite existence."[8] Here we encounter a deep and perhaps irreconcilable tension in the bioconservative position. The first aspiration is deeply and inextricably linked to our biological heritage and the finite contours of natural existence that it marks out for us. This is the aspiration consistent with Fukuyama's emphasis on the "species-typical" features of embodied human existence, and fellow bioconservative Michael Sandel's insistence on appreciation for the "gifts of our natural finitude."[9] It implies a commitment to not only embrace but to *preserve* the biological integrity of the human condition.

Kass frames his second aspiration, on the other hand, as a desire to *transcend* our finitude in the direction of something understood as higher, even if this requires some element of self-forgetting or self-denial. Now Kass has in mind the kind of religious self-denial in which one sacrifices one's natural desires in order to close the gap between, and establish a relationship with, a higher being that embodies the "good, the true, and the beautiful."[10] Yet why could this reading not be reconstructed to accord with the secular aspiration to *attain* a qualitatively different and higher form of being—that is, to 'deny' one's given, biological form in order to remake it into something recognizably more beautiful and good? Here the relationship sought with a higher being is not one of communion, but identity. That Kass would reject such hubris does not prevent his account of dignity from technically permitting this aspiration; in fact, he arguably invites it by celebrating the "god-like" quality of humans to "articulate a future goal and bring it into being by their own purposive conduct," along with our freedom to "quit the state of nature" and establish ourselves under laws of our own making.[11] Here a deep tension emerges between the claim that human dignity is inextricably linked to the natural givens of our biology, and the claim that it is inextricably linked to the conscious striving to transcend what is naturally given, in search of the beautiful, the true, or the good.

This tension is not merely abstract; it comes into play when we consider how bioconservatives might evaluate various concrete proposals for human enhancement. Imagine two possibilities. One is a hypothetical technique for germline modification and gene insertion to bioengineer humans with canine DNA, giving them olfactory capacities well beyond the species-typical range. With proper training, these capacities can greatly enhance the work of soldiers, airline security personnel, police officers, firefighters, and health workers. The second proposal is a biomechanically engineered, permanent subdermal implant that releases a neurochemical agent to stimulate species-typical feelings of self-esteem and satisfaction at the user's command. Present these proposals to two bioconservatives, one primarily motivated by respect for the integrity of humanity's biological givenness, and the other primarily motivated by respect for effortful human striving toward excellent activity and virtue. It is reasonable to think that they may come to very different judgments about which proposal poses a greater threat to human dignity.

Notice that the first proposal employs germline modifications of human DNA to alter our species-typical capacities in a manner that clearly fails to preserve the integrity and finitude of our biological givens. Yet for the humans altered with nonhuman DNA, as well as for others who depend upon their work, the changes would appear to enhance and expand, rather than reduce, their human potential for virtuous activity. Such enhanced humans could attend to a far greater range of morally significant phenomena—for example, previously undetectable olfactory markers of fear, anxiety, aggression, decomposition, contamination, infection, cancer, fire, or airborne toxins. They would thus be able to glean more information related to core human goods such as security, justice, and health. Moreover, learning to use their new sensory powers in these excellent ways would require just the same sorts of education, practice, and deliberate striving that professionals undertake *now* in order to refine their professional discernment and perception. Thus at least some radical human enhancements could not only be compatible with, but might even *enrich* our aspirations to moral self-cultivation, our attainment of virtue and the goods internal to practices, and our consequent flourishing. How could this offend human dignity?

Now consider the second proposal. Though the implant is an alien presence in the human body, it is no more so than a pacemaker or stent, neither of which alarm bioconservatives. Imagine also that the neurochemical agent released by the implant is biologically identical to a substance naturally produced in the human body. No alterations have been made here to our genetic heritage, nor would such an implant produce any experience beyond the 'species-typical' capacity of an unenhanced human. By one bioconservative standard, then, the change seems fairly benign. Yet the use of this implant, which produces on-demand feelings of contentment regardless of one's circumstances, could *radically* reduce the motivation for effortful human striving, and the achievements of excellence that often result from such efforts, by dampening the desire to transform one's life situation, and/or one's own character, into something more satisfying. By the *other* bioconservative standard, then, this proposed enhancement would appear to present a far more profound threat to human dignity than the former.

Most bioconservatives articulate a concern for biological integrity *and* for human striving.[12] However, the tension between them renders such accounts problematic, both conceptually and practically. *If* the human aspiration to cultivate ourselves is the root of our dignity, and if human enhancement can open up new paths of cultivation and higher states of cultivated excellence, then at least some imaginable enhancements could reinforce our dignity by removing biological obstacles to those higher states. Thus bioconservatism appears incoherent as long as these disparate moral intuitions are conflated and packaged together under the amorphous heading of 'human dignity'. Why has this tension not been purged from the bioconservative position?

Fukuyama's account of dignity provides an important clue. Beyond his unhelpful invocation of 'Factor X,' he tells us that human dignity is rooted in a coherent biological whole that is more than the sum of its parts, incorporating reason, consciousness, the capacity of moral choice, and "the distinctive gamut of human emotions."[13] Now, some higher animals also appear to possess consciousness, instrumental reason, and a rich emotional life. Why is there not also, say, 'chimp dignity' or 'elephant dignity'? Are the champions of human dignity motivated only by what transhumanist James Hughes labels "human racism," a special moral regard for humankind grounded in no relevant moral facts?[14] Fukuyama *could* avoid this objection by noting the one item on his list not widely acknowledged as appearing in other higher animals: the faculty of 'moral choice.' This is the capacity to reflect upon, select, and freely carry out a course of action guided by a cognized moral ideal. A person can say to herself, 'what kind of person do I want to become?' or even 'what kind of person *should* I become?' and can consciously act on the answers she finds in reflection. Perhaps elephants, gorillas, or whales meditate on things like this and act accordingly, but there is no compelling evidence to date that this is so. It might seem odd, then, that Fukuyama has not made *this* capacity the linchpin of his account of human dignity, particularly since it could save him from Hughes' charge of 'human racism.'

This is no oversight on Fukuyama's part, however. He does not make this argument because he rejects Kant's idea that moral choice *liberates* us from our natural inclinations.[15] If dignity arose *only* from the capacity of autonomous moral choice, as Kant believed, then respecting dignity would only require preserving that capacity, not other given elements of our biology. In fact, we might be *obligated* to ignore, modify, or override our natural heritage in the interests of our own moral and intellectual development. Such a view supports transhumanist claims that human enhancement has already been taking place for millennia by means of cultural enterprises such as art, philosophy, and education, and that enhancement is, in the words of transhumanist Ronald Bailey, the "highest expression of human dignity and human nature."[16] A celebration of moral choice as the locus of human dignity is *simply too amenable to deliberate projects of self-transformation* to sit well with the bioconservative position.[17]

It is worth noting that while bioconservatives tend to self-identify as humanists, their conflicted attitudes toward humanity's creative moral freedom embody a different strain of humanism than that found in many voices from the past. Consider a much older tradition of philosophical and religious humanism, captured in 1486 by Pico della Mirandola's *Oration on the Dignity of Man*, in which he imagines God giving this charge to Adam:

> We give you no fixed place to live, no form that is peculiar to you, nor any function that is yours alone. According to your desires and judgment, you will have and possess whatever place to live, whatever form, and whatever functions you yourself choose. All other things have a limited and fixed nature prescribed and bounded by our laws. You, with no limit or no bound, may choose for yourself the limits and bounds of your nature. We have placed you at the world's center so that you may survey everything else in the world. We have made you neither of heavenly nor of earthly stuff, neither mortal nor immortal, so that with free choice and dignity, you may fashion yourself into whatever form you choose. To you is granted the power of degrading yourself into the lower forms of life, the beasts, and to you is granted the power, contained in your intellect and judgment, to be reborn into the higher forms, the divine.[18]

For della Mirandola, it is this capacity for self-transformation that makes humanity deserving of respect: "Who could not help but admire this great shape-shifter? In fact, how could one admire anything else?. . . "[19] A broad range of philosophical, scientific, religious, and artistic perspectives have resonated with della Mirandola's vision, from Condorcet to Nietzsche to Teilhard de Chardin to Shaw. Its embodiment on the front lines of transhumanism is simply its newest expression. Why, then, do bioconservatives, who portray themselves as humanistic insofar as they wish to celebrate the special majesty of our species above all others, shrink from the newest invocation of this vision?[20]

For one thing, the prospects of 21st century technological convergence raise the stakes of this vision by several orders of magnitude. Classical and medieval possibilities of self-transformation were limited to those habits, passions, and aspirations that could be remade by unaided reason. Despite della Mirandola's evocation of unlimited creative freedom, our physical embodiment was, until very recently, still largely beyond the scope of our choice. Certainly, cultural traditions of body modification, meditative control of autonomic functions, and asceticism testify that humans have tested the boundaries of our corporeal givenness for millennia. The possibilities, however, were always quite limited. Today, through the potential of converging technologies, it appears that *any* part of us might someday be made responsive to our creative whims. Confronted with this dizzying power, many contemporary humanists understandably experience deep unease.

Yet della Mirandola insists that bodily transformation is no threat to human dignity so long as our reflective and aspirational nature endures:

> Bark does not make a plant a plant, rather its senseless and mindless nature does. The hide does not make an animal an animal, but rather its irrational but sensitive soul. . . . Who would not admire man, who is called by Moses and the Gospels "all flesh" and "every creature," because he fashions and transforms himself into any fleshly form and assumes the character of any creature whatsoever?[21]

That bioconservatives do not share this liberality with regard to the flesh is plain. Yet it is worth asking why. By Kass's logic it would seem that as long as we transform our minds and bodies according to a clear vision of the "good, the true, and the beautiful," the dignity of human aspiration to a higher form should remain intact. One can also enumerate a host of conventional modifications of the flesh, done for both medical and aesthetic reasons, to which most bioconservatives offer no objection. What aspect of the transhumanist program of modification presents a moral difference sufficient to justify the bioconservative reaction?

I suggest that there is no sufficient moral difference. Instead, resistance to human enhancement becomes persuasive only when framed not in terms of an *a priori* argument from dignity, but a contingent argument from *virtue*, one that appreciates the contemporary practical obstacles to acquiring that "clear vision" of the good needed to guide transhumanist aspirations to their successful realization.

10.2 Technomoral Humility, Wisdom, and the Argument from Hubris

Our contingent argument from virtue takes seriously one objection to the transhumanists that we have not yet considered: that they are guilty of technomoral *hubris*. This objection can take many forms. Arguments from hubris are often muddled by vague warnings about the dangers of 'playing God' or trying to 'become Gods'; but as C.A.J. Coady notes, such warnings are highly ambiguous. They can refer to specific theological concerns likely to be dismissed by nonreligious parties to the debate, but such phrases are frequently used in a secular context as shorthand for the limits of human virtue, both intellectual and moral.[22]

Arguments from hubris, at least as old as the myth of Daedalus and Icarus, can represent more than just blind conservatism or an irrational fear of change. They can also reflect enduring concerns about human aspirations to transcendence becoming excessive or pathological. Is the transhumanist project guilty of such excess? By joining the vision of transcendence with the promise of virtually unlimited technological powers, the transhumanist vision exponentially raises the moral and material stakes of human aspiration. For perhaps the first time in history, della Mirandola's remark about degrading ourselves to a subhuman form of life becomes a literal possibility, should we wield these new technologies unwisely. This danger calls forth a need for *technomoral humility*: the ability to make an honest and clear-headed assessment of the present limits of our ability to wield these powers wisely and well.

Among transhumanists, Nick Bostrom has been the most transparent about the considerable risks of human enhancement technologies. While optimistic that there is likely "more distance to rise than to fall," he acknowledges potentially grave risks of these technologies. There are non-negligible risks that we might foolishly or unwittingly use them to "clip the wings of (our) own souls," losing "hold of our ideals, our loves and hates, or our capacity to respond spontaneously with the full register of human emotion."[23] Perhaps through ignorant and misguided tinkering with the human genome we may even bring about our own extinction. Against the backdrop of these very real risks, the old argument from moral hubris, waved off by many transhumanists as the counsel of cowardly or unimaginative souls, acquires renewed force.

Remember that courage, in our case *technomoral* courage, is a mean between cowardice and rashness. If the libertarian transhumanists who champion unrestrained experimentation in this arena do not represent the vicious extreme of rashness, who does? Yet having rejected the bioconservative argument for *a priori* restraint, doesn't *that* view look now like the extreme of moral cowardice, or at least an unreasonable and even *anti-humanist* affinity for the status quo? Where, then can we find the appropriately virtuous mean with respect to human enhancement technologies? How do we find the delicate balance of moral attitude in which both technomoral humility *and* courage are embodied, and integrated in a holistic moral perspective? Someone who can achieve this delicate calibration and integration of multiple technomoral virtues is an exemplar of *technomoral wisdom*. Looking at the parties and stakeholders to the enhancement debate, how many such exemplars can we confidently identify at present? Who among them are the 21st century equivalents of the *phronimoi*, the *junzi*, or the *Sangha?*

But why, the transhumanists will argue, should we exercise cautious restraint now, when centuries of technosocial innovation have repeatedly demonstrated the triumph of futurists and 'imagineers' over the naysayers and spreaders of moral panic? It may seem that the new always wins out over the old, but if we ask ourselves *honestly* just how reliably our innovations have served us in a century marked by global economic, environmental, public health, and military disasters, we will admit that our recent record of human ingenuity has some shining successes; but arguably as many misfires and more than a few horror stories. The transhumanists are right that our time calls for bold action, not for clinging to the status quo, but that is precisely because the status quo is already destabilized by our past and ongoing technomoral failures. Before we push on blindly as we have, we ought to heed della Mirandola's warning that humanity's creative destiny is not to be taken lightly, but embraced only with the greatest possible virtue, lest we spoil our opportunity and allow ourselves to become "animals and senseless beasts":

> Above all, we should not make that freedom of choice God gave us into something harmful, for it was intended to be to our advantage. Let a holy ambition enter into our souls; let us not be content with mediocrity, but rather strive after the highest and expend all our strength in achieving it.[24]

Let us not forget that proposals to chemically tinker with our moral faculties, to replace neurons with silicon, or to modify the content and emotional tone of our memories, all carry some risk of rendering us, if not "animals and senseless beasts," at least something *even less* noble than our present selves. For example, it is tempting to embrace proposals to enhance the brains of soldiers so that they are less vulnerable than ordinary humans to post-traumatic stress. If successful, we could spare soldiers often crippling psychological and moral injuries strongly associated with increased risks of suicide, substance abuse, anxiety, rage, and depression. Such injuries tear apart individuals, families, and communities every day, around the globe. If we have the technological means to shield soldiers' minds from such injuries, it would seem inhumane not to use them.

And yet, to dampen the moral horrors of war for individual soldiers would also dampen these same horrors on the civic level, where decisions to make war are collectively made, endorsed, and funded. Historically, our wish to spare our sons and daughters the grave physical and mental harm that comes with war has served as one of the strongest impediments to casual warmaking. As with the robotic means of warfare discussed in chapter 9, lowering the psychological barriers to waging war could have devastating results around the globe.[25] How should a moral society, in a world where war is sometimes necessary to restrain genocide or similar evils, resolve this tension? How many political leaders, biomedical researchers, or transhumanist scholars can you name whom you are confident could resolve this profound moral dilemma wisely, at this very moment? What about your fellow citizens—how confident are you that a popular vote today would yield the right result? How confident are you that *you* can discern the right result?

In such a case, and many others like it that arise from various human enhancement proposals, any virtuous response has to begin from a stance of technomoral *humility*. There are few easy answers to be found, and the consequences of choosing wrongly are likely to be among the most devastating and potentially irreversible that we face. Yet we also need technomoral *courage* to act; even rejecting an enhancement is a choice for which we will be accountable. We cannot evade moral responsibility with passivity or self-imposed ignorance. In the example of enhancing soldiers' mental and emotional resilience, a virtuous response must meet demanding conditions: it must embody appropriate *empathy* for all those who suffer from war, actively take *care* of those for whose welfare we are most responsible, display *civility* in reaching a decision cooperatively with other local and global stakeholders, and attain a moral *perspective* in which the moral meaning of our global human and planetary situation is held in view. Any person or body actually capable of such a response would have to take an ambitious stance of *magnanimous* moral leadership with respect to the various medical, commercial, political, and other institutions with the power to produce and distribute such enhancements. All of this together requires the cultivation and exercise of exemplary *technomoral wisdom*.

Even if we adopt della Mirandola's religious verbiage as only a metaphor, how many of us today are filled with such a 'holy ambition,' or embody the deep individual or collective technomoral wisdom needed to justify and exercise it? It was precisely this challenge of steering biomedical enhancement in an era of moral confusion and apathy that led Hans Jonas to lament that, in a cynical postmodern age, "We need wisdom most when we believe in it least."[26] Nor was he the first to judge his own era ill-equipped for moral ambition. George Bernard Shaw apparently despaired of the 'Creative Evolution' celebrated in passages of *Man and Superman* ever being realized in an age given to moral laziness, waste, and senseless destruction.[27] Even Nietzsche's Zarathustra, hardly a voice of conservative caution, warned his disciples against liberating themselves from the constraints of society's given norms without a clear vision of the future for which they would liberate themselves, and the virtue to actually realize that vision:

> You call yourself free? Your dominant thought I want to hear, and not that you have escaped from a yoke. Are you one of those who had the right to escape from a yoke? There are some who threw away their last value when they threw away their servitude. Free from what? . . . But your eyes should tell me brightly: free *for* what?[28]

Historically, then, the argument from hubris is not owned by timid defenders of the status quo. It also reflects the insight that visionary human projects, especially those that involve great risk to ourselves and others, require *actual* visionaries—people with a clear vision and the considerable moral and intellectual virtue needed to realize that vision with others. Thus the greatest barrier to successful human enhancement is not the dubious technical means often proposed to achieve a posthuman future,[29] nor is it our inability to anticipate all of the effects or challenges we will face along the way—that is, the inherent *technosocial opacity* of the project.[30] Challenges will declare themselves and be overcome, or not, as our ingenuity and resources permit. No, of all the difficulties transhumanism faces, the real problem is *knowing what it is that we ought to wish for*. It is not that transhumanists simply wish for the wrong things. Rather, the libertarian philosophies that pervade the transhumanist community seem to preclude them from wishing for any clear ends at all, only the widespread availability of certain technological *means*, to be used however free individuals and groups see fit.

Consider Nietzsche's warning once more. We hear much from transhumanists about what they want to free us *from*: sickness, aging, death; the limitations of our bodily form; the tyrannies of entropy, space, and time. But do we hear as clearly what they want to free us *for*? They offer a host of possible scenarios for our consideration, but these form no comprehensible whole. We are told we might choose to constantly renew and rebuild our human bodies, or switch them out for non-biological or virtual ones; we might choose to adopt new bodies and personalities at will or form a global 'hive' mind. We might chase immortality or we might just choose to live to a modest age and pull the plug. We might use these technologies to make ourselves more emotionally stable, more emotionally free and capricious, or to enjoy emotions inconceivable in our current state; to create better lives for our children, or to shed biological reproduction as an outmoded relic; to make the Earth whole again, or to abandon the Earth for the stars.

The problem is not the unforeseeable technological possibilities, but a problem that defines our time and our world: in a neoliberal age, who among us has the courage and genuinely magnanimous moral leadership to point the way to a positive vision of the human future? Among the transhumanists we find many negative visions, by which I mean visions of liberation from constraint. In a posthuman future one can choose anything, any physical form or mental constitution, as long as the technological means and social opportunity exist. The posthuman future is one of free play, experimentation, a blank canvas upon which individuals can realize their personal visions of the good without interference from government or their fellow citizens.[31] But this vision itself has little or no positive substance; it is a moral ghost.[32]

Transhumanist and science fiction writer Ramez Naam tells us that, although we may not be able to imagine how our descendants will use human enhancement technology, "the lesson of history is that they will use the powers we provide them to make their world a better place."[33] One can debate the extent to which history really warrants this conclusion. Many would argue that we have used our technological powers to make the world better in certain respects, and much worse in others. Yet the statement is a confession that today's transhumanist visionaries *lack* a coherent moral vision and resist any call to provide one, while still assuring us that human enhancement technologies will indeed yield genuine *enhancements*—that is, that they will be conducive to human flourishing and the good life in general, despite the absence of any clear sense of what this involves. Transhumanists wish to let the inheritors of this technology "chart their own course."[34] But if we do not trust *ourselves* to reliably chart a wise course, what warrants such confidence in the ability of our descendants, the inheritors of our culture and our wisdom, to do so? What *could* warrant this confidence, *except a widespread human recommitment to the practice of moral self-cultivation, embodied in a new culture of global technomoral education?*

The standard transhumanist reply is to reject the demand that any one course be charted. Naam's book tells us that "We will not all opt for the same changes. . . . Different men and women, different communities, different ideologies will all select different goals to work toward. . . . Humanity will expand, splinter, and blossom."[35] Yet this claim betrays itself. We are told not merely that we will expand and splinter, but that we will "blossom"—this is a normative metaphor that presupposes flourishing in the achievement of some appropriate state or end. Certainly the metaphor is compatible with a wide *range* of such ends, but still there must be criteria for what qualifies as 'blossoming' and what does not. Not all change or growth in living systems amounts to flourishing. Some change is *rot. Some growth is malignant.* What justifies Naam's faith that the free and undirected pursuit of enhancement is more likely to promote human 'blossoming' than the alternatives? As he goes on, the incoherence becomes even starker. In the same passage he adds that though our future descendants might be so different "in ways we cannot imagine" that we would fail to recognize them, we may be reassured that nevertheless they "will have the traits most dear to us."[36] What could *possibly* ensure that result, given what has otherwise been said?

Similarly, famed inventor and futurist Ray Kurzweil, known even among transhumanists for extraordinarily exotic visions of the future involving cyborgs and uploaded minds, assures us that our transition to the next phase of existence will be guided by (unspecified) "human values." He tells us that our nonbiological descendants "will be even more humanlike than humans today" and will exhibit the "finer qualities of human thought to a far greater degree" than do we; that they will cherish knowledge above all while enjoying the "full range of emotional and spiritual experiences that humans claim to have."[37] It is unclear why a transhumanist calling for radically experimental transformations of human minds and bodies under conditions of profound technosocial opacity would anticipate, or even desire, such constancy of values and experience.

This tension recurs again and again in transhumanist writings. Ronald Bailey tells us that this century "will offer an ever-increasing menu of life plans and choices" which should allow people maximal freedom "to pursue the good and the true in their own ways," without the burden of any prespecified normative ideal.[38] Yet he tells us that "the future will see miracles, cures, ecological restoration, vivid new art forms, and a greater understanding of the wellsprings of human compassion."[39] Set aside the question of how we are to evaluate this as a positive outcome without some normative ideal of a good life. Even if we accept that the future so described is a good one, what justifies our confidence that it will be the actual terminus of the transhumanist project? Bailey assures us that "history has shown that with vigilance—not with blanket prohibitions—humanity can secure the benefits of science for posterity while minimizing the tragic results of any possible abuse."[40] Is this really what, say, the history of environmental policy in the modern era tells us? If not, then why should one be confident that 'ecological restoration' will be a priority of an enhanced human society? Imagine that new technology greatly reduces the burden of living in a degraded environment—say, by using nanoscale robots to repair cell damage from radiation and toxins, or by using clouds of linked nanobots to form a "utility fog" that cosmetically disguises unpleasant surroundings, or enabling colonization of other planets, or by providing a host of pharmacological tonics to compensate for the lost psychological benefits of pristine wilderness.[41] Might ecological restoration be among the *lowest* priorities of such a society?

Transhumanist writings seem to assume the unobtrusive operation in technosocial life of some sort of 'invisible moral hand,' an assumption that is neither made explicit nor defended. And yet, even without enhancement, our species already faces a number of grave perils, from extinction risks to 'merely' a miserable and hopeless existence for the vast majority of the human family. Nick Bostrom is right that stasis is probably not the safest bet.[42] If enhancement *can* boost our cognitive and moral resources to confront these perils, along with the inevitable challenges that will result from enhancement itself, then perhaps throwing our resources behind the development of such technologies is the least risky alternative. An 'enhanced' future might not be any better than ours, but then again, an unenhanced future might not be very good either. As Coady notes, conservatism can be its own kind of hubris.[43] Still, is a technologically aggressive but morally blind 'Hail Mary' pass *really* our best option?

Inaction is impossible; given the risks of humanity's current status, resisting innovation requires moral wisdom to justify it too, and if we lack the virtue to safely push ahead, we also lack the virtue to safely stay put. *We are at an impasse, and the cultivation of technomoral virtue is our only way forward.* While the technological means for human enhancement are already emerging, we lack the moral means today to reliably answer the question of whether and how it can be done ethically. We are not ready to answer, though as Hans Jonas warned us, we should have made ourselves ready long ago. The challenge we face today is not a moral dilemma; it is rather a moral *imperative,* long overdue in recognition, to collectively cultivate the technomoral virtues needed to confront this and many other emerging technosocial challenges wisely and well.

Does this mean that the only prudent course is to put all human enhancement research on indefinite hold until we manage to jump-start a massive global reinvestment in the practices of moral self-cultivation and education for technomoral virtue? Not necessarily. We have noted that forgoing any given technological development carries certain risks just as does pushing forward. Technomoral humility, courage, perspective, and wisdom all entail a skilled disposition not to *avoid* but to *confront* and intelligently *manage* the unavoidable moral risks of technosocial life. In their 2014 book *Unfit for the Future: The Need for Moral Enhancement,* philosophers Ingmar Persson and Julian Savulescu argue that forgoing one particular type of human enhancement research—in particular, research aimed at enhancing the *moral* capacities of human beings—may greatly amplify the existential risks to human flourishing *already* posed by the hitherto glacial pace of our species' collective moral development.

The technomoral failures of existing and past human generations have already placed the flourishing of future generations in grave jeopardy. How much more time can we safely allow ourselves to wise up by traditional means of moral self-cultivation, before we attempt more drastic measures to artificially accelerate our moral maturation as a species? Persson and Savulescu argue that we may have very little time, and that biomedical and pharmacological means of enhancing and extending our capacities for human empathy, justice, and other moral traits, if successful, could hasten our species' collective moral development, perhaps enough to greatly reduce the danger of our self-destruction.[44]

This will not be easy, since the lion's share of the planet's technological and scientific resources are controlled by liberal societies operating in cultures focused on short-term private gains, alienated from any shared conception of the ultimate moral goods to be realized by collective action in the long-term interests of the human family. Even the notion that there *are* such goods—as opposed to the immediate, subjective, and often arbitrary preferences of private individuals—is no longer readily granted in many such cultures. This ethical alienation from a vision of our own higher flourishing was described by the 20th-Century Spanish philosopher José Ortega y Gasset as a uniquely modern crisis that inhibits the intelligent and creative use of technical power.[45] This crisis must at least begin to be mitigated by new and vital programs of technomoral education and habituation before any part of the transhumanist vision can be wisely managed. Thus the active pursuit of human enhancement technologies, including moral enhancement, is probably indefensible without a concurrent cultural movement to invest significant resources and energies in the wider cultivation of technomoral virtues and leadership.

Such a commitment represents humanity's best hope for living, and living *well,* in the 21st century and beyond. The human family is not yet up to the moral task it is facing: finding a way to wisely manage *together* the increasing global complexity, instability, plurality, interdependence, rapid change, and growing opacity of our technosocial future. *This* book recommends a classical solution: an energetic (perhaps even desperate) collective effort to reinvest our cultures in the habits of moral self-cultivation and education for technomoral wisdom. Notably Persson and Savulescu, although they are proponents of human enhancement, endorse the very same effort. They write, "We are not envisaging that moral bioenhancement will ever reach a point at which traditional methods of moral education—or other social strategies like institutional redesign using incentives—will ever be redundant . . . we think that these methods will need to be used well and, indeed, that they should be employed more extensively than they are today."[46]

Given, however, that traditional methods of moral education and self-cultivation *alone* may not work well enough, or fast enough, to save us from the shortsighted and selfish habits that have already brought our planet to the edge of ruin, they conclude that biomedical technologies of moral enhancement, should they become available, may be necessary and prudent accelerants. They could amplify and sustain the impact of traditional moral practices and character education on our habits and virtues, especially the virtues of care and empathic concern. While the technical path to effective moral enhancement is not well marked, and the obstacles to it hard to foresee, Persson and Savulescu note that "the predicament of humankind is so serious that all possible ways out of it should be explored."[47]

Yet in the absence of new cultural reinvestments in moral education and practice adapted for contemporary technosocial life, the *wise* use of moral enhancement is unlikely, given that we need it precisely *because* moral wisdom among our species is in such dangerously short supply. Persson and Savulescu thus call for a new educational dedication to "what might be called a 'science-sophy,'"—that is, a new body of moral wisdom about appropriate and justified uses of scientific and technical power.[48] The account of technomoral wisdom developed here is one small step toward such a philosophy, one that must evolve *together* with emerging technologies in ways that will inevitably challenge existing neoliberal norms of technology development. Short of a long string of dumb luck or unearned grace, this path is our species' best, and likely only, hope for continued flourishing.

By philosophical *and* technological means, then, technomoral virtues such as self-control, courage, empathy, justice, care, flexibility, civility, and perspective must begin to be explicitly targeted for promotion by researchers, educators, parents, and policymakers. Such efforts will require private and public funding, sometimes at the expense of other possible paths of educational and technological development. Among the paths that must be sacrificed are human enhancements that would impede our cultivation of technomoral virtue: enhancements that would weaken our capacities for self-control and moderation, generate group solidarity even in the absence of justice or care, or reduce empathic concern by blunting our emotional response to suffering. Each of these enhancements could be desirable to many for psychological, commercial, or political reasons. Yet if freely but widely chosen in an unrestricted market, they would preclude the very flourishing that transhumanists promise us in a posthuman world.

Indeed, to insist upon neoliberal, free-market norms for human enhancement or any other new technosocial power is to assume either 1) that our descendants will, without any wise guidance from us, *magically* acquire the moral convictions we lack to build a future worth choosing; *or,* 2) that our progeny will chart their own arbitrary technosocial course more or less as we have, without the benefit of any greater moral wisdom or insight than we enjoy today. The latter world will be what it is, but cannot be expected to be any *better* than our own, and given the depleted resources and compromised environment that we will have left them, it is quite likely to be worse.

10.3 Technomoral Virtue and Contemporary Life: A Crisis of Moral Wishing

Before we close this chapter, let us reflect upon the words of José Ortega y Gasset. His 1939 essay "Meditación de la Técnica" (known in English as "Man the Technician") expresses his existentialist conception of human life as *autofabrication*: a project of self-realization, bringing into being "the aspiration we are."[49] For Ortega y Gasset as for later existentialists, the freedom of human choice means that a human person is not a *thing*, natural or otherwise, but "a project as such, something which is not yet but aspires to be."[50] To those who advocate relinquishment of the power to enhance ourselves, Ortega y Gasset might have replied that "in the very root of his essence man finds himself called upon to be an engineer. Life means to him at once and primarily the effort to bring into existence what does not exist offhand, to wit: himself . . . "[51]

Yet he tempers his enthusiasm for human creativity and freedom with a warning about technology and virtue. He notes that for a human being, flourishing is the only subjective necessity, and she will often place her organic life in peril in order to pursue it. Of course, conceptions of human flourishing are not fixed, but individually and culturally plastic. Yet this very plasticity exposes a problem at the heart of our technosocial aspirations. For technology is not an end in itself; nor is it merely a means to the preservation of organic life—were that its ultimate aim, advanced technology would be quite redundant, a duplication of more primitive mechanisms for survival that we already possess. Rather, technologies are instruments whose ultimate ends must exist outside the technological sphere. As Ortega y Gasset puts it, ". . . technology is, strictly speaking, not the beginning of things. It will mobilize its ingenuity and perform the task life is; it will—within certain limits, of course—succeed in realizing the human project. But it does not draw up that project; *the final aims it has to pursue come from elsewhere.* The vital program is pretechnical."[52]

The unresolved crisis of the 20th century, still with us in the 21st, is a crisis of meaning—the meaning of human excellence, of flourishing, of the good life. The transhumanist project pushes this crisis to its absolute limit. What is the good that radical life-extension will serve? More time, obviously. But time for what? Not simply to live, for I desire more than that to live *well.* What will maximal detachment from my bodily needs and limitations afford me? More freedom, certainly. But to echo Nietzsche: freedom *for* what? Freedom to *do* what? Ortega y Gasset tells us that our humanity rests entirely upon the "to do" of projected action, and hence "the mission of technology consists in releasing man for *the task of being himself.*"[53] But who do we want to be? Is our vision of this task any clearer than it was for our classical or Enlightenment predecessors? Or is it in fact more clouded than ever?

It is here that contemporary technosocial culture, especially in the wealthy nations driving global technology innovation, rings hollow at its core. We are showered with an endless and ever-deepening deluge of choices. The distinctions between the thousands of cellphones on Amazon, the millions of apps in the App Store, the hundreds of kinds of cereal in an American supermarket, or the dozens of gourmet coffee presses reviewed on a 'lifestyle blog,' are often virtually impossible for us to identify, much less evaluate. With such choice comes not happiness, but paralysis of the will, the haunting fear of a suboptimal selection. While Ortega y Gasset did not witness technosocial desire as it is today, his words still resonate for us:

> Desiring is by no means easy. The reader need only remember the particular quandary of the newly rich man. With all wish-fulfilling means at his command he finds himself in the awkward situation of not knowing how to wish. At the bottom of his heart he is aware that he wishes nothing, that he himself is unable to direct his appetite and to choose among the innumerable things offered by his environment. He has to look for a middleman to orient him. And he finds one in the predominant wishes of

> other people, whom he will entrust with wishing for him. Consequently, the first purchases of the newly rich are an automobile, a radio, and an electric shaver. As there are hackneyed thoughts, ideas which the man who thinks them has not thought originally and for himself but blindly and automatically, so there are hackneyed wishes which are but the fiction and the gesture of genuine desire.

Ortega y Gasset's point is not the easy condemnation of shallow consumerism. In his view, the latter is merely the shadow cast by a malady of far greater significance:

> If this happens in the realm of wishing with objects which are there and lie to hand before they are wished for, one may imagine how difficult the properly creative wish must be, the wish that reaches out for things yet nonexistent and anticipates the still unreal. . . . If a man is unable to wish for his own self because he has no clear vision of a self to be realized, he can have but pseudo wishes and spectral desires devoid of sincerity and vigor. It may well be that one of the basic diseases of our time is a *crisis of wishing* and that for this all our fabulous technical achievements seem to be of no use whatsoever.[54]

For Ortega y Gasset, the emergent technosocial crisis is a vacuity of moral imagination for life, and its chief symptom is an "appalling restlessness" that manifests itself in frenzied but directionless seeking.[55] This "crisis of wishing" is a culturally-induced deficiency of practical wisdom, the absence of authentically motivating visions of the appropriate ends of a human life. If Ortega y Gasset was right, then in the absence of some deliberate intervention, contemporary technosocial life is likely to be marked by a progressive paralysis of practical wisdom, in which our expanding technical knowledge of effective means receives less and less direction by meaningful desires and moral ends. Like nerve cells gradually cut off by a neurodegenerative disease from their directing impulses in the brain, a technosocial crisis of wishing would result in actions that appear increasingly spasmodic, uncoordinated, and lacking in purpose.

Ortega y Gasset's diagnosis is an empirical claim; it stands or falls based upon the social, psychological, and moral phenomena it predicts. The prediction is clear—that humans in the technologically developed world, powerful and marginalized alike, will increasingly find themselves rudderless and impotent to direct their own lives or to discover their own ends, and in desperation will gradually turn to the compulsive consumption of ends manufactured for them by mass technosocial culture, consumption that itself points to no further ends or aims, and in satisfying only 'spectral desires' leaves them ever searching for more. The reader may judge for herself the extent to which this diagnosis has been confirmed.[56]

My aim in this book has been to shine a light on this all too real disease: a widening cultural gap between the scope of our global technosocial power and the depth of our technomoral wisdom. Throughout the book, culminating with this chapter, I have suggested the only plausible first step toward a cure: to convene new institutions, communities, and cultural alliances in the service of global technomoral cultivation. This will require intense, cooperative, and sustained human efforts, many of them on a worldwide scale. Yet such efforts are not without precedent in a species well acquainted with fighting massive, protracted world wars for far less happy gains. A long-overdue commitment to the cultivation of technomoral habits and virtues may be the human family's only real chance for not merely continuing to live, but live *well* in this century and those to come. If we act now, while there is still time and hope, we may at last discover what kind of human, or posthuman, future is worth wishing for.

Epilogue

THE EMERGING TECHNOLOGIES explored in Part III underscore the immense challenge of predicting and securing the future of human flourishing in the 21st century and beyond. From biomedical enhancement to social robotics and artificial intelligence, technology is opening up a seemingly infinite number of new trajectories for our species, and those with whom we share our planet. Some of these possible futures look quasi-utopian. Most are the usual mixed bag of blessings and curses, but a few are truly catastrophic for humanity—and it is not only technophobes and neo-Luddites who are worried about those scenarios. Prominent voices among the technoscientific cognoscenti, from Stephen Hawking to Elon Musk to the worldwide community of climatologists, are already issuing urgent warnings concerning the risks posed by mass extinctions, dramatic climate shifts, artificial intelligence, and other potentially 'existential' threats to human survival and flourishing. Similar concerns haunt the popular imagination, as we can see mirrored in the exploding number of apocalyptic scenarios in film, television, and literature.

Are the voices that warn of us a coming crisis alarmist, or do we have actual reasons to worry? Some of the rhetoric is almost certainly overblown. Those who are predicting an imminent 'rise of the robots' or an 'AI singularity,' in which artificially intelligent beings decide to dispense with humanity or enslave us, in my view serve as an unhelpful distraction from the far more plausible but less cinematic dangers of artificial intelligence. These mostly involve unexpected interactions between people and software systems that aren't smart *enough* to avoid wreaking havoc on complex human institutions, rather than robot overlords with 'superintelligence' dwarfing our own.[1] Other existential risks, such as environmental devastation, nuclear or biological warfare, and massive cosmic impacts are very, very real, and more importantly, the magnitude of those risks depends greatly upon what we choose to do today with our technosocial power. Even with near-earth objects such as asteroids, the paths of which we do not presently control, the existential risk they pose can conceivably be mitigated by wise investment in science and technology, from improved detection systems to improvised devices to alter their course. Likewise, the existential risk of environmental devastation, resource depletion, and food chain collapse on a planetary scale will depend greatly on how we choose to confront the technosocial challenges of clean and renewable energy, ocean acidification, and a rapidly warming planet.

We may or may not be the first intelligent species in the universe to face such crucial tests. Unfortunately, there is reason to believe that if we are *not* the first, most of those who have gone before us have failed. I'm speaking here of something known as "Fermi's Paradox" and the hypothesis of the "Great Filter." Fermi's Paradox, posed by the physicist Enrico Fermi in 1950, refers to the curious absence of detectable signals from alien civilizations in a universe that appears to host so many suitable planets for intelligent life, and is, in the words of one contemporary astronomer, "bulging at the seams with ingredients for biology."[2] One commonly proposed solution to Fermi's Paradox is a hypothesis that is deeply disturbing, precisely because it is so intuitively plausible: the possibility that the universe is silent because technologically advanced species typically don't last very long before they self-destruct—for example, by draining and toxifying their nonrenewable resources through uncontrolled industry, or by developing weapons of planet-killing power. This hypothesis holds that humanity may be quickly approaching a "Great Filter," one that may have prevented other intelligent civilizations in the universe from surviving their own sudden explosions of technosocial innovation.[3]

A "Great Filter" need not be of our own making—some of the things that could tend to 'filter out' advanced species from the galaxy, such as large asteroids, are simply markers of our universe as a dangerous place. However, the scenario in which advanced species tend to bring a Great Filter upon themselves is plausible precisely because the only instance of a technological civilization that we have as a data point, our own, has already been endangered in precisely these ways. Historians of the Cold War have documented multiple occasions during that period in which the likely devastation of the human race by intercontinental nuclear bombardment was narrowly prevented, not by advanced technological controls and failsafes, but only by the extraordinary prudence, courage, and critical judgment of a few people, or in some cases, just one.[4] Later, international treaties successfully brought at least a temporary pause to the nuclear arms race, but the existential risk of a global nuclear holocaust endures, and on some accounts, is again increasing.[5] Unfortunately for us and for the planet, global environmental devastation, mass extinctions, and resource depletion cannot, unlike an impending nuclear launch, be called back by the virtue of one or two wise souls in key positions of responsibility. They can only be halted by the cooperative efforts of millions, even billions of humans willing to countenance the sacrifice of significant short-term positional gains in wealth and power in the interests of the long-term survival of the species. Nor is survival our sole concern. Even if our descendants have the good fortune to make it through the technosocial growing pains of the next few centuries, what are their chances of not merely living, but living *well*?

All of this underscores the claim articulated in this book—that better technical systems *alone* will not secure the future of human flourishing. Only with the broader and more intensive social cultivation of technomoral virtues such as wisdom, courage, and perspective can this aim be accomplished. Regardless of how high or low one estimates the existential risks to humanity to be, reason compels us to lower that risk if we can. Yet it would be the height of foolishness to attempt to lower that risk by a reactionary campaign against further scientific and technical developments, for these innovations in knowledge and practice offer us new ways of preserving and enhancing human flourishing, not just tools to destroy it. Consider the significant natural risks to humanity posed by asteroids, massive radiation bursts from the sun, and the mutation of dangerous flu strains; then ask yourself if you want your children to forego the technoscientific innovations that will be necessary to manage and mitigate those future dangers.

In any case, José Ortega y Gasset was right; the human family may be many things, but in essence we are a family of engineers. We engineer *ourselves*, in various modes: science and craftsmanship, but also humanistic education, art, literature, music, spiritual practice, physical exertion, and of course, the practice of philosophy. Over the millennia these modes have produced many new tools and many visions of the future; today it happens to be the tools of the digital era that shape most powerfully our sense of what is possible for the human family. Yet this book concerns the one tool that humanity *must* finally master, if we and those with whom we share our world are to have any solid hope of living well in the 21st century and beyond. This tool is not an artifact of silicon but of cultural wisdom; its repeated invention is documented in the histories of moral practice that are the legacy of philosophical cultures around the globe, including those explored in this book. Now we must use it to hammer out a new culture of technomoral virtue, in which human individuals, families, communities, and institutions consciously work to inculcate the specific moral skills and capabilities that intelligent life needs in order to responsibly direct and wisely manage the use of technoscientific power.

If we estimate the likelihood of humanity successfully meeting these challenges by looking at our recent record, we will be given pause—indeed, many of the environmental problems we face today are the result of a record of exceedingly poor management of our technoscientific powers, in ways that repeatedly underestimate their risks and discount the value of the future, prioritizing modest short-term gains, often for the wealthy few, over long-term harms to the home that we and future generations must all share. Fortunately, humans are not doomed to repeat the mistakes of the past. Our cognitive powers of reflection, imagination, creativity, and judgment, not to mention our capacities for moral ambition, empathy, and hope, have already enabled us to resist many of our most cruel and self-destructive impulses. We still live in a world where fear, hatred, and violence based on race, sex, gender, religion, and nationality are rampant—but we no longer live in a world where these evils are uncontroversial, and in many countries both the laws and the will of the majority now oppose them. We still live in a world with nuclear weapons—but no longer one in which these instruments of the apocalypse proliferate with impunity. We still live in a world with poverty, disease, and famine. But we no longer live in a world without tools to fight them, and scientific, political, and economic advances can deliver even more powerful tools in the future. We also know that humans are capable of massive and sustained collaboration in immensely complex and challenging tasks. Consider the construction of the Large Hadron Collider in Switzerland, the largest machine on the planet, built over ten years by a cooperative effort of 10,000 scientists from more than one hundred different countries. Is it beyond our imagination to think that cooperative efforts of this kind, on even greater scales, are possible if the stakes are the survival of meaningful human life on this planet?

Even if most other intelligent civilizations in the universe have failed to meet this challenge, we need not despair at our prospects of doing so. Our species enjoys a long, rich, and diverse cultural history of successfully harnessing and disseminating our creative power, our capacity for philosophical insight, our nobler ambitions, and our sheer will to flourish, with astounding results—from space flight, to vanquishing polio and smallpox, to building global networks and institutions dedicated to higher learning and moral and spiritual practice. So there is cause for much hope—but not much comfort, for there is a lot of work for us to do if we are to equip ourselves for the existential challenges that lie ahead. And as we all know, we humans are chronic procrastinators, especially when the hard work that has to be done today will only pay off—if it pays off at all—in a distant tomorrow. This may be the hardest weakness for us to overcome. *Can* we cultivate the will, and the wisdom, to do today what is needed for tomorrow?

For human beings, nothing is written in stone; yet in a perplexing irony, it is often those most welcoming of technosocial innovation who succumb to the false belief that present patterns of moral, economic, and political practice are permanent fixtures, rather than what they are—malleable cultural habits with a long history of adapting to changing social conditions.[6] While the funhouse mirrors of internet comment boards, nationalistic politics, religious extremism, and corporatized mass media produce a distorted image of the human being as incorrigibly selfish, petty, violent, paranoid, and foolhardy, most humans have always made an effort to be something far better than that, using millennia-old habits of moral and intellectual cultivation encoded in virtually every stable culture. Updating these codes for technosocial life in the 21st century, and renewing our attention to their importance for our survival and flourishing on 'Spaceship Earth,' is not an idealistic fantasy—it is a practical necessity staring us in the face, and a task for which our history offers important cultural resources to be reclaimed and renewed, as I have suggested in this book. As important as they are, the engineers of the 21st century who fashion code for machines are not as critical to the human mission as those who must fashion, test, and disseminate *technomoral* code for humans—new habits and practices for living well with emerging technologies.

Thus instead of seeking solace in the post-apocalyptic fantasies of the cultural present, which express a yearning for a global calamity that will press the 'restart button' for humanity and erase our past mistakes along with our triumphs, I urge the reader to embrace a far more courageous hope and ambition, one embedded in another popular entertainment vision of the future—*Star Trek*. In the words of its philosophical and creative voice, Gene Roddenberry:

> *Star Trek* speaks to some basic human needs: that there is a tomorrow—it's not all going to be over with a big flash and a bomb; that the human race is improving; that we have things to be proud of as humans.[7]

In the original series, the humans of the 23rd century repeatedly reference a critical leap in *moral* development made by 21st–22nd century humans; a *cultural* transition that enabled their narrow escape from self-destruction in internecine wars fueled by new technoscientific powers. In the future Roddenberry envisioned, humanity passes through its Great Filter not by inventing warp drives and transporters, nor by enduring a global apocalypse that erases our weakened cultures and broken institutions, but by consciously cultivating the technomoral virtues needed to improve them: the self-control, courage, empathy, civility, perspective, magnanimity, and wisdom to make humanity worthy of its greatest technoscientific aspirations. Such a future has not been promised to us; but it is the only future worth wanting.

Notes

INTRODUCTION

1. Hawking et al. (2014).
2. See the July 28, 2015 open letter on autonomous weapons at the website of the Future of Life Institute (www.futureoflife.org/ai-open-letter/) with which Tegmark, Hawking, Musk, Russell, and Wilczek are now associated. The website states their aim as working to "mitigate existential risks facing humanity," while "focusing on potential risks from the development of human-level artificial intelligence." Their broader mission: "To catalyze and support research and initiatives for safeguarding life and developing optimistic visions of the future, including positive ways for humanity to steer its own course considering new technologies and challenges."
3. Much contemporary philosophy of technology reflects this view; see Verbeek (2011) for one influential account of how technologies and human social practices are co-constitutive. In this book the adjectives *technosocial* and *technomoral* reflect this perspective.
4. de Waal (2007), 23.
5. The reference here is to the widespread emissions software fraud by Volkswagen publically exposed in September 2015.
6. The ubiquity of the opposite complaint is highlighted by Turkle (2011), 167.
7. Post-industrial visions of a technocratic future are found in the works of Henri Saint-Simon, Auguste Comte, Thorstein Veblen, and John Dewey, to name just a few.
8. See Packer (2013).
9. Ray Kurzweil (2005) is the radical standard-bearer for the former, techno-optimist camp of futurists; Bill Joy (2000) is often cited as the contemporary voice of techno-pessimism. Garreau (2005) gives a thorough account of these competing visions, along with several alternatives.
10. See Allenby and Sarewitz (2011) for a related claim.

11. Kant ([1785] 1997). The rule has another well-known formulation, which forbids treating rational beings as mere means to our own ends. Though more helpful in certain cases, this too fails to address the technomoral complexities of a world in which, for example, an intelligent, autonomous robot or software agent challenges our concept of a 'rational being.'
12. Persson and Savulescu (2012), 46.
13. Lewis et al. (2014).
14. Mill ([1861] 2001), 23–25.
15. Although I reject Martha Nussbaum's view that the label of "virtue ethics" is so broad as to be vacuous, in this book I take her advice that we who wish to make use of the rich practical resources of virtue traditions not get wrapped up in the art of defining our unique kind, and get on with the business of "figuring out what we ourselves want to say." Nussbaum (1999), 163.
16. Mine, then, is an explicitly *pluralistic* account—one that willingly sacrifices some theoretical unity in exchange for practical cash-value in addressing collective problems of technomoral wisdom. It is also a human-centered account, insofar as I take humans to be the only agents presently capable of deliberating together about the good life; yet an ability to attend and respond to *all* forms of moral worth, including that of nonhumans and the environment, is explicitly identified with the technomoral virtues of empathy, care, and perspective in Chapter 6.
17. These virtues do not function *exclusively* for this purpose; individually, they each serve other ethical functions in the various cultural traditions for which they are meaningful. Yet when cultivated *together* they may also serve a global function in helping us to address a global human problem of 21st century life: namely, how to flourish in a condition of acute technosocial opacity and increasing existential risk.

CHAPTER 1

1. For an excellent analysis of the historical trajectory of this concept in classical ethical discourse, see Annas (1993). I have chosen to bypass the debate about whether classical ethics engages the specifically *moral* domain of modern ethics (see Anscombe 1958 and Williams 1985). However, I share Annas' view (1993, 120–131) that it does.
2. See Boodberg (1953).
3. Keown (1992), 49.
4. Aristotle (1984), *Nicomachean Ethics* (hereafter *NE*), 1103a20-35. This, of course, allows us to distinguish human *virtues of character* from the type of virtue described earlier as possessed by an excellent knife. Furthermore, as noted by Foot (1978), virtue of character entails an agent's habitual *willing* of the good in a sense that involves moral discernment and is not narrowly instrumental. Hence physical health is not a virtue of character even though it can to a significant degree be deliberately cultivated as a rationally chosen state of human excellence.

5. See Aristotle, *NE* 1103a15-35; 1106b15-25; 1144b25. The integration of cognitive, affective, and desiderative states in virtuous dispositions is discussed extensively by McDowell (1998).
6. See Aristotle, *NE* Book VI for his account of practical wisdom and its role in ethics.
7. See McDowell (1998), 39. Complete or 'full' virtue is an aspirational ideal rather than a minimum standard for ethical behavior; thus we may properly refer to *degrees* of virtue, and to its gradual *cultivation* rather than its achievement.
8. *NE* 1106b20.
9. There are ongoing debates among ethicists about the extent to which the virtues can be successfully integrated; how attainable virtue is for the ordinary person; to what extent the cultivation of virtue requires external goods or luck; whether and under what circumstances a person's virtue can coexist with vice; and how to distinguish clearly *moral* virtues such as justice from apparently nonmoral virtues such as cleverness, creativity, and perseverance. For an overview of these debates, see Crisp (1996); Crisp and Slote (1997); French et al. (1988); and Statman (1997). Some of these debates will be directly engaged in this book, while others fall outside the scope of our inquiry. Despite the open questions, most virtue ethicists agree upon the *general* conceptual profile of the term 'virtue' presented above.
10. Cafaro (1997) rebuts the charge that virtue ethics is narrowly agent-focused. Regarding the conceptual linkage between virtue and happiness/flourishing, some virtue theorists deny this *eudaimonist* implication—for example, Michael Slote (1992). Such theorists explicitly acknowledge, however, that in doing so they depart from a core principle of the classical virtue ethical tradition.
11. Aristotle, *NE* 1098b30. Controversies over the translation of *eudaimonia* as 'happiness' have resulted in the term 'human flourishing' being preferred by some; I will use both terms depending on context. On the distinct advantages of each, see Hursthouse (1999), 9–10.
12. Insufficient because, as Aristotle notes at *NE* 1101a, humans are vulnerable to devastating misfortunes from which not even great virtue can provide complete protection, although virtuous persons will bear such misfortunes better and preserve more of their happiness than will the vicious.
13. Nussbaum (1999) rejects this standard description of virtue ethics as a distinct "third approach" in ethics; while she is right to note the rich diversity of contemporary views that fall under this label, her assertion that the use of rule or decision principles in virtue ethics is indistinguishable from that of other approaches is unconvincing.
14. See Louden (1984) and Schneewind (1990). The canonical form of the latter objection appears in Henry Sidgwick's influential *The Methods of Ethics* (1884), in which he claims that at best, Aristotle's ethics "only indicates the *whereabouts* of Virtue. It does not give us a method for finding it" (375).
15. For a careful survey of the first wave of philosophical responses to Anscombe's call, see Pence (1984); also French et al. (1988).

16. MacIntyre (1984); McDowell (1998); Hursthouse (1999); and Annas (2011) are examples of prominent neo-Aristotelian accounts. Baier (1985); Becker (1998); and Swanton (2003) offer accounts of the second sort, while Slote (1992) and Driver (2001) are among the best-known adherents of the third strategy.
17. Hursthouse (1999) provides an overview of and response to such criticisms; see Harman (2000) and Doris (2002) for the situationist critique of the empirical validity of character traits; excellent responses to the situationist challenge are made by Kupperman (2001); Kamtekar (2004); and Annas (2005).
18. See Miller (2014) for a helpful discussion of this point.
19. Badhwar (2014a).
20. For a review of contemporary developments in Confucian virtue ethics, see Tiwald (2010). For examples of these developments, see Yearley (1990); Ivanhoe (2002a); Van Norden (2007); Yu (2007); and Angle (2009). For examples of contemporary work on Buddhist virtue ethics, see Keown (1992); Whitehill (2000); Harvey (2000); Cooper and James (2005); and Bommarito (2014).
21. See Tiwald (2010), 57–58 on the opportunities presented by comparative virtue ethics.
22. Wong (2013).
23. See Wallach and Allen (2009); Lin, Abney, and Bekey (2012).
24. See Annas (1993) and Hursthouse (1999).
25. Ibid.
26. Such claims are made by O'Neill (1990) and Sherman (1997), and applied to computer ethics by Grodzinsky (1999); yet ineliminable differences remain that I will argue constrain Kantian deontology's application to emerging technological practices.
27. The story is recounted in Makos (2012).
28. Yearley (2002), 256.
29. See *NE* Book II Chapters 8–9, see also chapter 5, section 6 in this volume.
30. Nussbaum (1990), 73–74.
31. *NE* 1094b16.
32. See *NE* Book VI, Chapter 13, especially 1144b20-30. See also MacIntyre (1984), 150–152. This commitment to the dependent status of rational principles of ethics distinguishes Aristotle's view from those who grant the virtues an essential place in ethics, but regard their practical content as derivable from fixed ethical principles; see for example O'Neill (1996).
33. Indeed, in their book *The Techno-Human Condition*, Braden Allenby and Daniel Sarewitz (2011) argue that consequentialist and deontological ethics are crippled by the unprecedented and irremediable ignorance of the future that marks this condition—what I have termed *acute technosocial opacity*. Despite their correct diagnosis, Allenby and Sarewitz fail to explicitly acknowledge that the solution they call for is, in fact, not a "reinvented Enlightenment" ethic (187) but a technosocial *virtue* ethic heavily indebted to classical conceptions of practical wisdom.

34. See Bainbridge and Roco (2006), also Khushf (2007), and Nordmann (2004).
35. See Ess (2010a); Vallor (2010; 2012); Volkman (2010); Wong (2012); and Couldry (2013a and 2013b).
36. See anthologies by Scharff and Dusek (2003); Kaplan (2004); and Hanks (2010) for an overview. See also Mitcham (1994).
37. For critical perspectives on first-wave philosophy of technology, and the thesis of technological determinism in particular, see Feenberg (1999); Achterhuis (2001); and Verbeek (2005).
38. Jonas (1979), 28.
39. Jonas (1984), 21.
40. Jonas (1979), 28.
41. Jonas (1984).
42. Ibid., 12.
43. Ibid., 6.
44. Jonas's account is not beyond criticism—for example, he vastly underestimates the extent to which Aristotle saw virtue as a matter of political (and hence collective) concern rather than private interest, and the extent to which management of novel and unpredictable circumstances is an essential component of *phronēsis*. Still Jonas is correct that Aristotle's whole account *as given* is not adequate to our contemporary situation.
45. Jonas (1984), chapter 1.
46. Ibid., 26.
47. Ibid., 203.
48. Borgmann (1984).
49. Ibid., especially chapters 9 and 18.
50. Ibid., chapter 25.
51. Ibid., 233.
52. See Feenberg (1999); Achterhuis (2001); Ihde (2004); and Verbeek (2005).
53. See Brey (2004); Tavani (2005); and Moor (2008). Luciano Floridi's *The Ethics of Information* (2013) is perhaps the best-known candidate for a new technosocial ethic, although as a principle-based and universalist framework it is vulnerable to many of the same objections as other deontological systems, as detailed by Volkman (2010). For a more enthusiastic take on the relationship between information ethics and an Aristotelian ethic of flourishing, see Bynum (2006).
54. See the debate between Ihde (2010) and Scharff (2012) about whether philosophy of technology needs strong normativity.
55. Such a person would *also* probably be recognized as vicious in other historical and cultural settings; hence a new framework for technosocial virtues will not be wholly discontinuous with traditional conceptions of virtue, differing mainly in its prioritization and interpretation of the virtues. This is also why such a framework can resonate even with those who remain culturally wedded to, say, traditional Confucian or Buddhist values.

56. See Oakley and Cocking (2001); Sandler and Cafaro (2005); Cooper and James (2005); Walker and Ivanhoe (2007); and Couldry (2013b).

CHAPTER 2

1. To give just one example, the important Aristotelian virtue of *megalopsuchia,* which entails a (deserved) sense of moral superiority to others, is thought by many to be inconsistent with modern liberal norms. But see Kristjánsson (1998), who claims that this in fact tells against modern liberal conceptions of morality. See also chapter 6, Section 11 in this volume.
2. MacIntyre (1984), 159.
3. See MacIntyre's powerful critique of Aristotle on this point in *Dependent Rational Animals* (1999).
4. Closely related views are found in the *NE's* likely predecessor, the *Eudemian Ethics*, as well as his *Politics* and the *Magna Moralia*, the authorship of which is contested. For contemporary analyses of Aristotle's ethics, see Annas (2011) and Hursthouse (1999).
5. For excellent English-language analyses of Confucian virtue ethics, see Ivanhoe (2000) and (2002a); Van Norden (2002) and (2007); and Yu (2007).
6. This is not entirely the case for neo-Confucianism, a robustly influential school of Chinese thought marked by strong Buddhist influences that significantly transformed the classical core. However, our focus in this book will be on the classical form of Confucianism.
7. Yearley (1990), 71.
8. Yu (2007), 82–83.
9. On 'semblances of virtue,' see Yearley (1990), 19, 124–125.
10. See Yu (2007), 101–102, 124–125 for a fuller discussion of the contrasts between Aristotelian and Confucian ethics.
11. See Yearley (1990), 38–41 and Yu (2007), 150–152.
12. See Yearley (1990).
13. For English-language accounts of Buddhist virtue ethics, see Keown (1992); Harvey (2000); Cooper and James (2005); and Wright (2009).
14. Buddhism's denial of an enduring, substantive self as a metaphysical entity remains entirely compatible with the existence of an ethical self whose habits and beliefs can be cultivated over time.
15. Gethin (1998), 82.
16. See Wright (2009) on the importance of character cultivation in Buddhism.
17. Cooper and James (2005), 83.
18. The qualification 'might' here is essential, as the virtuous person's decision would result from the application of practical wisdom to the salient features of the particular situation, as opposed to being decided at a higher level of abstraction by uncritically applying general principles such as 'always honor your father and king' or 'show equal compassion for the suffering of strangers.' One could imagine such

a dilemma with special circumstances that motivated a virtuous Buddhist to honor kin preference, or a virtuous Confucian to oppose his corrupt king's cruelty to citizens of a neighboring kingdom.

19. See Kristjánsson (1998).
20. On Confucianism as a virtue ethic, see Ivanhoe (2002a), 4–10; Van Norden (2007), 15–23; and Yu (2007), 226 ff. 5, where he notes that "the virtue ethics approach to Confucianism has become prevailing in recent years." For similar responses with respect to Buddhism, see Keown (2002), 193–227; Harvey (2000), 49–51; and Cooper and James (2005), 85–89.
21. See Van Norden (2007), 17, 20 on the roots of this distinction and its use in comparative ethics.
22. Ibid., 33, 37–59. Van Norden's analysis serves his comparison of Confucian and Aristotelian ethics, but is intended to encompass all virtue ethical accounts.
23. Of course, MacIntyre is primarily concerned with the *Western* moral tradition and with recovering its Aristotelian and Thomistic heritage. My task here is more ecumenical, and although it is beyond our scope to defend this aim against MacIntyre's narrower commitments, I believe that the possibility and reasonableness of such a defense is implicit in the arguments of this chapter.
24. MacIntyre (1984), 187.
25. Ibid.
26. It goes without saying that global technosocial activity has also produced significant harms, and that some of these alleged goods (e.g., efficiency) have been identified by critics of modern technology such as Jacques Ellul (1964) and Herbert Marcuse (1964) as injurious to human freedom and/or happiness.
27. MacIntyre (1984), 191.
28. Ibid., 188.
29. Ibid., 190.
30. Ibid., 218–220.
31. Ibid., 191.
32. See Yearley (1990) for a comparative treatment of courage in Confucian and Western thought.
33. Peterson and Seligman (2004), Chapter 2.
34. Van Norden (2002), 7.
35. See Singer (2011); Pinker (2011); and Floridi (2013) for predictions of such expansion.
36. *NE* 1155a20; Kongzi (Confucius), *Analects* 13:19 and 15:5; and Cooper and James (2005), 108, where they argue that Buddhism's locus of moral concern, while global in scope, remains the human family. It should be noted that I have not ruled out a technosocial practice that pursues goals framed in nonanthropocentric terms; however, it remains the case that while nonhumans can and must be *subjects* of moral concern, at present only human agents can constitute a moral practice and tradition in the sense defined by virtue ethics.

37. Those who have explored such possibilities (Ess 2006, 2009, 2010a; Hongladarom and Ess 2007; Hongladarom 2008; Capurro 2008; Floridi 2006) often frame their goal as a global, cross-cultural, or intercultural *information* ethic. I suggest that this label may unduly, if unintentionally, limit the scope of our attention to information technologies as narrowly construed (excluding, say, technologies for biomedical enhancement), and as abstracted from their social context—hence my choice to reframe the goal as a global *technosocial* ethic.
38. Ess (2006), 216.
39. Ibid., 218.
40. Ibid.
41. Ibid., ff. 37; also Ess (2010a), 112.
42. Ess (2014).

CHAPTER 3

1. MacIntyre (1984), 276.
2. As noted in chapter 2, this global community is composed of members of other communities bound by more deeply shared and local concerns, practices, and values; thus a technomoral virtue ethic cannot replace the existing plurality of local conceptions of the good life and moral excellence. At best it can supervene upon them in a manner that addresses the need for more globally coherent human responses to moral dilemmas involving technosocial practices.
3. See Ess (2010a), 115.
4. Cavell (1990), 57.
5. Van Norden (2007), 50–51. Van Norden claims that Aristotle's view of moral habituation is nearer to the perspective of Mengzi's Confucian rival, Xunzi, who held that ritual practice corrects and radically reshapes human nature. In all likelihood Aristotle's view lies somewhere in between the two Confucian alternatives, as he argues that virtues arise "neither by nature . . . nor contrary to nature." (*NE* 1103a25) See also Liu and Ivanhoe (2002) on Mengzi's moral philosophy.
6. *Anguttara Nikāya* 10.2; translation from Cooper and James (2005), 82.
7. In Aristotelian and Buddhist ethics, moral sages may seek contemplative lives largely detached from practical problems of the sort with which we are concerned. In the Confucian tradition, the sage (*shengren*) functions as a regulative ideal rather than a demonstrable example like the *junzi*, though even a true Confucian sage would be fully engaged in political life. However, since technomoral virtues must enable cooperative ethical discourse concerning global problems of the sort described in chapter 2 and Part III of this book, we will be more interested in what classical virtue traditions can tell us about cultivating excellence in ordinary human beings than in what these traditions say about exceedingly rare specimens of moral sagehood.
8. See Harvey (2000), 39.

9. See Lockwood (2013) for a nuanced account of habituation in Aristotle's ethics.
10. *NE* 1105b8.
11. *NE* 1105b17.
12. *NE* 1179b5-10.
13. *NE* 1104b10.
14. *NE* 1104b15-1105a15.
15. *NE* 1104b1-3.
16. The apparent circularity of this definition has been much discussed in the literature, but the circle is generally regarded by virtue ethicists as nonvicious (no pun intended). In any classical account, a virtuous person is understood as responsive to the salient moral features of his or her practical environment (see chapter 5); virtue is not defined internally (or externally) to the agent, but by a 'fit' between the possibilities for human flourishing afforded to the agent by the moral environment and the intelligence and effectiveness with which he or she enacts those possibilities. In practice, however, the not-yet-virtuous person is unable to reliably perceive the moral possibilities in the environment or discern how best to seize them; hence the necessity of relying upon moral exemplars.
17. *NE* 1141a 25-1141b20. There is much dispute among Aristotle scholars concerning the *phronimos* or man of practical wisdom (*phronēsis*); for example, over whether there are different types of *phronimoi*, whether they manifest practical wisdom in varying degrees, how rare they might be, and why Aristotle believed that women and slaves naturally lack the rational faculties needed to become a *phronimos.*
18. *NE* 1120a24.
19. *NE* 1180a1.
20. See chapter 6, Part 2 for a discussion of this virtue.
21. *NE* 1179b15-25. Compare with Kongzi, *Analects* 7:26.
22. Van Norden (2007), 106 ff. 58.
23. Slingerland (2003), 200. Hereafter references to the *Analects* of Kongzi (Confucius) appear in the following style: *Analects* 17:2. Translations generally follow Slingerland (2003); occasional minor departures from Slingerland's translation are my own, except where otherwise noted.
24. Van Norden (2007), 338. See also *Analects* 8:2.
25. Ibid., 107.
26. Yearley (1990), 41.
27. *Analects* 12:1. Here 'humane,' rendered as 'Goodness' by Slingerland (2003), is a translation of *ren* or '*ren*-minded,' a term we shall have cause to discuss in the following sections.
28. Ibid., 12:15; 13:13; see also Yu (2007) 116–117 on the similar role of habit in Aristotle and Kongzi. This notion of correction is, however, in tension with Mengzi's account of self-cultivation as developing our natural sprouts of human goodness.
29. Ibid., 6:29; 13:21; 11:16; 17:8; 20:2.
30. Ibid., 12:19.

31. Ibid., 4:17. See also *Analects* 4:1, 4:25, and 7:22. As with debates about the *phronimoi*, it is unclear whether Kongzi regarded the ideal of the perfected *junzi* as practically achievable. At a minimum it seems clear that a perfected *junzi* would be, like the sage (*shengren*), exceedingly rare. Still, given that Kongzi describes himself as constantly emulating virtuous persons (*Analects* 7:22), despite regarding his own time as remarkably corrupt (*Analects* 7:26), it is safe to say that he believed that moral exemplars worthy of emulation were common enough, even if they were not yet fully perfected.
32. Ibid., 1:12.
33. Ibid., 9:14. See Ivanhoe (2002b) on the role of joy in ritual practice as motivating self-cultivation.
34. Harvey (2000), 48.
35. Cooper and James (2005), 48.
36. Ibid.
37. Buddhists distinguish between pleasures and joy; the former involve the mental and physical cravings that are inimical to the cultivated state, while the latter can be consistent with equanimity and relinquishment of the fruits of action (*karma*). A related distinction appears in Confucian thought, where Kongzi says of the Way that "One who knows it is not the equal of one who loves it, and one who loves it is not the equal of one who takes joy in it." (*Analects* 6:20) See also Ivanhoe (2002b), 224, on the distinction in Mengzi between moral joy and pleasure. In all three classical traditions, however, the cultivated person finds a profound type of happiness in acting well, whereas the uncultivated person may do good acts only with reluctance, discomfort, or displeasure.
38. Harvey (2000), 37, 68.
39. Ibid., 39.
40. Ibid., 41.
41. Ibid.
42. Ibid., also Keown (1992), 44–56. See also Yu (2007), 148–154 on the complex and mutually reinforcing interactions of ethical wisdom, habit, choice, and feeling in Aristotelian and Confucian ethics.

CHAPTER 4

1. Van Norden (2007), 320. See Hursthouse (1999), 224–226, and 240 for her criticisms of Kantian and utilitarian forms of the impersonal, universal, or neutral moral point of view. For a detailed account of the emergence in Stoic virtue ethics of the demand for moral impartiality or neutrality, and the debate between this view and Aristotle's, see Annas (1993), chapter 12. It is worth noting that the Stoic position merely radicalizes the demand of virtue ethics for the moral extension of natural relational concerns, as described later in this chapter. Thus Stoicism is still rooted in a relational understanding of the self, rather than a monadic one.

2. See Yu (2007), 210–212 for a detailed discussion of the Confucian self as "relational" and its close affinity with the political self in Aristotle's ethics. Yu further defends these views from the objection that a relational self would lack moral independence or in some other way be dissolved into the social context. See also Shun and Wong (2004) on the relational self in Confucian ethics.
3. As MacIntyre notes in *Dependent Rational Animals*, reliable knowledge of oneself is made possible only by "those social relationships which on occasion provide badly needed correction for our own judgments. When adequate self-knowledge is achieved, it is always a shared achievement" (1999, 95).
4. For a detailed contemporary defense of key aspects of this account, see MacIntyre (1999); classical virtue traditions generally assume this viewpoint rather than explicitly defending it. As we will see in later chapters, even the more contemplative and solitary states of practice found in Buddhist and Aristotelian ideals of philosophical enlightenment must be understood as relational.
5. Aristotle (1984), *Politics* 1253a1-5. Hereafter *P.*
6. *NE* 1130a9.
7. *NE* 1134b10-17. See also Aristotle's *P* 1262a1-1262b35, where he rejects Plato's proposal in the *Republic* to collapse the fine-grained structure of kinship and other traditional human relations.
8. *NE* 1156a1-5.
9. *NE* 1169b13.
10. See *NE* 1158b10-30.
11. *NE* 1160a1-8. See also *NE* 1158b10-35; 1161a1-1162a30.
12. *NE* 1171b30-1172a15.
13. *NE* 1162a25. Even in this case Aristotle thinks that man and woman will display different kinds of excellence, but sufficiently complementary that their enjoyment of virtue can be shared.
14. *P* 1253a19.
15. *Analects* 1:2.
16. Yu (2007), 33 and 77; also Van Norden (2007), 117–118.
17. *Analects* 14:42.
18. See Yu (2007), 94 where he argues to this effect.
19. See *Analects* 4:18 and 2:7.
20. Ibid., 2:21. See also Mengzi, *Mencius* 7A15, and chapter 5 in this volume on moral extension.
21. Ibid., 1:6.
22. Ibid., 12:22.
23. Ibid., 1:16.
24. Van Norden (2007), 115. See also *Analects* 13:28.
25. See *Mencius* 3A5, 7A26, and 7B26; also Wong (2002), 203–204; Van Norden (2007) offers extensive analysis of the debate between Mohists and Ruists on the subject of graded love.

26. *Mencius* 7A26 and 6A5. See also chapter 5 herein, on prudential judgment.
27. *Mencius* 7A26.
28. Harvey (2000), 104–109; Cooper and James (2005), 95–96.
29. In Cooper and James (2005), 133, where they explore the implications of Buddhist relational understanding for environmental ethics.
30. In Harvey (2000), 29.
31. See Ibid., 109-118. See also chapter 5 herein on moral attention.
32. Ibid., 12.
33. Ibid., 51.
34. See ibid., 108–109, on the developmental progression from acting virtuously toward those for whom it is easiest to do so, to radiating outward to more and more difficult personal contexts until one's virtuous practice is properly extended to all. The process involves many of the features of moral habituation described in the previous chapter, especially the gradual overcoming of affective and cognitive obstacles to acting well.
35. See Keown (1992), 150–163; Harvey (2000), 134–140; and Cooper and James (2005), 76. Further discussion of this point can be found in chapter 5 concerning prudential judgment.
36. Plato, *Apology* 38a.
37. *NE* 1109b1-5.
38. *NE* 1114a5.
39. For more detailed discussion of the history of self-examination as a technique of self-care in Western thought, see Foucault (1988); Cavell (1990); and Nehamas (1998).
40. *NE* 1170a1-3; 1171b30-1172a7.
41. *NE* 1128b15-35. This disdain of moral humility is one reason why Aristotle's virtue of magnanimity (*megalopsuchia*) has attracted much modern critical scorn; but see Kristjánsson (1998) for a defense of magnanimity as a contemporary virtue, see also chapter 6, section 12 in this volume.
42. See *Analects* 7:32, 7:33, 9:7; also Van Norden (2007), 19–20 on Confucian humility.
43. Ibid., 2:4. See also 7:16 and 9:15.
44. Ibid., 1:14, 4:17, 9:24, 15:14.
45. Ibid., 15:29. (Chai translation). See also *Mencius* 4A4; 4A19.
46. Ibid., 5:27.
47. Ibid.,14:32 (Waley translation); see also 15:19 and *Mencius* 4B28.
48. Ibid., 12:16, 4:7, 4:17, 7:22, 13:13, 15:20.
49. *Mencius* 7A5 and 4A19.
50. See *Analects* 4:6, 6:12, 7:30; and *Mencius* 6A11, 7B21.
51. *Analects* 1:4. See also 19:5, and *Xunzi* (2003), p. 15.
52. This is not to say that it need *only* be this, of course; simply that any ethical philosophy, even one with metaphysical and/or religious underpinnings, has practical worldly action as its central subject matter.

53. Harvey (2000), 36–37.
54. An exception would be certain forms of Pure Land Buddhism that allow for immediate enlightenment by something akin to grace. See ibid., 143.
55. *NE* 1105a30-1105b5.
56. *NE* 1140b1-10.
57. *NE* 1105b15.
58. More will be said about this virtue in chapter 6, section 12.
59. This issue is discussed throughout the *NE*, but especially in Books IX and X. See also Annas (1993), chapter 12.
60. *NE* 1168b30, *NE* 1169a5-10.
61. *NE* 1098a18.
62. *Analects* 7:8.
63. See Van Norden (2007) 227, on this remark in *Mencius* 2A2. See also Nivison (1996).
64. As Van Norden (2007) notes on page 228–229, it is unclear whether Mengzi saw these environmental factors as crucial to *childhood* moral development, as is made explicit in the Greek virtue tradition.
65. *Mencius* 7A42. Translation by Dobson in Mencius (1963), 189, there listed as 7.18. Translations of 7A42 vary widely.
66. See Van Norden (2007), 231–232.
67. *Mencius* 2A2.
68. *Analects* 9:19. See also 4:5.
69. *Xunzi* (2003), 23.
70. Ibid., 19.
71. *Analects* 8:7. See also Yu (2007), 168 on the resonance with this point in Aristotle's view, and 177–182 on self-completion (*cheng*) in Confucian thought.
72. *Mencius* 4B14.
73. See Harvey (2000) 61, 91–92, on the distinction between the mundane path of the Buddhist layperson or 'householder' and the supramundane path of the monastic *Sangha*.
74. In Harvey (2000) 11, from Tatz 1986, 47.
75. Ibid., 37, 80.
76. See Cooper and James (2005), 60–61; and *Analects* 14:45.
77. Cooper and James (2005), 94.
78. Ibid., 95.
79. Ibid.
80. See *Analects* 16:11; 18:7; 18:8.
81. Cooper and James (2005), 95.
82. Although see Van Norden (2007), 231, on the use of *si* to focus attention away from sensual desires.
83. Harvey (2000), 11.

CHAPTER 5

1. *NE* 1142a25-30.
2. *NE* 1142a10-20; 1143b10-15.
3. McDowell (1998).
4. As both McDowell (1998) and Nussbaum (1990) have noted, this element of moral practice is missing from deontological or consequentialist accounts of ethics in which sufficient action-guidance can be obtained simply from applying standing rules or principles of morality to the present facts of the case. For such accounts ignore the way in which moral attention selects from among all the available facts those that most acutely call for a moral response; a selection which, as we will see in the next section, may require a modification, nonstandard implementation, or even suspension of a moral rule or principle that would otherwise apply.
5. Nussbaum (1990), 38.
6. Ibid.
7. Ibid., 40–42.
8. Waley (1989), 45. The Chinese term rendered in Pinyin as *si* is written as *ssu* by Waley, using the traditional Wade-Giles romanization.
9. *Mencius* 4A17, 5A2, and 2B7, respectively.
10. See Yearley (1990), 62–70, on the relationship between *zhi* and *si* (Like Waley, Yearley also uses Wade-Giles romanization, rendering these as *chih* and *ssu*).
11. See Van Norden (2007), 234–245.
12. Angle (2014), 173.This is part of a larger debate about how much automaticity is present in virtue.
13. See Keown (1992), 206–217 on the comparison between Buddhist and Aristotelian psychology.
14. Ibid., 70–82.
15. Cooper and James (2005), 98.
16. Ibid., 100.
17. *NE* 1143b6-9.
18. *NE* 1141b12-15.
19. *NE* 1142b1-30.
20. *NE* 1143a19-24.
21. *NE* Book VI, Chapter 12.
22. Van Norden (2007), 123. Here he notes one key difference from Aristotle's account: that *zhi* cannot, as does the intellectual virtue of *phronēsis*, represent complete virtue, since for Kongzi this also requires the virtue of *ren* (humaneness). On the other hand, Van Norden acknowledges (124–125) that *zhi* seems to entail an appreciation or love of *ren*. In that case the parallel is nearer, since Aristotle also requires the man of *phronēsis* to have a love of moral virtue. The remaining difference is that for Aristotle moral virtue is centered more on acting justly toward others

than on acting benevolently toward them. How this will translate into different moral behavior is an interesting question, but beyond the scope of our inquiry.

23. *Mencius* 5B1; see Van Norden (2007), 277 for his interpretation of this passage as highlighting skill in means-ends deliberation.
24. Yu (2007), 94.
25. Yearley (1990), 19–23. A related phrase in Confucian moral thought is the "village honest man," who is thought to be well cultivated but in fact is simply a pale imitation of genuine virtue.
26. See Harvey (2000), 134–140.
27. Keown (1992), 151.
28. Ibid. See also Cooper and James (2005), 63–64, and 76, where they note that the presence of the doctrine of skillful means is a powerful argument for classifying Buddhism as a virtue ethical tradition.
29. Harvey (2000), 135.
30. See Becker, (1986).
31. See *NE* Book V.
32. *NE* 1155a26.
33. *NE* 1166a20. See Vallor (2012) on the role of empathy (the sharing of joys and pains) in Aristotelian virtue friendship, and its implications for friendship in the age of new social media.
34. *NE* 1166a30-35.
35. *NE* 1161b17-35. He suggests that this is why a mother's love is stronger than that of a father (i.e., since on his view mother and child are literally of common matter, whereas the father contributes only form).
36. *NE* 1171b20.
37. *NE* Book VIII, Chapter 10.
38. *Mencius* 1A7, 1B5.
39. Ibid.
40. See Nivison (1980); Ivanhoe (2002a); Wong (2002); Yu (2007), 119–120.
41. *Mencius* 1B5.
42. *Mencius* 6A18. Translation from Chan (1963), 60.
43. Wei-Ming (1984), 381–382.
44. Harvey (2000), 64.
45. Cooper and James (2005), 40, 61.
46. Ibid., 61.
47. Ibid.
48. Yu (2007), 120.
49. This strategy resonates with much contemporary literature on the narrative exercise of moral imagination and its role in moral motivation, as well as contemporary work in moral psychology on the cognitive, affective, and perceptual underpinnings of empathy. See Slote (2007); Batson (2011); and Trivigno (2014).

CHAPTER 6

1. See Snow and Trivigno (2014) on the psychology of virtue; see also chapter 10 herein for a discussion of the possibility of biomedical enhancement of these capacities.
2. Nussbaum (1988).
3. See Peterson and Seligman (2004) for an interesting attempt to give a comprehensive taxonomy of six "core virtues" and a list of supporting "character strengths" that facilitate each virtue.
4. Yu (2007), 177.
5. See Ess (2010b) and Ess and Thorseth (2011).
6. See 2015 Gallup poll: http://www.gallup.com/poll/185927/americans-trust-media-remains-historical-low.aspx?
7. Frankfurt (2005).
8. Aquinas (1975), *Summa Contra Gentiles*, Book IV, Ch. Lv.
9. See Grenberg (2005).
10. See Floridi and Taddeo (2014).
11. Bostrom (2002).
12. Wright (2009), 135. Wright notes that contemporary Buddhist practice is likely to show an increasing concern for social and political justice, given its increasingly global reach.
13. Oremus (2013).
14. It also indicates a vicious deficiency of other technomoral virtues, especially *care, civility,* and *empathy.*
15. *NE* 1115b15-20. Minor variation from the Barnes translation is mine.
16. See Yearley (1990), 114–116.
17. For an extensive discussion of these 'semblances' of courage, see Yearley (1990), 124–129, rooted in Aristotle's account in *NE* Book Three, Chapter VIII as well as Aquinas's analyses of that chapter.
18. See Yearley (1990), 130–132, rooted in Aquinas' *Summa Theologiae*, Part 2-2, Question 123.
19. Ibid., 134–135.
20. See *Analects* 9:29, 14:28, and 17:23.
21. See Slingerland (2003), 96 on Miao Xie's commentary on *Analects* 9:29.
22. Yearley (1990), 146.
23. *Mencius* 6A10, translation from Nivison (1996). See also 2A2.
24. It is worth noting that while models of heroic martial courage are still culturally influential in many parts of the world, the modern transformation of military practice by long-range weapons and automated systems has placed considerable pressure on this conception (Singer 2009; Vallor 2013a). The predicted expansion of drone and robotic warfare will only increase this pressure; both globally and locally humans will be challenged to find new ways to understand martial courage, or else to question whether and to what extent this form of courage continues to warrant its historical privilege.

25. Significant portions of this section are adapted from Vallor (2012).
26. Batson (2011), 20. See also Goleman (2013), 98.
27. See Trivigno (2014) for a discussion of this.
28. See *NE* 1106a4, also Vallor (2012).
29. *NE* 1166a8.
30. *NE* 1106a4
31. *EE* 1240a36
32. *NE* 1166b18. See also *NE* 1167a5-20, where Aristotle describes as a precondition of friendship the finding in each other of some excellence or worth to admire, which under certain conditions evolves from mere goodwill into the rare love and empathy between two people that typifies virtue friendship (NE 1171a5-15).
33. *NE* 1166b34
34. *NE* 1167a2; *NE* 1167a10
35. MacIntyre (1999), 116.
36. Ibid., 7, 64. However, *misericordia* as defined by Aquinas and MacIntyre is narrower than empathy as I, and Aristotle, define it: while *misericordia* is "grief or sorrow over someone else's distress . . . just insofar as one understands the other's distress as one's own" (Ibid., 125), empathy includes, in addition, the disposition to rejoice in the joy of another (NE 1166a8). See Vallor (2012) on the importance of this oft-neglected component of empathy. An exception to this pattern of one-sided neglect is the Buddhist virtue of *muditā* (sympathetic joy).
37. Or in the case of a sociopath, in either sense.
38. See Slingerland (2003), 242, also Ivanhoe (2002a), 2–3.
39. Indeed, the character for *shu* is composed of the characters for *xin* (heart-mind) and *ru* (comparing). The analogical dimension of *shu* prevents it from meaning mere pity or sympathy; I must take on the other's state of mind or suffering in some way that makes me see the other's condition as *like* my own. Seeing another's suffering simply as a deplorable, sad state of affairs, even one I feel duty-bound to remedy, is not enough. Hence I have chosen to conceive of *shu* as nearer to 'empathy' than 'sympathy.'
40. *Analects* 15:24 and 4:15.
41. *Mencius* 7A4. Translation adapted from Bloom (2002), 93.
42. *Mencius* 2A6. Translation adapted from Ihara (1991), 51.
43. See Bloom (2002), 75–76.
44. This indiscriminate extension of empathy was opposed by classical Confucians, who, as we saw in chapter 5, vigorously defended the notion of moral affection as properly graded by kinship and political ties. Philip J. Ivanhoe notes that neo-Confucian scholars like Wang Yangming (1472–1529) later tried to reduce this tension with Buddhism, by reinterpreting Confucianism as cultivating a natural "sensitivity" or feeling for all parts of the universe as one body: "Those who don't feel this unbounded connection, who feel no compassion for a world in pain, are like people with paralyzed limbs who cannot feel an injury inflicted upon their own

hands and feet." Still, Ivanhoe notes that this view is "completely alien" to the classical Confucian *ethos*; see (2002a), 28–29.

45. Wright (2009), 30.
46. Ibid., 31.
47. Turkle (2011, 2015) is perhaps the most widely influential advocate of this claim.
48. See Twenge et al. (2008); Konrath et al. (2011); and Turkle (2015).
49. Wright (2009), 18–19.
50. See Li (1994, 2000); Star (2002); and Yuan (2002) regarding the close proximity and important distinctions between the concepts of *ren* and care.
51. See Held (2006) and Slote (2007).
52. See Slote (2007); Held (2006); and McLaren (2001) for arguments for and against regarding care ethics as a form of virtue ethics.
53. Tronto (1993).
54. See Vallor (2011b).
55. See MacIntyre (1999) on Aristotle's failure to fully acknowledge the ethical import of human dependence and vulnerability; see *NE* 1169b13 on the importance of having friends to benefit.
56. Walzer (1974) names it as one of five key civic virtues, along with loyalty, service, tolerance, and participation.
57. *NE* 1155a22.
58. *NE* 1171b32.
59. Schwarzenbach (1996), 102.
60. Ibid., 105.
61. In certain US communities, one hears this in the grammatically odd but socially important expression "He (or she)'s good people."
62. Schwarzenbach (1996), 106.
63. Ibid., 109–110.
64. Ibid., 113.
65. Ibid., 122.
66. Ibid. She notes that in modern translations this term is more commonly rendered as 'attention,' 'discipline,' or 'control,' but suggests that 'care' is a more appropriate translation.
67. *NE* 1125b30-1126a30.
68. *Analects* 9:4; 17:8; 15:37; 17:24; 19:11.
69. Ibid. 17:7.
70. Ibid. 14:34.
71. Wright (2009), 95.
72. Śāntideva, *Bodhicaryavatara*, Chapter 6, verse 2, quoted in Wright (2009), 101.
73. Bommarito (2014), 273.
74. Ibid.
75. Ibid., 274.
76. Ibid., 277. Bommarito does not, however, appear to regard moral perspective as itself a virtue, for he claims that it "need not be consciously or intentionally developed."

(Ibid., 274). I would respectfully disagree, though the argument cannot effectively be made here.

77. Exactly how the virtues can be successfully integrated, and how the person of practical wisdom responds when virtues appear in a given situation to be placed in tension or conflict, is the subject of much discussion by contemporary virtue ethicists. Some dismiss the hypothesis of the 'unity of the virtues,' others continue to embrace it as a regulative ideal. Yet while important, these theoretical concerns lie beyond the scope of our present inquiry.

CHAPTER 7

1. See Baym (2011).
2. boyd (2014).
3. See, for example, Google executives Eric Schmidt and Jared Cohen's *The New Digital Age* (2014), although even they acknowledge the need for human wisdom in guiding the technological future.
4. See Diener (2009).
5. See Badwhar (2014b) for a thoughtful defense of objective theories of well-being.
6. Vallor (2010) and (2012).
7. Selinger (2013).
8. Sheldon (2008); Pierce (2009); and Laghi et al. (2013).
9. Bilton (2014).
10. Selinger (2013).
11. See Vallor (2012) on the incompatibility of effortless connection and moral friendship.
12. Wong (2013).
13. Nissenbaum (2010); Ess (2013).
14. See Kuss and Griffiths (2011). The scare quotes serve to distinguish the colloquial notion from the technical term that entails increasing physical tolerance for and dependence on a chemical substance.
15. Carr (2010), 200.
16. Young and Nabuco de Abreu (2011).
17. Doshi (2014).
18. Ibid.
19. Mill [1861] (2001), 10–11.
20. Ibid.
21. Morozov (2013), 20–21.
22. Vallor (2015), 118.
23. Hern (2014).
24. Ophir, Nass, and Wagner (2009); Carr (2010); Wang and Tchernev (2012).
25. Goleman (2013), 100–104.
26. Immordino-Yang et al., (2009).
27. Goleman (2013), 115.

28. Batson (2011), 49.
29. Goleman (2013), 88–89.
30. Ibid., 81.
31. Kaplan (1995); Berman et al. (2008).
32. See Walji (2014) and Oremus (2014).
33. See Keen (2007); also Mar et al. (2011). Mark Coeckelbergh (2007) has written extensively on the use of narrative to enrich moral imagination and other moral capabilities in technological contexts.
34. Shirky (2008). Regrettably, Daniel Goleman has also fallen victim to this pernicious illusion, claiming that "This collective intelligence, the sum total of what everyone in a distributed group can contribute, promises maximal focus, the summation of what multiple eyes can notice" (2013, 21). This is the illusion that knowledge is simply the aggregation of disembodied factoids rather than the integration of facts into a comprehensive and responsive worldview.
35. Floridi (2014), Section 4.6.
36. Chou and Edge (2012); Kross et al. (2013); Krasnova et al. (2013)
37. Davenport et. al. (2014).
38. Gross (2014).
39. *Mencius* 7A42. Translation by Dobson in Mencius (1963), 189; there listed as 7.18.
40. Schmidt and Cohen (2013), 121.
41. Ibid., 34
42. Shirky (2008), 20–21.
43. Morozov (2013), 169, 344.
44. See also Wheeler (2011), 195.
45. Bakardjieva (2011), 76.
46. Consider, for example, the dramatic YouTube cell-phone video uploaded by Minnesota resident Chris Lollie depicting his arrest and Tasering by police, which along with others like it, received widespread social and traditional media coverage in the weeks after the Ferguson protests. See Friedersdorf (2014a).
47. Dewey (2014).
48. Pariser (2011); and Sunstein (2007).
49. Lotan (2014).
50. Honan (2014).
51. Bilton (2013).
52. See http://www.reddit.com/wiki/reddiquette.
53. Borgmann (2004), 63–64.
54. Hampton et al. (2014); Couldry (2012), 127; Couldry, Livingstone and Markham (2007), 179–184.
55. Hampton et. al. (2014).
56. Ibid., 4.
57. Ibid., 25.
58. Cohen (2014).
59. Tsukayama (2014).

60. See Verbeek (2011).
61. See Couldry (2012, 2013a, 2013b).
62. Couldry (2013b), 49.
63. Couldry (2013a), 17.
64. Jonas (1984), 21.

CHAPTER 8

1. Foucault (1995).
2. See Mann, Nolan and Wellman (2003); and Ganascia (2010).
3. Brin (2014).
4. Olmstead v. United States, 277 U.S. 438 (1928).
5. Schneier (2008).
6. Brin (2014).
7. Morozov (2013), 320–321.
8. See video at http://www.huffingtonpost.com/2009/12/07/google-ceo-on-privacy-if_n_383105.html. March 18, 2010, updated May 25, 2011.
9. For an illuminating illustration of this point, see Friedersdorf (2014b).
10. Hardy (2014).
11. Morozov (2013), 243.
12. Hartzog and Selinger (2014).
13. Nyhan and Reifler (2010); Nyhan, Reifler and Ubel (2013); Nyhan, Reifler, Richey and Freed (2014).
14. Perry et. al. (2013).
15. Brin (1996).
16. See Morozov (2013), 213–214.
17. Plato, *Apology*, 37e-38a.
18. *Mencius* 2A2. Translation from Ivanhoe and Van Norden (2001), 126.
19. Kelly (2011).
20. This and the following section of chapter 8 are adapted from Vallor (2013b).
21. *Mencius* 7A5.
22. Foucault (1988), 18.
23. *Mencius* 2A2.
24. Cavell (1990), 57.
25. See Latour (1999) on the use of speedbumps and other environmental artifacts as social controls.
26. Morozov (2013), 121–124.

CHAPTER 9

1. See Ford (2015); Kaplan (2015); Markoff (2015); Osborne and Frey (2013).
2. Brynjolfsson and McAfee (2014).

3. Lin, Abney, and Bekey (2012).
4. Ibid. See also Sullins (2011).
5. Turkle (2011); Scheutz (2012).
6. Consider the unfortunate fate of HitchBOT, the Canadian robot dismembered in 2015 by vandals in Philadelphia; also the online storm of visceral repugnance directed toward the 2015 Boston Dynamics video demonstration of researchers kicking their robot dog 'Spot.' See also Whitby (2012), 240.
7. See Turkle (2011) on the power and resilience of such projections, not only for children but for rational, educated adults who simultaneously affirm that they 'know' that the system is 'just' a machine.
8. Kant, for example, famously argued against animal cruelty on these grounds.
9. U.S. Army ARL September 27, 2011: "Supersizing robotic system capabilities: Army's collaborative effort with industry, academia break open new space for autonomous warfare." http://www.arl.army.mil/www/?article=682.
10. Arkin (2010).
11. See Vallor (2015).
12. Heyns (2013), 10.
13. See Sparrow (2009) and Sullins (2010).
14. Overly (2012).
15. For competing views on these questions, see Sharkey (2012) and Arkin (2009).
16. *NE* 1115b15-20. As noted in chapter 6, while Aristotle regards martial or physical courage as foundational, Mengzi privileges moral courage. Our account assumes martial or physical courage as historically prior, growing out of the basic duties of moral care to protect family and extended kin; but takes moral courage to be this virtue's mature and ultimate form.
17. While Olsthoorn (2007) suggests that only moral courage should be seen as essential to military virtue, Sparrow (2013), 90, holds that "for the foreseeable future," military virtue seems to presuppose both forms of courage.
18. Thompson (2011).
19. Otto and Webber (2013); Matthews (2014); and Kirkpatrick (2015).
20. Mengzi 6A10.
21. See Arquilla (2011).
22. 'Bugsplat' is the name of the U.S. Defense Department's software program for calculating civilian deaths from airstrikes.
23. Vallor (2013a), 180.
24. Vallor (2013a), 11.2.
25. See Sparrow (2013) and (2015).
26. Lanier (2010).
27. Parts of this section and remaining sections of Chapter Nine are adapted from Vallor (2011b).
28. Sharkey and Sharkey (2012b).
29. GeckoSystems, Investors: Press Release, November 4, 2013. http://www.geckosystems.com/investors/press_releases/20131104_GeckoSystems_%20Updates_Shareholders.php.

30. See Sparrow and Sparrow (2006); Sharkey and Sharkey (2012a); Borenstein and Pearson (2010); Coeckelbergh (2010); and Turkle (2011) for excellent articulations of these concerns.
31. Coeckelbergh (2009).
32. Ibid., 217.
33. Ibid., 219.
34. Carebot ethicists commonly state or imply that this is the primary, if not the sole ethical concern; see Borenstein and Pearson (2012), 251.
35. Sharkey and Sharkey (2012a), 28. See also Borenstein and Pearson (2010), 283–285; and Coeckelbergh (2010), 183.
36. Borenstein and Pearson (2010), 286.
37. Ibid., 284.
38. Sparrow and Sparrow (2006), 152.
39. Nussbaum (2006), 102.
40. Borenstein and Pearson (2010), 284.
41. Sharkey and Sharkey (2012a), 29. See also Borenstein and Pearson (2010).
42. Aristotle *NE* 1167a10.
43. Sparrow and Sparrow (2006), 154.
44. See chapter 6 on the distinction between empathy as a basic human capacity and as a cultivated virtue.
45. Noddings (1984), 14.
46. Jennifer Parks, in her feminist analysis of the prospect of carebots, endorses a related view of care ethicist Annette Baier: that we only become "skilled in the arts of personhood through relationships of dependency that we share with others." (Parks 2010), 111.
47. Noddings (1984), 16.
48. Ibid., 14.
49. Ibid., 24.
50. Ibid., 12, 50.
51. Mengzi 6A8.
52. Noddings (1984), 51.
53. Sparrow and Sparrow (2006); Sharkey and Sharkey (2012a).
54. van Wynsberghe (2013), 424.

CHAPTER 10

1. This chapter is adapted from Vallor (2011a).
2. The use of 'enhancement' as a conceptual distinction from 'therapeutic' uses of technology has been rightly judged as problematic by many scholars engaged in the debate over human enhancement, and it is far from obvious that this distinction can be drawn in a consistent, nonarbitrary, and morally significant way. That said, it is widely employed in the relevant literature. See Allhoff et al., (2009).
3. Bostrom (2005a); Savulescu et al. (2011); Hughes (2004); Harris (2007). For reasons of space, this passes over two key distinctions: the first is between the strong

transhumanist program for enhancement and more modest enhancement goals that stop short of radical alteration of the human species. The second is the distinction between 'transhuman' and 'posthuman' philosophies; some transhumanists explicitly call for a posthuman future, that is, one in which humanity has been surpassed. Others, like Kurzweil (2005), reject the notion of leaving our humanity behind, while simultaneously regarding the nature of our humanity as almost infinitely malleable. Finally, there are uses of 'posthuman' in literary theory, gender, and culture studies (see Haraway 1991) that do not map neatly onto the transhumanist conception of posthumanity. Both distinctions are important but fluid and contested.

4. Kass (2008); Sandel (2007); Fukuyama (2002).
5. For fine surveys of the debate, see Allhoff et al. (2009), Savulescu and Bostrom (2009) and Buchanan (2011).
6. Fukuyama (2002), 149.
7. Kass (2008).
8. Ibid.
9. Sandel (2007).
10. Ibid.
11. Ibid.
12. Bioconservatives frequently associate transhumanism with a world in which "there'd be nobody left to worry about work, or challenge, or satisfaction, or sweat" (McKibben 2003, 100). Abandoning the argument from striving in favor of strict respect for the gifts of nature helps little; it avoids incoherence, but at the price of seeming to suggest that human dignity would be best served by eschewing medical treatment of congenital disability or disease; the engineering of peaceful and egalitarian social arrangements; or the educational cultivation of new artistic and intellectual capacities.
13. Fukuyama (2002), 169, 171.
14. Hughes (2004), 78.
15. Fukuyama (2002) 119–120, 151.
16. Bailey (2005), 6i. See also Naam (2005), 228; Bostrom (2005a); Kurzweil (2005). As Kurzweil argues: ". . . the essence of being human is not our limitations—although we do have many—it's our ability to reach beyond our limitations. We didn't stay on the ground. We didn't even stay on the planet. And we are already not settling for the limitations of our biology" (2005, 311).
17. The President's Council on Bioethics (2003), 146–149. Kass (2008) states elsewhere: "the dignity of rational choice pays no respect at all to the dignity we have through our loves and longings—central aspects of human life understood as a grown-togetherness of body and soul. Not all of human dignity consists in reason or freedom."
18. della Mirandola (1994).
19. Ibid.
20. Indeed, Kass declines the label 'ethicist' in favor of describing himself as an "old-fashioned humanist" (Wilkinson, 2008). It is reasonable to ask *why* his humanism departs so sharply from della Mirandola's.

21. della Mirandola (1994).
22. Coady (2009), 165.
23. Bostrom (2008b), 191. See also Bostrom (2005b).
24. della Mirandola (1994).
25. See Lin et al. (2014) on this point.
26. Jonas (1984), 21.
27. Gibbs (1991), 81–83.
28. Nietzsche (1968), 175.
29. Stock (2002), 22.
30. McKibben (2003), 92.
31. See Bostrom (2005b) and Bailey (2005), 241. Some with sympathies for the transhumanist vision accept the necessity of some government regulation to prevent grave ethical abuses, but even these thinkers generally defend the liberal ideal, wishing to avoid the promotion of any explicit moral vision for the human species (Agar 2004).
32. Even Hughes (2004), who rejects the libertarian principles of many transhumanists, acknowledges the absence of a shared moral or political vision among them (xv), and limits his characterization of its goals to essentially negative aims: release from the natural limits of body, age, intelligence, and suffering. Yet these fail to satisfy Aristotle's demand for a moral vision beyond the mere pursuit of proximal ends such as health and strength (1140a28).
33. Naam (2005), 229.
34. Ibid., see also Agar (2004), 87.
35. Naam (2005), 233.
36. Ibid.
37. Kurzweil (2005), 372-377.
38. Bailey (2005), 51, 245.
39. Ibid., 246.
40. Ibid. See also Bostrom (2008a), 112, and (2008c), 5.
41. See Kurzweil (2005), 310, and Hall (1993).
42. Bostrom (2008b).
43. Coady (2009), 180.
44. Persson and Savulescu (2014).
45. Ortega y Gasset (2002).
46. Persson and Savulescu (2012), 121.
47. Ibid., 123.
48. Ibid.,131.
49. Ortega y Gasset (2002), 113.
50. Ibid.
51. Ortega y Gasset (2002), 116.
52. Ibid., 119, emphasis added.
53. Ibid., 118, emphasis added.
54. Ibid., 120-121, emphasis added.

55. Ibid.
56. Even if this diagnosis were a form of cultural hypochondria, a delusion of moral incapacity, the delusion itself may, as is often the case in hypochondriacs, produce the very incapacity it imagines.

EPILOGUE

1. See Bostrom (2014).
2. Astronomer and 'exoplanet-hunter' Geoffrey Marcy made this claim at a news conference for the Breakthrough Listen initiative backed by Russian philanthropist Yuri Milner; the project provides new funding for the so-far fruitless international effort known as SETI (Search for Extraterrestrial Intelligence). The quote and its relevance to the Fermi Paradox is mentioned in Overbye (2015).
3. The 'Great Filter' concept is credited to Hanson (1998); Nick Bostrom (2008d) has recently employed it in his discussions of the existential risks posed by emerging technologies, especially artificial intelligence.
4. A recent Chatham House Report of The Royal Institute of International Affairs (Lewis et al. 2014) lists thirteen such cases since 1962; of these, the most widely known are the events of the Cuban Missile Crisis as well as the false alert of incoming U.S. ICBM's received by a Soviet satellite early warning station in 1983, in which disaster was averted only by the actions of a single Soviet commander, Lieutenant Colonel Stanislav Yevgrafovich. Yevgrafovich broke protocol by failing to report the alert to his superiors, gambling that it was a false alarm produced by a technical glitch (indeed, sunlight had confused the satellite's sensors). The authors of the Chatham House report express concern that the risk of more such close calls (and by extension, the risk that one of them will not be 'close' but deadly) is rising due to global political instability. They also note that "individual decision-making, often in disobedience of protocol and political guidance, has on several occasions saved the day." This underscores this book's claim that human flourishing is most crucially contingent upon the cultivation of technomoral virtues such as wisdom and courage, in the absence of which technosocial innovations are as likely to be destructive as constructive of our flourishing.
5. Ibid.
6. The contrary attitude is lampooned by the Twitter wag @stopthecyborgs, who defines a transhumanist as one who "believes that death can be defeated but that western neoliberalism is unchallengeable." Award-winning computer scientist Kentaro Toyama (2015) documents this delusion in his book *Geek Heresy: Rescuing Social Change from the Cult of Technology*, where he advances the core thesis reinforced by this book: that finding ways to develop *better people*, not just better technology, holds the key to the future of human flourishing.
7. Roddenberry Interview (20 September 1988), included in *Star Trek: The Next Generation* Season 5, DVD 7, "Mission Logs: Year Five," "A Tribute to Gene Roddenberry," 0:26:09.

References

NOTE: The following classical texts used most commonly in the book are cited as follows, in the standard manner for classical scholarship:

Aristotle: All references are from the (1984) Revised Oxford Translation edited by Jonathan Barnes unless otherwise noted, and are cited with abbreviated title and standard Bekker numbers. Example: the *Nicomachean Ethics*, Book V, Chapter 4, Section 3 would be abbreviated as *NE* 1132a1-7. The *Eudemian Ethics* is abbreviated as *EE*, the *Politics* as *P*.

Kongzi (non-Latinized from of Confucius): All references are from the *Analects*, Slingerland (2003) translation unless otherwise noted, and cited with book and section number. Example: *Analects* Book Nine, Section 4 would be abbreviated as *Analects* 9:4.

Mengzi: (non-Latinized form of Mencius): All references are from Mencius (1970), D.C. Lau translation unless otherwise noted, cited with the alphanumerical code for the book, part and chapter. Example: *Mencius* Book 2, Part A. Chapter 2 would be abbreviated as *Mencius* 2A2.

Achterhuis, Hans, ed. 2001. *American Philosophy of Technology: The Empirical Turn.* Translated by Robert P. Crease. Bloomington, IN: Indiana University Press.

Agar, Nicholas. 2004. *Liberal Eugenics: In Defence of Human Enhancement.* Malden, MA: Blackwell.

Allenby, Brad and Daniel Sarewitz. 2011. *The Techno-Human Condition.* Cambridge, MA: MIT Press.

Allhof, Fritz, Patrick Lin, James Moor, and John Weckert. 2009. "Ethics of Human Enhancement: 25 Questions and Answers." Report for US National Science Foundation. www.humanenhance.com/NSF_report.pdf.

Angle, Stephen C. 2009. *Sagehood: The Contemporary Significance of Neo-Confucian Philosophy.* Oxford: Oxford University Press.

Angle, Stephen C. 2014. "Seeing Confucian 'Active Moral Perception' in Light of Contemporary Psychology." In *The Philosophy and Psychology of Character and Happiness*, edited by Nancy E. Snow and Franco V. Trivigno, 163–180. New York: Routledge.

Annas, Julia. 1993. *The Morality of Happiness*. Oxford: Oxford University Press.

Annas, Julia. 2005. "Comments on John Doris' *Lack of Character*." *Philosophy and Phenomenological Research* 71 (3): 636–642.

Annas, Julia. 2011. *Intelligent Virtue*. New York: Oxford University Press.

Anscombe, G.E.M. 1958. "Modern Moral Philosophy." *Philosophy* 33 (124): 1–19.

Aquinas, St. Thomas. 1975. *Summa Contra Gentiles. Book Four, Salvation*. Translated by Charles J. O'Neil. Notre Dame, IN: University of Notre Dame Press.

Aristotle. 1984. *The Complete Works of Aristotle: Revised Oxford Translation*. Edited by Jonathan Barnes. Princeton: Princeton University Press.

Arkin, Ronald C. 2009. *Governing Lethal Behavior in Autonomous Robots*. Boca Raton, FL: CRC.

Arkin, Ronald C. 2010. "The Case for Ethical Autonomy in Unmanned Systems." *Journal of Military Ethics* 9 (4): 332–341.

Arquilla, John. 2011. *Insurgents, Raiders and Bandits: How Masters of Irregular Warfare Have Shaped Our World.* Lanham, MD: Ivan R. Dee.

Badhwar, Neera K. 2014a. "Reasoning about Wrong Reasons, No Reasons, and Reasons of Virtue." In *The Philosophy and Psychology of Character and Happiness*, edited by Nancy E. Snow and Franco V. Trivigno, 35–53. New York: Routledge.

Badhwar, Neera K. 2014b. "Objectivity and Subjectivity in Theories of Well-Being." *Philosophy and Public Policy Quarterly* 32 (1): 23–28.

Baier, Annette. 1985. *Postures of the Mind: Essays on Mind and Morals*. Minneapolis, MN: University of Minnesota Press.

Bailey, Ronald. 2005. *Liberation Biology: The Scientific and Moral Case for the Biotech Revolution*. Amherst, NY: Prometheus.

Bainbridge, William Sims and Roco, Mihail C., eds. 2006. *Managing Nano-Bio-Info-Cogno Innovations: Converging Technologies in Society*. Dordrecht, The Netherlands: Springer.

Bakardjieva, Maria. 2011. "The Internet in Everyday Life: Exploring the Tenets and Contributions of Diverse Approaches." In *The Handbook of Internet Studies*, edited by Mia Consalvo and Charles Ess, 59–82. Malden, MA: Blackwell.

Batson, C. Daniel. 2011. *Altruism in Humans.* New York: Oxford University Press.

Baym, Nancy. 2011. "Social Networks 2.0." In *The Handbook of Internet Studies*, edited by Mia Consalvo and Charles Ess, 384–405. Malden, MA: Wiley-Blackwell.

Becker, Lawrence C. 1986. *Reciprocity*. London: Routledge & Kegan Paul.

Becker, Lawrence C. 1998. *A New Stoicism*. Princeton, NJ: Princeton University Press.

Berman, Marc G., John Jonides, and Stephen Kaplan. 2008. "The Cognitive Benefits of Interacting with Nature." *Psychological Science* 19 (12): 1207–1212.

Bilton, Nick. 2013. "Knowing Where to Focus the Wisdom of Crowds." Bits, *New York Times*, April 22. http://bits.blogs.nytimes.com/2013/04/22/knowing-where-to-focus-the-wisdom-of-the-crowds/.

Bilton, Nick. 2014. "Steve Jobs Was a Low-Tech Parent." Disruptions, *New York Times*, September 10. http://www.nytimes.com/2014/09/11/fashion/steve-jobs-apple-was-a-low-tech-parent.html.

Bloom, Irene T. 2002. "Mengzian Arguments on Human Nature (Ren Xing)." In *Essays on the Moral Philosophy of Mengzi*, edited by Xiusheng Liu and Philip J. Ivanhoe, 64–100. Indianapolis: Hackett.

Bommarito, Nicolas. 2013. "Modesty as a Virtue of Attention." *Philosophical Review* 122 (1): 93–117.

Bommarito, Nicolas. 2014. "Patience and Perspective." *Philosophy East and West* 64 (2): 269–286.

Boodberg, Peter A. 1953. "The Semasiology of Some Primary Confucian Concepts." *Philosophy East and West* 2 (4): 317–332.

Borenstein, Jason and Pearson, Yvette. 2010. "Robot Carers: Harbingers of Expanded Freedom for All?" In *Special Issue: Robot Ethics and Human Ethics*. Edited by Anthony Beavers. *Ethics and Information Technology* 12 (3): 277–288.

Borenstein, Jason and Pearson, Yvette. 2012. "Robot Caregivers: Ethical Issues Across the Human Lifespan." In *Robot Ethics*, edited by Patrick Lin, Keith Abney, and George Bekey, 251–265. Cambridge, MA: MIT Press.

Borgmann, Albert. 1984. *Technology and the Character of Contemporary Life*. Chicago: University of Chicago Press.

Borgmann, Albert. 2004. "Is the Internet the Solution to the Problem of Community?" In *Community in the Digital Age*, edited by Andrew Feenberg and Darin Barney, 53–67. Lanham, MD: Rowman and Littlefield.

Bostrom, Nick. 2002. "Existential Risks: Analyzing Human Extinction Scenarios and Related Hazards." *Journal of Evolution and Technology* 9 (1): 1–30.

Bostrom, Nick. 2005a. "In Defense of Posthuman Dignity." *Bioethics* 19 (3): 202–214.

Bostrom, Nick. 2005b. "Transhumanist Values." *Review of Contemporary Philosophy* 4 (1–2): 87–101.

Bostrom, Nick. 2008a. "Why I Want to be a Posthuman When I Grow Up." In *Medical Enhancement and Posthumanity*, edited by Bert Gordijn and Ruth Chadwick, 107–136. Dordrecht, The Netherlands: Springer.

Bostrom, Nick. 2008b. "Dignity and Enhancement." In *Human Dignity and Bioethics: Essays Commissioned by the President's Council on Bioethics*, 173–207. Washington, DC: US Independent Agencies and Commissions.

Bostrom, Nick. 2008c. "Letter from Utopia." *Studies in Ethics, Law and Technology* 2 (1): 1–7.

Bostrom, Nick. 2008d. "Where are They? Why I Hope the Search for Extraterrestrial Life Finds Nothing." *MIT Technology Review*, May/June issue.

Bostrom, Nick. 2014. *Superintelligence: Paths, Dangers, Strategies*. Oxford: Oxford University Press.

boyd, danah. 2014. *It's Complicated: The Social Lives of Networked Teens*. New Haven: Yale University Press.

Bradbury, Ray. 1951. "The Veldt." In *The Illustrated Man*. By Ray Bradbury, 9–27. New York: Doubleday.

Brey, Philip. 2004. "Disclosive Computer Ethics." In *Readings in Cyberethics*, edited by Richard Spinello and Herman Tavani, 55–66. 2nd ed. Sudbury, MA: Jones and Bartlett.

Brey, Philip. 2007. "Is Information Ethics Culture-Relative?" *International Journal of Technology and Human Interaction* 3 (3): 12–25.

Brin, David. 1996. "The Transparent Society." *Wired* 4 (12) http://www.wired.com/1996/12/fftransparent/. Accessed September 4, 2014.

Brin, David. 1998. *The Transparent Society: Will Technology Force Us to Choose Between Privacy and Freedom?* Cambridge, MA: Perseus.

Brin, David. 2014. "In Defense of a Transparent Society." http://www.davidbrin.com/tsdefense.html. Accessed September 4, 2014.

Brynjolfsson, Erik and Andrew McAfee. 2014. *The Second Machine Age: Work, Progress and Prosperity in a Time of Brilliant Technologies*. New York: W.W. Norton.

Buchanan, Allen. 2011. *Better Than Human: The Promise and Perils of Enhancing Ourselves*. New York: Oxford University Press.

Bynum, Terrell W. 2006. "Flourishing Ethics." *Ethics and Information Technology* 8 (4): 157–173.

Cafaro, Philip. 1997. "Virtue Ethics (Not Too) Simplified." *Auslegung: A Journal of Philosophy* 22 (1): 49–67.

Capurro, Rafael. 2008. "Intercultural Information Ethics: Foundations and Applications." *Journal of Information, Communication and Ethics in Society* 6 (2): 116–126.

Carr, Nicholas. 2010. *The Shallows: What the Internet is Doing to Our Brains*. New York: W.W. Norton.

Cavell, Stanley. 1990. *Conditions Handsome and Unhandsome: The Constitution of Emersonian Perfectionism*. Chicago: University of Chicago Press.

Chai, Ch'u and Winberg Chai, eds. and trans. 1965. *The Sacred Books of Confucius and other Confucian Classics*. New Hyde Park, NY: University Books.

Chan, Wing-Tsit, trans. 1963. *A Source Book in Chinese Philosophy*. Princeton: Princeton University Press.

Chou, Hui-Tzu and Nicholas Edge. 2012. "They are Happier and Having Better Lives Than I am: The Impact of Using Facebook on Perceptions of Others' Lives." *Cyberpsychology, Behavior, and Social Networking* 15 (2): 117–121.

Coady, C.A.J. 2009. "Playing God." In *Human Enhancement*, edited by Julian Savulescu and Nick Bostrom, 155–180. Oxford: Oxford University Press.

Coeckelbergh, Mark. 2007. *Imagination and Principles: An Essay on the Role of Imagination in Moral Reasoning*. New York: Palgrave Macmillan.

Coeckelbergh, Mark. 2009. "Personal Robots, Appearance and Human Good: A Methodological Reflection on Roboethics." *International Journal of Social Robotics* 1 (3): 217–221.

Coeckelbergh, Mark. 2010. "Health Care, Capabilities, and AI Assistive Technologies." *Ethical Theory and Moral Practice* 13 (2): 181–190.

Cohen, Jodi S. 2014. "'Civility' a Divisive Issue in U. of I. Faculty Decisions." *Chicago Tribune*, August 29. http://www.chicagotribune.com/news/local/ct-salaita-ayers-backlash-20140829-story.html#page=1.

Cohen, Julie. 2012. *Configuring the Networked Self: Law, Code and the Play of Everyday Practice*. New Haven, CT: Yale University Press.

Cooper, David E. and Simon P. James. 2005. *Buddhism, Virtue and Environment*. Burlington, VT: Ashgate.

Couldry, Nick. 2012. *Media, Society, World: Social Theory and Digital Media Practice*. Malden, MA: Polity.

Couldry, Nick. 2013a. "Why Media Ethics Still Matters." In *Global Media Ethics:Problems and Perspectives,* edited by Stephen J.A. Ward, 13–29. Malden, MA: Wiley Blackwell.

Couldry, Nick. 2013b. "Living Well and Through Media." In *Ethics of Media,* edited by Nick Couldry, Mirca Madianou, and Amit Pinchevski, 39–56. New York: Palgrave Macmillan.

Couldry, Nick, Sonia Livingstone, and Tim Markham. 2007. *Media Consumption and Public Engagement: Beyond the Presumption of Attention*. New York: Palgrave Macmillan.

Crisp, Roger, ed. 1996. *How Should One Live? Essays on the Virtues.* New York: Oxford University Press.

Crisp, Roger, and Michael Slote, eds. 1997. *Virtue Ethics.* Oxford: Oxford University Press.

Davenport, Shaun W., Shawn M. Bergman, Jacqueline Z. Bergman, and Matthew E. Fearrington. 2014. "Twitter Versus Facebook: Exploring the Role of Narcissism in the Motives and Usage of Different Social Media Platforms." *Computers in Human Behavior* 32: 212–220.

de Waal, Frans. 2007. *Chimpanzee Politics: Power and Sex Among Apes*. Baltimore: Johns Hopkins University Press.

della Mirandola, Giovanni Pico. 1994. *Oration on the Dignity of Man, Paragraphs 1–7*. Translated by Richard Hooker. www.fordham.edu/halsall/med/oration.html. Accessed December 9, 2009.

Dewey, Caitlin. 2014. "Hackers Have the Names and Social Security Numbers of Ferguson Police. But Should They Share Them?" *Washington Post,* August 14. http://www.washingtonpost.com/news/the-intersect/wp/2014/08/14/hackers-have-the-names-and-social-security-numbers-of-ferguson-police-but-should-they-share-them/.

Diener, Ed. 2009. "Overview of Subjective Well-Being Scales," http://internal.psychology.illinois.edu/~ediener/scales.html. Accessed April 2, 2016.

Doris, John. 2002. *Lack of Character: Personality and Moral Behavior*. Cambridge, UK: Cambridge University Press.

Doshi, Suhail. 2014. "Mixpanel: How Addictive is Your App?" *Re/Code*, March 6. http://recode.net/2014/03/06/mixpanel-how-addictive-is-your-app/.

Driver, Julia. 2001. *Uneasy Virtue*. New York: Cambridge University Press.

Ellul, Jacques. 1964. *The Technological Society*. New York: Knopf.

Ess, Charles. 2006. "Ethical Pluralism and Global Information Ethics." *Ethics and Information Technology* 8 (4): 215–226.

Ess, Charles. 2009. "Floridi's Philosophy of Information and Information Ethics: Current Perspectives, Future Directions." *The Information Society* 25 (3): 159–168.

Ess, Charles. 2010a. "The Embodied Self in a Digital Age: Possibilities, Risks, and Prospects for a Pluralistic (Democratic/Liberal) Future?" *Nordicom Information* 32 (3–2): 105–118.

Ess, Charles. 2010b. "Trust and New Communication Technologies: Vicious Circles, Virtuous Circles, Possible Futures." *Knowledge, Technology and Policy* 23 (3–4): 287–305.

Ess, Charles. 2013. *Digital Media Ethics*. 2nd ed. Cambridge, UK: Polity.

Ess, Charles. 2014. "Selfhood, Moral Agency and the Good Life in Mediatized Worlds? Perspectives from Medium Theory and Philosophy." In *Mediatization of Communication*, edited by Knut Lundby, 617–640. Berlin: de Gruyter.

Ess, Charles and May Thorseth, eds. 2011. *Trust and Virtual Worlds*. New York: Peter Lang.

Feenberg, Andrew. 1999. *Questioning Technology*. London: Routledge.

Floridi, Luciano. 2006. "Four Challenges for a Theory of Informational Privacy." *Ethics and Information Technology* 8 (3): 109–119.

Floridi, Luciano. 2008. "Information Ethics: A Reappraisal." *Ethics and Information Technology* 10 (2–3): 189–204.

Floridi, Luciano. 2013. *The Ethics of Information*. Oxford: Oxford University Press.

Floridi, Luciano, ed. 2014. *The Onlife Manifesto: Being Human in a Hyperconnected Era*. New York: Springer.

Floridi, Luciano and Mariarosaria Taddeo, eds. 2014. *The Ethics of Information Warfare*. New York: Springer.

Foot, Philippa. 1978. *Virtues and Vices*. Oxford: Blackwell.

Ford, Martin. 2015. *Rise of the Robots: Technology and the Threat of a Jobless Future*. New York: Basic Books.

Foucault, Michel. 1988. *Technologies of the Self: A Seminar with Michel Foucault*. Edited by Luther H. Martin, Huck Gutman, and Patrick H. Hutton. Amherst: University of Massachusetts Press.

Foucault, Michel. 1995. *Discipline and Punish: The Birth of the Prison*. Translated by Alan Sheridan. New York: Vintage.

Foucault, Michel. 1998. *Technologies of the Self: A Seminar with Michel Foucault*. Edited by Luther H. Martin, Huck Gutman, and Patrick H. Hutton. Amherst, MA: University of Massachusetts Press.

Frankfurt, Harry G. 2005. *On Bullshit*. Princeton: Princeton University Press.

French, Peter A., Theodore UehlingJr., and Howard Wettstein, eds. 1988. *Ethical Theory: Character and Virtue*. Midwest Studies in Philosophy 13. Notre Dame, IN: University of Notre Dame Press.

Friedersdorf, Conor. 2014a. "Man Arrested While Picking Up His Kids: 'The Problem is I'm Black.'" *The Atlantic*, August 29. http://www.theatlantic.com/national/archive/2014/08/the-problem-is-im-black/379357/.

Friedersdorf, Conor. 2014b. "This Man Has Nothing to Hide: Not Even His Email Password." *The Atlantic*, August 26. http://www.theatlantic.com/politics/archive/2014/08/this-man-has-nothing-to-hide/379041/.

Fukuyama, Francis. 2002. *Our Posthuman Future: Consequences of the Biotechnology Revolution*. New York: Farrar, Straus and Giroux.

Ganascia, Jean-Gabriel. 2010. "The Generalized Sousveillance Society." *Social Science Information* 49 (3): 1–19.

Garreau, Joel. 2005. *Radical Evolution: The Promise and Peril of Enhancing our Minds, Our Bodies—and What It Means to be Human*. New York: Doubleday.

Gethin, Rupert. 1998. *The Foundations of Buddhism*. Oxford: Oxford University Press.

Gibbs, A.M. 1991. "Shaw and Creative Evolution." In *Irish Writers and Religion*, edited by Robert Welch, 75–88. Savage, MD: Barnes and Noble.

Goleman, Daniel. 2013. *Focus: The Hidden Driver of Excellence*. New York: Harper Collins.

Grenberg, Jeanine. 2005. *Kant and the Ethics of Humility*. New York: Cambridge University Press.

Grodzinsky, Frances. 1999. "The Practitioner Within: Revisiting the Virtues." *Computers and Society* 29 (1): 9–15.

Gross, Doug. 2014. "Meet Sobrr, the Anti-Facebook App." *CNN*, August 21. http://www.cnn.com/2014/08/21/tech/mobile/sobrr-app/.

Hall, John Storrs. 1993. "Utility Fog: A Universal Physical Substance." In *Vision 21: Interdisciplinary Science and Engineering in the Era of Cyberspace*, 115–126. Proceedings of a Symposium Cosponsored by the NASA Lewis Research Center and the Ohio Aerospace Institute and Held in Westlake, Ohio, March 30–31, 1993. Westlake, OH: NASA Conference Publication 10129.

Hampton, Keith, Lee Rainie, Weixu Lu, Maria Dwyer, Inyoung Shin, and Kristen Purcell. August 2014. "Social Media and the 'Spiral of Silence.'" *Pew Research Center*, Washington, DC. http://www.pewinternet.org/2014/08/26/social-media-and-the-spiral-of-silence/.

Hanks, Craig, ed. 2010. *Technology and Values*. Malden, MA: Wiley Blackwell.

Hanson, Robin. 1998. "The Great Filter: Are We Almost Past It?" http://mason.gmu.edu/~rhanson/greatfilter.html. Accessed August 30, 2015.

Haraway, Donna. 1991. *Simians, Cyborgs and Women*. New York: Routledge.

Hardy, Quentin. 2014. "How Urban Anonymity Disappears When All Data is Tracked." *New York Times*, Bits, April 19. http://bits.blogs.nytimes.com/2014/04/

19/how-urban-anonymity-disappears-when-all-data-is-tracked/?_php=true&_type=blogs&_r=0.

Harman, Gilbert. 2000. "The Nonexistence of Character Traits." *Proceedings of the Aristotelian Society* 100 (2): 223–226.

Harris, John. 2007. *Enhancing Evolution: The Ethical Case for Making Better People*. Princeton, NJ: Princeton University Press.

Hartzog, Woodrow and Evan Selinger. 2014. "Two Reasons Why Extreme Social Surveillance Doesn't Replace Privacy." *Forbes*, September 1. http://www.forbes.com/sites/privacynotice/2014/09/01/two-reasons-why-extreme-social-surveillance-doesnt-replace-privacy/.

Harvey, Peter. 2000. *An Introduction to Buddhist Ethics: Foundations, Values, and Issues*. Cambridge, UK: Cambridge University Press.

Hawking, Stephen, Stuart Russell, Max Tegmark, and Frank Wilczek. 2014. "Transcendence Looks at the Implications of Artificial Intelligence - But Are We Taking AI Seriously Enough?" *The Independent*, May 1. http://www.independent.co.uk/news/science/stephen-hawking-transcendence-looks-at-the-implications-of-artificial-intelligence-but-are-we-taking-9313474.html.

Held, Virginia. 2006. *The Ethics of Care: Personal, Political and Global*. Oxford: Oxford University Press.

Hern, Alex. 2014. "OKCupid Experiments are Standard 'Scientific Methods,' Says Founder." *The Guardian*, August 4. http://www.theguardian.com/technology/2014/aug/04/okcupid-dating-ethics-facebook-experiments.

Heyns, Christof. 2013. *Report of the Special Rapporteur on Extrajudicial, Summary or Arbitrary Executions*. United Nations General Assembly, Human Rights Council. Office of the High Comissioner. http://www.ohchr.org/en/issues/executions/pages/srexecutionsindex.aspx. Accessed November 21, 2013.

Honan, Mat. 2014. "I Liked Everything I Saw on Facebook for Two Days. Here's What It Did To Me." *Wired*, August 11. http://www.wired.com/2014/08/i-liked-everything-i-saw-on-facebook-for-two-days-heres-what-it-did-to-me/.

Hongladarom, Soraj. 2008. "Floridi and Spinoza on Global Information Ethics." *Ethics and Information Technology* 10 (2–3): 175–187.

Hongladarom, Soraj and Ess, Charles, eds. 2007. *Information Technology Ethics: Cultural Perspectives*. Hershey, PA: Idea Group.

Hughes, James. 2004. *Citizen Cyborg: Why Democratic Societies Must Respond to the Redesigned Human of the Future*. Cambridge, MA: Westview.

Hursthouse, Rosalind. 1999. *On Virtue Ethics*. New York: Oxford University Press.

Ihara, Craig. 1991. "David Wong on Emotions in Mencius." *Philosophy East and West* 41 (1): 45–54.

Ihde, Don. 2004. "Has the Philosophy of Technology Arrived? A State-of-the-Art Review." *Philosophy of Science* 71 (1): 117–131.

Ihde, Don. 2010. *Heidegger's Technologies: Postphenomenological Perspectives*. New York: Fordham University Press.

Immordino-Yang, Mary H., Andrea McColl, Hanna Damasio, and Antonio Damasio. 2009. "Neural Correlates of Admiration and Compassion." *Proceedings of the National Academcy of Sciences* 106 (19): 8021–8026.

Ivanhoe, Philip J. 2000. *Confucian Moral Self-Cultivation*. 2nd ed. Indianapolis, IN: Hackett.

Ivanhoe, Philip J. 2002a. *Ethics in the Confucian Tradition: The Thought of Mengzi and Wang Yangming*. Indianapolis, IN: Hackett.

Ivanhoe, Philip J. 2002b. "Confucian Self-Cultivation and Mengzi's Notion of Extension." In *Essays on the Moral Philosophy of Mengzi*, edited by Xiusheng Liu and P.J. Ivanhoe, 221–241. Indianapolis: Hackett.

Ivanhoe, Philip J. and Bryan W. Van Norden, eds. 2001. *Readings in Classical Chinese Philosophy*. 2nd ed. Indianapolis: Hackett.

Jonas, Hans. 1979. "Toward a Philosophy of Technology." *Hastings Center Report* 9 (1): 34–43. http://greatbooksojai.com/Hans-Jonas-Toward-a-Philosophy-of-Technology.pdf.

Jonas, Hans. 1984. *The Imperative of Responsibility: In Search of an Ethics for the Technological Age*. Chicago: University of Chicago Press.

Joy, Bill. 2000. "Why the Future Doesn't Need Us." *Wired* 8 (4). http://www.wired.com/2000/04/joy-2/

Kamtekar, Rachana. 2004. "Situationism and Virtue Ethics on the Content of Our Character." *Ethics* 114 (3): 458–491.

Kaplan, David M., ed. 2004. *Readings in the Philosophy of Technology*. Lanham, MD: Rowman and Littlefield.

Kaplan, Jerry. 2015. *Humans Need Not Apply: A Guide to Wealth and Work in the Age of Artificial Intelligence*. New Haven, CT: Yale University Press.

Kaplan, Stephen. 1995. "The Restorative Benefits of Nature: Toward an Integrative Framework." *Journal of Environmental Psychology* 15 (3): 169–182.

Kant, Immanuel. [1785] 1997. *Groundwork of the Metaphysics of Morals*. Translated by Mary Gregor. Cambridge, UK: Cambridge University Press.

Kass, Leon. 2003. "Ageless Bodies, Happy Souls." *The New Atlantis* Spring (1): 9–28.

Kass, Leon. 2008. "Defending Human Dignity." In *Human Dignity and Bioethics: Essays Commissioned by the President's Council on Bioethics*, 297–332. Washington DC: The President's Council on Bioethics.

Keen, Suzanne. 2007. *Empathy and the Novel*. Oxford: Oxford University Press.

Kelly, Kevin. 2011. "The Quantifiable Self." *The Technium*, June 26. http://kk.org/thetechnium/2011/06/the-quantifiabl/.

Keown, Damien. 1992. *The Nature of Buddhist Ethics*. London: Macmillan.

Khushf, George. 2007. "The Ethics of NBIC Convergence." *The Journal of Medicine and Philosophy: A Forum for Bioethics and Philosophy of Medicine* 32 (3): 185–196.

Kirkpatrick, Jesse. 2015. "Drones and the Martial Virtue of Courage." *Journal of Military Ethics*. 14 (3–4): 202–219.

Konrath, Sara, Edward H. O'Brien, and Courtney Hsing. 2011. "Changes in Dispositional Empathy in College Students Over Time: A Meta-Analysis." *Personality and Social Psychology Review* 15 (2): 180–198.

Krasnova, Hanna, Helena Wenninger, Thomas Widjaja, and Peter Buxmann. 2013. "Envy on Facebook: A Hidden Threat to Users' Life Satisfaction?" *11th Annual Conference on Wirtschaftsinformatik, Leipzig, Germany March 2013*. 92–108.

Kristjánsson, Kristján. 1998. "Liberating Moral Traditions: Saga Morality and Aristotle's *Megalopsychia*." *Ethical Theory and Moral Practice* 1 (4): 397–422.

Kross, Ethan, Phillippe Verduyn, Emre Demiralp, Jiyoung Park, David Seungjae Lee, Natalie Lin, Holly Shablack, John Jonides, and Oscar Ybarra. 2013. "Facebook Use Predicts Declines in Subjective Well-Being in Young Adults." *PLoS ONE* 8 (8): e69841. doi:10.1371/journal.pone.0069841. http://journals.plos.org/plosone/article?id=10.1371/journal.pone.0069841.

Kupperman, Joel J. 2001. "The Indispensability of Character." *Philosophy* 76 (2): 239–250.

Kurzweil, Ray. 2005. *The Singularity is Near: When Humans Transcend Biology*. New York: Penguin.

Kuss, Daria J., and Mark D. Griffiths. 2011. "Online Social Networking and Addiction: A Review of the Psychological Literature." *International Journal of Environmental Research and Public Health* 8 (9): 3528–3552.

Laghi, Fiorenzo, Barry H. Scheider, Irene Vitoroulis, Robert J. Coplan, Roberto Baiocco, Yair Amichai-Hamburger, Natasha Hudek, Diana Koszycki, Scott Miller, and Martine Flament. 2013. "Knowing When Not to Use the Internet: Shyness and Adolescents' On-line and Off-line Interactions with Friends." *Computers in Human Behavior* 29 (1): 51–57.

Lanier, Jaron. 2010. *You Are Not a Gadget: A Manifesto*. New York: Knopf.

Latour, Bruno. 1999. *Pandora's Hope: Essays on the Reality of Science Studies*. Cambridge, MA: Harvard University Press.

Lewis, Patricia, Heather Williams, Susan Aghlani, and Benoît Pelopidas. 2014. "Too Close for Comfort: Cases of Near Nuclear Use and Options for Policy." Chatham House Report of the Royal Institute of International Affairs, April 28, 2014. http://www.chathamhouse.org/publications/papers/view/199200?dm_i=1TY5,2EIQH,BHZJ2P,8Q9SA,1.

Li, Chenyang. 1994. "The Confucian Concept of *Jen* and the Feminist Ethics of Care: A Comparative Study." *Hypatia* 9 (1): 70–89.

Li, Chenyang. 2000. "Confucianism and Feminist Concerns: Overcoming the Confucian 'Gender Complex.'" *Journal of Chinese Philosophy* 27 (2): 187–199.

Lin, Patrick, Keith Abney, and George Bekey, eds. 2012. *Robot Ethics*. Cambridge, MA: MIT Press.

Lin, Patrick, Max Mehlman, Keith Abney, Shannon French, Shannon Vallor, Jai Galliott, Michael Burnam-Fink, Alexander R. LaCroix, Seth Schuknecht. 2014. "Super Soldiers (Part 2): The Ethical, Legal, and Operational Implications." In *Global Issues and Ethical Considerations in Human Enhancement Technologies*, edited by Steven J. Thompson, 139–160. Hershey, PA: IGI Global.

Liu, Xiusheng, and Philip J. Ivanhoe, eds. 2002. *Essays on the Moral Philosophy of Mengzi.* Indianapolis: Hackett.

Lockwood, Thornton C. 2013. "Habituation, Habit and Character in Aristotle's *Nicomachean Ethics*." In *A History of Habit from Aristotle to Bourdieu*, edited by Tom Sparrow and Adam Hutchinson, 19–36. Lanham, MD: Lexington.

Lotan, Gilad. 2014. "Israel, Gaza, War and Data: Social Networks and the Art of Personalizing Propaganda." *Medium*, August 4. https://medium.com/i-data/israel-gaza-war-data-a54969aeb23e.

Louden, Robert. 1984. "On Some Vices of Virtue Ethics." *American Philosophical Quarterly* 21 (3): 227–236.

MacIntyre, Alasdair. 1984. *After Virtue*. 2nd ed. Notre Dame, IN: University of Notre Dame Press.

MacIntyre, Alasdair. 1999. *Dependent Rational Animals: Why Human Beings Need The Virtues*. Chicago: Open Court.

Makos, Adam. 2012. *A Higher Call.* New York: Berkley Books.

Mann, Steve, Jason Nolan, and Barry Wellman. 2003. "Sousveillance: Inventing and Using Wearable Computing Devices for Data Collection in Surveillance Environments." *Surveillance and Society* 1 (3): 331–355.

Mar, Raymond A., Keith Oatley, Maja Djikic, and Justin Mullin. 2011. "Emotion and Narrative Fiction: Interactive Influences Before, During and After Reading." *Cognition and Emotion* 25 (5): 818–833.

Marcuse, Herbert. 1964. *One-Dimensional Man: Studies in the Ideology of Advanced Industrial Society*. Boston: Beacon.

Markoff, John. 2015. *Machines of Loving Grace: The Quest for Common Ground Between Humans and Robots*. New York: HarperCollins.

Matthews, Michael D. 2014. "Stress Among UAV Operators: Post-Traumatic Stress Disorder, Existential Crisis, or Moral Injury?" *Ethics and Armed Forces* 1 (1): 53–57.

McDowell, John. 1998. *Mind, Value and Reality*. Cambridge, MA: Harvard University Press.

McKibben, Bill. 2003. *Enough: Staying Human in an Engineered Age*. New York: Times Books.

McLaren, Margaret. 2001. "Feminist Ethics: Care as a Virtue." In *Feminists Doing Ethics*, edited by Peggy DesAutels and Joanne Waugh, 101–118. Lanham, MD: Rowman and Littlefield.

Mencius. 1963. *Mencius*. Translated by W.A.C.H. Dobson. Toronto: University of Toronto Press.

Mencius. 1970. *Mencius*. Translated by D.C. Lau. London: Penguin.

Mill, John Stuart. [1861] 2001. *Utilitarianism*. 2nd ed. Indianapolis, IN: Hackett.

Miller, Christian B. 2014. "The Real Challenge to Virtue Ethics from Psychology." In *The Philosophy and Psychology of Character and Happiness,* edited by Nancy E. Snow and Franco V. Trivigno, 15–34. New York: Routledge.

Mitcham, Carl. 1994. *Thinking Through Technology: The Path Between Engineering and Philosophy*. Chicago: University of Chicago Press.

Moor, James. 2008. "Why We Need Better Ethics for Emerging Technologies." In *Information Technology and Moral Philosophy*, edited by Jeroen van den Hoven and John Weckert, 26–39. Cambridge: Cambridge University Press.

Morozov, Evgeny. 2011. *The Net Delusion: The Dark Side of Internet Freedom*. New York: PublicAffairs.

Morozov, Evgeny. 2013. *To Save Everything, Click Here: The Folly of Technological Solutionism*. New York: PublicAffairs.

Naam, Ramez. 2005. *More than Human: Embracing the Promise of Biological Enhancement*. New York: Broadway Books.

Nehamas, Alexander. 1998. *The Art of Living: Socratic Reflections from Plato to Foucault*. Berkeley: University of California Press.

Nietzsche, Friedrich. 1968. *The Portable Nietzsche*. Edited and translated by Walter Kaufmann. New York: Viking.

Nissenbaum, Helen. 2010. *Privacy in Context: Technology, Policy, and the Integrity of Social Life*. Stanford, CA: Stanford University Press.

Nivison, David S. 1980. "Mencius and Motivation." *Journal of the American Academy of Religion* 47 (3): 417–432.

Nivison, David S. 1996. *The Ways of Confucianism: Investigations in Chinese Philosophy*. Edited by Bryan W. Van Norden. Peru, IL: Open Court.

Noddings, Nel. 1984. *Caring: A Feminine Approach to Ethics and Moral Education*. Berkeley: University of California Press.

Nordmann, Alfred. 2004. *Converging Technologies: Shaping the Future of European Societies*.Luxembourg: EUR-OP, 2004. www.ntnu.no/2020/final_report_en.pdf. Accessed March 24, 2016.

Nussbaum, Martha C. 1988. "Non-Relative Virtues: An Aristotelian Approach." *Midwest Studies in Philosophy* 13 (1): 32–53.

Nussbaum, Martha C. 1990. *Love's Knowledge: Essays on Philosophy and Literature*. Oxford: Oxford University Press.

Nussbaum, Martha C. 1999. "Virtue Ethics: A Misleading Category?" *Journal of Ethics* 3 (3): 163–201.

Nussbaum, Martha C. 2006. *Frontiers of Justice: Disability, Nationality, Species Membership*. Cambridge, MA: Harvard University Press.

Nyhan, Brendan, and Jason Reifler. 2010. "When Corrections Fail: The Persistence of Political Misperceptions." *Political Behavior* 32 (2): 303–330.

Nyhan, Brendan, Jason Reifler, and Peter A. Ubel. 2013. "The Hazards of Correcting Myths about Health Care Reform." *Medical Care* 51 (2): 127–132.

Nyhan, Brendan, Jason Reifler, Sean Richey, and Gary L. Freed. 2014. "Effective Messages in Vaccine Promotion: A Randomized Trial." *Pediatrics* 133 (4): e835–842.

Oakley, Justin, and Dean Cocking, eds. 2001. *Virtue Ethics and Professional Roles*. Cambridge, UK: Cambridge University Press.

Olsthoorn, Peter. 2007. "Courage in the Military: Physical and Moral." *Journal of Military Ethics* 6 (4): 270–279.

O'Neill, Onora. 1990. *Constructions of Reason: Explorations of Kant's Practical Philosophy*. Cambridge, UK: Cambridge University Press.

O'Neill, Onora. 1996. *Towards Justice and Virtue: A Constructive Account of Practical Reasoning*. Cambridge, UK: Cambridge University Press.

Ophir, Eyal, Clifford Nass, and Anthony D. Wagner. 2009. "Cognitive Control in Media Multi-Taskers." *Proceedings of the National Academy of Sciences*, 106 (37): 15583–15587.

Oremus, Will. 2014. "Take the 'No Ice-Bucket' Challenge." *Slate*. August 12. http://www.slate.com/blogs/future_tense/2014/08/12/icebucketchallenge_you_don_t_need_an_ice_bucket_to_donate_to_als_research.html.

Ortega y Gasset, José. 2002. *Toward a Philosophy of History*. Translated by Helene Weyl. Urbana and Chicago: University of Illinois.

Osborne, Michael and Carl Benedikt Frey. 2013. "The Future of Employment: How Susceptible are Jobs to Computerisation?" Working paper of Oxford Martin Programme on the Impacts of Future Technology, 1–72. http://www.oxfordmartin.ox.ac.uk/downloads/academic/The_Future_of_Employment.pdf. Accessed August 20, 2015.

Otto, Jean and Bryant Webber. 2013. "Mental Health Diagnoses and Counseling Among Pilots of Remotely Piloted Aircraft in the United States Air Force." *Medical Surveillance Monthly Report* 20 (3): 3–8.

Overbye, Dennis. 2015. "The Flip Side of Optimism About Life on Other Planets." Out There. Space and Cosmos. *The New York Times*, August 3.

Overly, Steven. 2012. "At Vecna Technologies, Low Funding for BEARs Leads to Building Bots." Capital Business. *The Washington Post*, April 8.

Packer, George. 2013. "Change the World: Silicon Valley Transfers its Slogans—and Its Money—To The Realm of Politics." A Reporter at Large. *The New Yorker*, May 27.

Parks, Jennifer. 2010. "Lifting the Burden of Women's Care Work: Should Robots Replace the Human Touch?" *Hypatia* 25 (1): 100–120.

Pariser, Eli. 2011. *The Filter Bubble: What the Internet is Hiding From You*. New York: Penguin.

Pence, Gregory E. 1984. "Recent Work on Virtues." *American Philosophical Quarterly* 21 (4): 281–297.

Perry, Walter L., Brian McInnis, Carter C. Price, Susan Smith, and John S. Hollywood. 2013. *Predictive Policing: The Role of Crime Forecasting in Law Enforcement Operations*. Santa Monica, CA: RAND Corporation. http://www.rand.org/pubs/research_reports/RR233.html. Accessed April 3, 2016.

Persson, Ingmar and Julian Savulescu. 2012. *Unfit for the Future: The Need for Moral Enhancement*. Oxford: Oxford University Press.

Peterson, Christopher and Martin E.P. Seligman. 2004. *Character Strengths and Virtues: A Handbook and Classification*. New York: Oxford University Press.

Pierce, Tamyra. 2009. "Social Anxiety and Technology: Face-to-face Communication versus Technological Communication among Teens." *Computers in Human Behavior* 25 (6): 1367–1372.

Pinker, Steven. 2011. *The Better Angels of Our Nature*. New York: Penguin.

The President's Council on Bioethics. 2003. *Beyond Therapy*. Bioethics Research Library. Georgetown University. Washington, DC. http://hdl.handle.net/10822/559341.

Rainie, Lee, and Barry Wellman. 2012. *Networked: The New Social Operating System*. Cambridge, MA: MIT Press.

Sandel, Michael. 2007. *The Case Against Perfection: Ethics in the Age of Genetic Engineering*. Cambridge, MA: Harvard University Press.

Sandler, Robert, and Philip Cafaro, eds. 2005. *Environmental Virtue Ethics*. Lanham, MD: Rowman and Littlefield.

Savulescu, Julian, and Nick Bostrom, eds. 2009. *Human Enhancement*. Oxford: Oxford University Press.

Savulescu, Julian, Ruud ter Meulen, and Guy Kahane, eds. 2011. *Enhancing Human Capacities*. Oxford: Wiley-Blackwell.

Scharff, Robert C. 2012. "Empirical Technoscience Studies in a Comtean World: Too Much Concreteness?" *Philosophy and Technology* 25 (2): 153–177.

Scharff, Robert C., and Val Dusek, eds. 2003. *Philosophy of Technology: The Technological Condition*. Malden, MA: Blackwell.

Scheutz, Matthias. 2012. "The Inherent Dangers of Unidirectional Emotional Bonds Between Humans and Social Robots." In *Robot Ethics,* edited by Patrick Lin, Keith Abney, and George Bekey, 205–222. Cambridge, MA: MIT Press.

Schmidt, Eric and Jared Cohen. 2014. *The New Digital Age: Transforming Nations, Businesses, and Our Lives*. New York: Vintage.

Schneewind, J.B. 1990. "The Misfortunes of Virtue." *Ethics* 101 (1): 42–63.

Schneier, Bruce. 2008. "The Myth of the Transparent Society." *Wired*, March 6. http://archive.wired.com/politics/security/commentary/securitymatters/2008/03/securitymatters_0306.

Schwarzenbach, Sibyl. 1996. "On Civic Friendship." *Ethics* 107 (1): 97–128.

Selinger, Evan. 2013. "Facebook Home Propaganda Makes Selfishness Contagious." *Wired*, April 22. http://www.wired.com/2013/04/facebook-home-ads-make-selfishness-contagious/.

Sharkey, Noel. 2012. "Killing Made Easy: From Joysticks to Politics." In *Robot Ethics,* edited by Patrick Lin, Keith Abney, and George Bekey, 111–128. Cambridge, MA: MIT Press.

Sharkey, Amanda, and Noel Sharkey. 2012a. "Granny and the Robots: Ethical Issues in Robot Care for the Elderly." *Ethics and Information Technology* 14 (1): 27–40.

Sharkey, Noel, and Amanda Sharkey. 2012b. "The Rights and Wrongs of Robot Care." In *Robot Ethics,* edited by Patrick Lin, Keith Abney, and George Bekey, 267–282. Cambridge, MA: MIT Press.

Sheldon, Pavica. 2008. "The Relationship Between Unwillingness-to-Communicate and Students' Facebook Use." *Journal of Media Psychology* 20 (2): 67–75.

Sherman, Nancy. 1993. "The Virtues of Common Pursuit." *Philosophy and Phenomenological Research* 53 (2): 277–299.

Sherman, Nancy. 1997. *Making a Necessity of Virtue: Aristotle and Kant on Virtue.* Cambridge,UK: Cambridge University Press.

Shirky, Clay. 2008. *Here Comes Everybody: The Power of Organizing without Organizations.* New York: Penguin.

Shun, Kwong-loi and David B. Wong, eds. 2004. *Confucian Ethics: A Comparative Study of Self, Autonomy, and Community.* Cambridge, UK: Cambridge University Press.

Sidgwick, Henry. 1884. *The Methods of Ethics.* 3rd ed. London: Macmillan and Co.

Singer, Peter. 2011. *The Expanding Circle: Ethics, Evolution and Moral Progress.* Princeton, NJ: Princeton University Press.

Singer, Peter W. 2009. *Wired for War: The Robotics Revolution and Conflict in the 21st Century.* New York: Penguin.

Slote, Michael. 1992. *From Morality to Virtue.* New York: Oxford University Press.

Slote, Michael. 2007. *The Ethics of Care and Empathy.* New York: Routledge.

Slingerland, Edward. 2003. *Confucius: Analects.* Indianapolis: Hackett.

Snow, Nancy E., and Franco V. Trivigno, eds. 2014. *The Philosophy and Psychology of Character and Happiness.* New York: Routledge.

Sparrow, Robert. 2009. "Building a Better WarBot: Ethical Issues in the Design of Unmanned Systems for Military Applications." *Science and Engineering Ethics* 15 (2): 169–187.

Sparrow, Robert. 2013. "War Without Virtue?" In *Killing by Remote Control*, edited by Bradley Jay Strawser, 84–105. New York: Oxford University Press.

Sparrow, Robert. 2015. "Drones, Courage, and Military Culture." In *Routledge Handbook of Military Ethics*, edited by George R. Lucas, Jr., 380–394. New York: Taylor and Francis.

Sparrow, Robert, and Linda Sparrow. 2006. "In the Hands of Machines? The Future of Aged Care." *Minds and Machines* 16 (2): 141–161.

Star, Daniel. 2002. "Do Confucians Really Care? A Defense of the Distinctiveness of Care Ethics: A Reply to Chenyang Li." *Hypatia* 17 (1): 77–106.

Statman, Daniel, ed. 1997. *Virtue Ethics: A Critical Reader.* Washington, DC: Georgetown University Press.

Stock, Gregory. 2002. *Redesigning Humans: Our Inevitable Genetic Future.* Boston: Houghton Mifflin.

Sullins, John P. 2010. "RoboWarfare: Can Robots Be More Ethical Than Humans on the Battlefield?" *Ethics and Information Technology* 12 (3): 263–275.

Sullins, John P. 2011. "Introduction: Open Questions in Roboethics." *Philosophy and Technology* 24 (3): 233–238.

Sunstein, Cass R. 2007. *Republic.com 2.0.* Princeton, NJ: Princeton University Press.

Swanton, Christine. 2003. *Virtue Ethics: A Pluralistic View.* Oxford: Oxford University Press.

Tatz, Mark. 1986. *Asanga's Chapter on Ethics with the Commentary of Tsong-Kha-Pa: The Basic Path to Awakening.* Lewiston, NY: Edwin Mellen.

Tavani, Herman T. 2005. "The Impact of the Internet on our Moral Condition: Do We Need a New Framework of Ethics?" In *The Impact of the Internet on our Moral Lives*, edited by R.J. Cavalier, 215–237. Albany: SUNY Press.

Thompson, Loren. 2011. "How to Waste $100 Billion: Weapons That Didn't Work Out." *Forbes*, December 19.

Tiwald, Justin. 2010. "Confucianism and Virtue Ethics: Still a Fledgling in Chinese and Comparative Philosophy." *Comparative Philosophy* 1 (2): 55–63.

Toyama, Kentaro. 2015. *Geek Heresy: Rescuing Social Change from the Cult of Technology.* New York: PublicAffairs.

Trivigno, Franco V. 2014. "Empathic Concern and the Pursuit of Virtue." In *The Philosophy and Psychology of Character and Happiness*, edited by Nancy E. Snow and Franco V. Trivigno, 113–132. New York: Routledge.

Tronto, Joan. 1993. *Moral Boundaries: A Political Argument for an Ethic of Care.* New York: Routledge.

Tsukayama, Hayley. 2014. "Twitter Vows to 'Improve Our Policies' after Robin Williams' Daughter is Bullied Off the Network." The Switch. *Washington Post*, August 13.

Turkle, Sherry. 2011. *Alone Together: Why We Expect More from Technology and Less from Each Other.* New York: Basic Books.

Turkle, Sherry, 2015. *Reclaiming Conversation: The Power of Talk in a Digital Age.* New York: Penguin.

Twenge, Jean M., Sara Konrath, Joshua D. Foster, W. Keith Campbell, and Brad J. Bushman. 2008. "Egos Inflating Over Time: A Cross-Temporal Meta-Analysis of the Narcissistic Personality Inventory." *Journal of Personality* 76 (4): 875–902.

Vallor, Shannon. 2010. "Social Networking Technology and the Virtues." *Ethics and Information Technology* 12 (2): 157–170.

Vallor, Shannon. 2011a. "Knowing What to Wish For: Human Enhancement Technology, Dignity and Virtue." *Techné: Research in Philosophy and Technology* 15 (2): 137–155.

Vallor, Shannon. 2011b. "Carebots and Caregivers: Sustaining the Ethical Ideal of Care in the 21st Century." *Philosophy and Technology* 24 (3): 251–268.

Vallor, Shannon. 2012. "Flourishing on Facebook: Virtue Friendship and New Social Media." *Ethics and Information Technology* 14 (3): 185–199.

Vallor, Shannon. 2013a. "Armed Robots and Military Virtue." In *The Ethics of Information Warfare*, edited by Luciano Floridi and Mariarosaria Taddeo, 169–185. New York: Springer.

Vallor, Shannon. 2013b. "Examined Lives." *The Philosophers' Magazine* 63 (4): 91–98.

Vallor, Shannon. 2015. "Moral Deskilling and Upskilling in a New Machine Age: Reflections on the Ambiguous Future of Character." *Philosophy of Technology* 28 (1): 107–124.

van Wynsberghe, Aimee. 2013. "Designing Robots for Care: Care Centered Value-Sensitive Design." *Science and Engineering Ethics* 19 (2): 407–433.

Van Norden, Bryan W., ed. 2002. *Confucius and the Analects: New Essays*. New York: Oxford University Press.

Van Norden, Bryan W. 2007. *Virtue Ethics and Consequentialism in Early Chinese Philosophy*. New York: Cambridge University Press.

Verbeek, Peter-Paul. 2005. *What Things Do: Philosophical Reflections on Technology, Agency and Design*. University Park: Pennsylvania State University Press.

Verbeek, Peter-Paul. 2011. *Moralizing Technology: Understanding and Designing the Morality of Things*. Chicago: University of Chicago Press.

Volkman, Richard. 2010. "Why Information Ethics Must Begin with Virtue Ethics." *Metaphilosophy* 41 (3): 380–401.

Waley, Arthur, trans. 1989. *The Analects of Confucius*. New York: Vintage Books.

Walji, Moneeza. 2014. "Slacktivism: Truth or Cynicism?" *Canadian Medical Association Journal Blog*, August 28. http://cmajblogs.com/tag/als-ice-bucket-challenge/.

Walker, Rebecca and Philip J. Ivanhoe, eds. 2007. *Working Virtue: Virtue Ethics, and Contemporary Moral Problems*. Oxford: Oxford University Press.

Wallach, Wendell, and Colin Allen. 2009. *Moral Machines: Teaching Robots Right from Wrong*. New York: Oxford University Press.

Walzer, Michael. 1974. "Civility and Civic Virtue in Contemporary America." *Social Research* 41 (4): 593–611.

Wang, Zheng, and John M. Tchernev. 2012. "The Myth of Media Multitasking: Reciprocal Dynamics of Media Multitasking, Personal Needs, and Gratifications." *Journal of Communication* 62 (3): 493–513.

Wei-Ming, Tu. 1984. "Pain and Suffering in Confucian Self-Cultivation." *Philosophy East and West* 34 (4): 379–388.

Wheeler, Deborah. 2011. "Does the Internet Empower? A Look at the Internet and International Development." In *The Handbook of Internet Studies*, edited by Mia Consalvo and Charles Ess, 188–211. Malden, MA: Blackwell.

Whitby, Blay. 2012. "Do You Want a Robot Lover? The Ethics of Caring Technologies." In *Robot Ethics*, edited by Patrick Lin, Keith Abney, and George Bekey, 233–248. Cambridge, MA: MIT Press.

Whitehill, James. 2000. "Buddhism and the Virtues." In *Contemporary Buddhist Ethics*, edited by Damien Keown, 17–36. Richmond, UK: Curzon.

Wilkinson, Francis. 2008. "The Discover Interview: Leon Kass." *Discover* 29 (2): 62.

Williams, Bernard. 1985. *Ethics and the Limits of Philosophy*. London: Fontana.

Wong, Pak-Hang. 2012. "Dao, Harmony and Personhood: Toward a Confucian Ethics of Technology." *Philosophy and Technology* 25 (1): 67–86.

Wong, Pak-Hang. 2013. "Confucian Social Media: An Oxymoron?" *Dao* 12 (3): 283–296.

Wright, Dale Stuart. 2009. *The Six Perfections: Buddhism and the Cultivation of Character*. Oxford: Oxford University Press.

Xunzi. 2003. *Basic Writings*. Translated by Burton Watson. New York: Columbia University Press.

Yearley, Lee H. 1990. *Mencius and Aquinas: Theories of Virtue and Conceptions of Courage*. Albany, NY: SUNY Press.

Yearley, Lee H. 2002. "An Existentialist Reading of Book Four of the *Analects*." In *Confucius and the Analects, New Essays*, edited by Bryan Van Norden, 237–274. New York: Oxford University Press.

Young, Kimberly S. and Christiano Nabuco de Abreu. 2011. *Internet Addiction: A Handbook and Guide to Evaluation and Treatment*. Hoboken, NJ: Wiley and Sons.

Yu, Jiyuan. 2007. *The Ethics of Confucius and Aristotle: Mirrors of Virtue*. New York: Routledge.

Yuan, Lijun. 2002. "Ethics of Care and Concept of *Jen*: A Reply to Chenyang Li." *Hypatia* 17 (1): 107–130.

Index

Printed in the USA/Agawam, MA
January 15, 2019

695463.013